WIRELESS MULTIMEDIA COMMUNICATION SYSTEMS

Design, Analysis, and Implementation

WIRELESS MULTIMEDIA COMMUNICATION SYSTEMS

Design, Analysis, and Implementation

K.R. RAO

ZORAN S. BOJKOVIC

BOJAN M. BAKMAZ

CRC Press
Taylor & Francis Group
Boca Raton London New York

CRC Press is an imprint of the
Taylor & Francis Group, an **informa** business

CRC Press
Taylor & Francis Group
6000 Broken Sound Parkway NW, Suite 300
Boca Raton, FL 33487-2742

© 2014 by Taylor & Francis Group, LLC
CRC Press is an imprint of Taylor & Francis Group, an Informa business

No claim to original U.S. Government works

Printed on acid-free paper
Version Date: 20140501

International Standard Book Number-13: 978-1-4665-6600-2 (Hardback)

Visit the Taylor & Francis Web site at
http://www.taylorandfrancis.com

and the CRC Press Web site at
http://www.crcpress.com

Dedication

To our families

Contents

Preface

Wireless multimedia communication systems (WMCSs) have become one of the most important mediums for information exchange and the core communication environment for business relations as well as for social interactions. This is a rapidly evolving field with growing applications in science and engineering. Millions of people all over the world use WMCSs for a plethora of daily activities. As we move further into the information age, this rapid advancement will continue in all aspects of WMCSs technologies.

Wireless technologies affect research fields in several disciplines. At the same time, wireless research has also benefited from ideas developed in other fields. How to effectively utilize the interdisciplinary research ideas to develop more efficient wireless technologies opens a new thread of research and design. After many years of research activities related to the optimization of the physical layer of radio networks, the focus is shifting in part toward system and applications level engineering and optimization for an improved performance of wireless multimedia communications. New topics are introduced, most of them not only to incrementally improve current systems but also to provide superior capacity, spectral efficiency, and flexible operation to next generation wireless systems (NGWSs). This can be achieved by extending the system bandwidth and pushing up the transmission rate by coordinating radio resources across adjacent cells for reduced interference in dense reuse of cellular networks, and by introducing mesh networking components and smart antenna technology for spatial multiplex–based transmission.

BOOK OBJECTIVES

The objective of this book is not only to familiarize the reader with advanced topics in WMCSs but also to provide underlying theory, concepts, and principles related to design, analysis, and implementation—presenting a number of specific schemes that have proven to be useful. In this way, coverage of a number of significant advances in their initial steps in this field is included, together with a survey of both the principles and practice. Current WMCSs and architectural concepts must evolve to cope with complex connectivity requirements. This is vital to the development of present-day wireless technologies as a natural step to all progression in general research about WMCSs. In this respect, the intent is to highlight the technological challenges, not just to provide information that will help to understand the processes.

This book is aimed at graduate students. Advanced senior students in engineering should also be able to grasp this topic. Also, the book is intended for both an academic and professional audience. It can serve as a reference book to proceed further toward the algorithms and applications. The annotated references provide a quick survey of the topics for the enthusiasts who wish to pursue the subject matter in greater depth.

ORGANIZATION OF THE BOOK

In this book, we follow up our efforts to highlight the rapidly evolving technologies of WMCSs from the point of view of design, analysis, and implementation. With the widespread adoption of technologies such as heterogeneous networks, cognitive radio, wireless mesh and sensor networks, smart grid, embedded Internet, etc., WMCSs have become an essential component of various media.

In most cases, the book is limited to a reasonable size, but at the same time provides the reader with the power and practical utility of the WMCSs. A large number of figures (139), tables (38), examples (45), and more than 600 references, together with acronyms and subject index, are also included.

The book is composed of 10 chapters. Chapter 1 addresses next generation wireless technologies. The first part summarizes the specifications for wireless networking standards that have emerged to support a broad range of applications. Because standards mediate between the technology and the application, they provide an observation point from which to understand the current and future technological opportunities. The chapter continues with internetworking in heterogeneous wireless environment. Heterogeneity is directly associated with the fact that no single radio access technology (RAT) is able to optimally cover all the wireless communications scenarios. After that, cooperative communications, such as multiple-input multiple-output (MIMO) technologies, are presented. Next, this chapter deals with cooperation techniques from a networking perspective. Finally, congestion control protocol is invoked. It has been emphasized that wireless distributed computing (WDC) exploits wireless connectivity to share processing-intensive tasks among multipath devices.

Chapter 2 surveys cognitive radio (CR) networks. Because CR technology can significantly help spectrum utilization by exploiting some of the parts that are unused by licensed users, it is rapidly gaining popularity and inspiring numerous applications. In fact, CR is the driver technology that enables next generation wireless networks to use the spectrum in a dynamic manner. This chapter starts with a CR system concept. Next, dealing with CR-related applications is performed. Spectrum sensing is also included. After that, the importance of multihop CR networks is presented. Access control in distributed CR networks concludes the chapter.

The purpose of Chapter 3, which relates to mobility management in heterogeneous wireless systems, is to survey recent research in future generation WMCSs. In a heterogeneous environment, mobility management represents the basis for providing continuous network connectivity to mobile users roaming between access networks. Future generations in mobility management will enable different wireless networks to interoperate with each other to ensure seamless mobility and global portability of multimedia services. Mobility management affects the whole protocol stack, from the physical, data link, and network layers up to the transport and application layers. The rest of this chapter is organized as follows. First, primary and auxiliary mobility management services are briefly introduced. After that, current and perspective mobility management protocols are presented, whereas special attention is provided to mobile IP solutions.

Chapter 4 seeks to contribute to network selection in heterogeneous wireless environment. It starts with a synopsis of the corresponding handovers. In addition, the Media Independent Handover (IEEE 802.21), designed for vertical handover, is briefly described. After that, definition and systematization of the decision criteria for network selection is provided. Next, the influences of users' preferences together with services' requirements are envisaged. Finally, the existing studies on network selection using cost function, multiple attribute decision making, fuzzy logic, artificial neural networks, and others, are systematically discussed.

Chapter 5 focuses on wireless mesh networks (WMNs). The wireless mesh environment is envisaged to be one of the key components in the converged networks of the future, providing flexible high-bandwidth backhaul over large geographical areas. With the advances in wireless technologies and the explosive growth of the Internet, designing efficient WMNs has become a major task for network operators. Joint design and optimization of independent problems such as routing and link scheduling have become one of the leading research trends in WMNs. From this point of view, a cross-layer approach is expected to bring significant benefits to achieve high system utilization. After the introduction of the WMN architecture and its fundamental characteristics, the rest of the chapter is related to routing protocols, as well as fair scheduling techniques in WMNs.

Chapter 6 concentrates on wireless multimedia sensor networks (WMSNs). The design of these networks depends significantly on the applications. In recent years, extensive research has opened challenging issues for the deployment of WMSNs. To highlight these issues, a survey on WMSNs including different types of nodes, architecture, and factors important for designing such networks are provided. The internal architectures of multimedia sensor devices are also included. Different perspective solutions for all layers of the communication protocol stack, together with some cross-layer approaches, are presented. Next, the convergence of mobile and sensor systems is analyzed and a brief overview of WMNSs applications is provided. Finally, research issues concerning problems related to the WSN's automated maintenance closes the chapter.

The increasing use in application areas requires the provision of properties at the network or application layer, such as privacy of communications, nonrepudiation of communication members' actions, authentication of parties in application, and high availability of links and services among several others. Security technologies enable the provision of several of these properties at the required level. Chapter 7 starts with general security issues including security attacks in wireless networks. Then, security requirements are presented. Next, the security aspects in emerging mobile networks are analyzed. Security of cognitive radio and wireless mesh networks is also outlined. Some security aspects concerning wireless multimedia sensor networks conclude the chapter.

With the increasing interest from both the academic and industrial communities, Chapter 8 describes the developments in communication technology in the smart grid (SG). This is an electrical system that uses information, two-way secure communication technologies, and computational intelligence in an integrated fashion across the entire spectrum of the energy system from the generation to the end points of electricity consumption. Huge amounts of data related to monitoring and control will be transmitted across SG wireless communication infrastructures, with intensive interference and increasing competition over the limited and crowded radio spectrum for the existing wireless networking standards. The design of the communication network associated with the SG involves detailed analysis of requirements, including the selection of the most suitable technologies for each case study, and architecture for the resultant heterogeneous system. After a key requirement and establishing standards for the SG, this chapter deals with the main component of this system, together with the corresponding communication architecture. A brief description of the effective load control is presented. The significance of wireless mesh networking as well as heterogeneous network integration to coordinate the SG functions are also invoked. Next, the importance of smart microgrid network is emphasized. Finally, an outline of the SG demand response concludes the chapter followed by a discussion.

Chapter 9, which is devoted to the evolution of embedded Internet, invokes the novel paradigm named the Internet of Things (IoT). From a wireless communications perspective, the IoT paradigm is strongly related to the effective utilization of wireless sensor networking. Differences in traffic characteristics, along with the energy constraints and the specific features of the IoT communication environment will be the stimulus for research activities, modeling, and protocol design at both the network and transport layers. Also, the term mobile crowdsensing, which refers to a broad range of community sensing paradigms, is coined. A semantic Web of things is presented as a service infrastructure. Then, machine-to-machine technology is emphasized for the development of IoT communications platforms with high potential to enable a wide range of applications. Finally, the interconnection of nanotechnology devices with classic networks and the Internet defines a new paradigm, which is referred to as the Internet of nano-things.

Although future networks will require a multimedia transport solution that is more aware of a delivery network's requirements, the future Internet (FI) is expected to be more agile, scalable, secure, and reliable. Rapidly emerging applications with different requirements and implications for the FI's design pose a significant set of problems and challenges. The book ends with Chapter 10, which starts with the main principles for FI architecture. Then, delivery infrastructure for next-generation services is presented. Next, information-centric networking is invoked and analyzed through the most representative approaches. The concept of scalable video delivery is briefly presented as a key for efficient streaming in FI. Also, media search and retrieval in the FI are outlined. FI self-management scenarios conclude the chapter.

A major problem during the preparation of this book was the rapid pace of development both in software and hardware relating to the book's topics. We have tried to keep pace by including the latest research. In this way, it is hoped that the book is timely and that it will appeal to a wide audience in the engineering, scientific, and technical community. Also, we believe that this book will provide readers with significant tools and resources when working in the field of wireless multimedia communications systems.

Special thanks go to reviewers who dedicated their precious time in providing numerous comments and suggestions. Their careful reading for improvement of the text, together with numerous comments with constructive criticism and enthusiastic support, have been valuable.

In closing, we hope that this book should provide not only knowledge but also inspiration to researchers and developers of wireless multimedia communication systems including design and implementation. Of course, much work remains about achieving higher quality technologies at low cost.

K. R. Rao
Z. S. Bojkovic
B. M. Bakmaz

List of Acronyms

2G: Second generation
3G: Third generation
3GPP: Third Generation Partnership Project
4G: Fourth generation
5G: Fifth generation
6LNTP: 6LoWPAN network time protocol
6LoWPAN: IPv6 in low-power wireless personal area networks
AAA: Authentication, authorization, and accounting
ABC&S: Always best connected and served
ABS: Advanced base station
ACDM: Algebraic channel decomposition multiplexing
ACO: Ant colony optimization
ACP: Adaptive congestion protocol
ADA: Advanced distributed automation
AF: Amplify-and-forward
AGC: Automatic gain control
AHP: Analytical hierarchy process
AIMD: Additive-increase multiplicative-decrease
AKA: Authentication and key agreement
AMI: Advanced metering infrastructure
AMR: Access mesh router
AMS: Advanced mobile station
ANDSF: Access network discovery and selection function
ANN: Artificial neural network
AODV: Ad hoc on-demand distance vector
AODV-MR: AODV-multiradio
AODV-ST: AODV-spanning tree
AP: Access point
API: Application programming interface
AR: Access router
ARQ: Automatic repeat-request
ASAR: Ant based service-aware routing
ASN: Access service node
BAN: Building area network
BATMAN: Better approach to mobile ad hoc networking
BER: Bit error rate
BGP: Border gateway protocol
BPSK: Binary phase shift keying

BS: Base station
BU: Binding update
BUack: Binding update acknowledgement
BWA: Broadband wireless access
CBCA: Carrier-based coverage augmentation
CCC: Common control channel
CCDS: SG control center at distributed substation
CCN: Content-centric networking
CCRN: Cooperative cognitive radio networking
CDMA: Code-division multiple access
CDN: Content delivery network
CF: Cost function
CHGW: Cognitive home gateway
CIS: Consumer information system
CK: Ciphering key
C-MAC: Cognitive MAC
CMOS: Complementary metal oxide semiconductor
CN: Corresponding node
CNGW: Cognitive NAN gateway
CoA: Care-of address
CoAP: Constrained application protocol
COM-MAC: Clustered on-demand multichannel MAC
CoRE: Constrained RESTful environments
CPC: Cognitive pilot channel
CPE: Customer premise equipment
CPS: Common part sublayer
CQI: Channel quality indicator
CR: Cognitive radio
CRBS: Cognitive radio base stations
CRN: Cognitive radio network
CRS: Cognitive radio system
CS: Convergence sublayer
CSCF: Call session control function
CSG: Closed subscriber group
CSI: Channel state information
CSMA/CA: Carrier sense multiple access with collision avoidance
CSN: Connectivity service network
CTS: Clear-to-send
DAG: Directed acyclic graph
DAO: Destination advertisement object
DAP: Data aggregation point
DAR: Dynamic address reconfiguration
DCF: Distribution coordination function

DC-MAC: Decentralized cognitive MAC
DCT: Discrete cosine transform
DDNS: Distributed domain name system
DDoS: Distributed denial of service
DER: Distributed energy resource
DES: Data encryption standard
DF: Decode-and-forward
DFT: Discrete Fourier transform
DGR: Directional geographical routing
DMM: Distributed mobility management
DMS: Distribution management system
DNF: Denoise-and-forward
DNS: Domain name system
DODAG: Destination oriented DAG
DONA: Data-oriented network architecture
DoS: Denial of service
DOSS-MAC: Dynamic open spectrum sharing MAC
DR: Demand response
DRESS-WS: Distributed resource-based simple Web service
DSA: Dynamic spectrum access
DSCC: Differentiated services based congestion control
DSM: Dynamic spectrum management
DSONPM: Dynamic self-organizing network planning and management
DSR: Dynamic source routing
DS-UWB: Direct spectrum UWB
DVC: Distributed video coding
DySPAN-SC: Dynamic Spectrum Access Networks Standards Committee
E²E: End-to-end
E²R: End-to-end reconfigurability
EAP-AKA: Extensible authentication protocol-AKA
EAQoS: Energy aware QoS
ECMA: European Computer Manufacturers Association
EDGE: Enhanced data rates for GSM evolution
eNB: Enhanced node B
ENT: Effective number of transmission
EPC: Evolved packet core
ePDG: Evolved packet data gateway
EPS: Evolved Packet System
ESTCP: Explicitly synchronized TCP
ETSI: European Telecommunications Standards Institute
ETT: Expected transmission time
ETX: Expected transmission count

E-UTRAN: Evolved-Universal Terrestrial Radio Access Network
Ex-OR: Extremely opportunistic routing
FA: Foreign agent
FBack: Fast binding acknowledgment
FBU: Fast binding update
FCC: Federal Communications Commission
FCS: Feedback control system
FDD: Frequency-division duplex
FDMA: Frequency division multiple access
FEC: Forward error correction
FFD: Full function device
FI: Future Internet
FIB: Forwarding information base
FMIPv6: Fast MIPv6
FoV: Field of view
FPGA: Field-programmable gate array
FPMIPv6: Fast handovers for PMIPv6
GEAMS: Geographic energy-aware multipath stream
GGSN: Gateway GPRS support node
GIS: Geographic information system
GoS: Grade of service
GPRS: General packet radio service
GPS: Global positioning system
GPSR: Greedy perimeter stateless routing
GRA: Gray relational analysis
GSA: Global mobile Suppliers Association
GSM: Global system for mobile communications
GW: Gateway
HA: Home agent
HAck: Handover acknowledge
HAN: Home area network
HARQ: Hybrid automatic repeat request
HC-MAC: Hardware-constrained MAC
HDTV: High definition TV
HEC: Home energy controller
HeNB: Home enhanced node B
HetNets: Heterogeneous networks
HEVC: High-efficiency video coding
HI: Handover initiate
HMIPv6: Hierarchical MIPv6
HoA: Home-of address
HR-WPAN: High-rate WPAN
HSDPA: High-speed downlink packet access

HSPA: High-speed packet access
HSS: Home subscriber server
HSUPA: High-speed uplink packet access
HTTP: Hypertext transfer protocol
ICMP: Internet control message protocol
ICT: Information and communication technologies
IEC: International Electrotechnical Commission
IEEE: Institute of Electrical and Electronics Engineers
IEEE SCC: IEEE Standards Coordinating Commitee
IETF: Internet Engineering Task Force
IH-AODV: Improved hierarchical AODV
IK: Integrity key
IKEv2: Internet key exchange version 2
IMS: IP multimedia subsystem
IMSI: International mobile subscriber identity
IMT: International mobile telecommunications
INEA: Initial network entry authentication
InP: Infrastructure provider
IoNT: Internet of nanothings
IoT: Internet of things
IP: Internet protocol
IPSec: IP security
IR-UWB: Impulse radio UWB
ISGAN: International Smart Grid Action Network
ISIM: IMS subscriber identity module
ISM: Industrial, scientific, and medical
ISO: International Organization for Standardization
ISP: Internet service provider
ITU: International Telecommunication Union
JRRM: Joint radio resource management
JTCP: Jitter-based TCP
LIPA: Local IP address
LLC: Logical link control
LLN: Low-power and lossy network
LMA: Local mobility anchor
LMD: Localized mobility domain
LNA: Low-noise amplifier
LoS: Line-of-sight
LRCC: Load repartition–based congestion control
LR-WPAN: Low-rate WPAN
LTE: Long term evaluation
LTE-A: LTE advanced
M2M: Machine-to-machine

MAC: Medium access control
MADM: Multiple attributes decision making
MAG: Mobile access gateway
MAHO: Mobile-assisted handover
MANET: Mobile ad hoc network
MAP: Mesh access point/mobility anchor point
MB-OFDM: Multiband OFDM
MBWA: Mobile broadband wireless access
MC: Mesh client
MCDM: Multiple criteria decision making
MCM: Multicarrier modulation
MCS: Mobile crowdsensing
MD: Multiple description
M-DTSN: Multimedia distributed transport for sensor networks
METX: Multicast ETX
MEW: Multiplicative exponential weighting
MG: Mesh gateway
MGS: Master-gateway station
MICS: Media independent command services
MIES: Media independent event service
MIH: Media independent handover
MIHF: Media independent handover function
MIIS: Media independent information services
MIMO: Multiple-input multiple-output
MIMO-POWMAC: MIMO power-controlled MAC
MIPv4: Mobile IP version 4
MIPv6: Mobile IP version 6
MIR: Mobile IP reservation protocol
MISO: Multi-input single-output
MitM: Man-in-the-middle
ML: Minimum loss
MLME: MAC layer management entity
mMAG: Moving MAG
MME: Mobility management entity
MMESH: Multipath MESH
MN: Mobile node
MNN: Mobile network node
MNP: Mobile network prefix
MODM: Multiple objective decision making
MORE: MAC independent opportunistic routing and encoding
MP: Mesh point
MPH: Multimedia processing hub
MPP: Mesh portal

MR: Mesh router
MR²-MC: Multiradio, multirate multichannel
MRDT: Multiradio distributed tree
MR-LQSR: Multiradio link quality source routing
MRS: Mesh-relay station
MS: Mobile station
MSAP: Mesh station with access point
mSCTP: Mobile SCTP
MSE: Mean squared error
MT: Mobile terminal
MTC: Machine type communication
MTU: Maximum transmission unit
MU-MIMO: Multi-user MIMO
NADV: Normalized ADVance
NAHO: Network-assisted handover
NAN: Neighborhood area network
NAP: Network access provider
NC: Network coding, network coverage
NCHO: Network-controlled handover
NDN: Named data networking
NDO: Named data object
NEMO: Network mobility
NetInf: Network of information
NETLMM: Network-based localized mobility management
NGN: Next generation network
NGWS: Next generation wireless systems
NIST: National Institute of Standards and Technology
NRS: Name resolution service
NSP: Network service provider
NTP: Network time protocol
OAM: Operation, administration, and maintenance
ODMRP: On-demand multicast routing protocol
OFDM: Orthogonal frequency division multiplexing
OLSR: Optimized link state routing
OMA: Open mobile alliance
O-MAC: Opportunistic MAC
OPC: Out-of-band pilot channel transmitter
OQPSK: Orthogonal QPSK
OSA-MAC: Opportunistic spectrum access MAC
OSM: Operator spectrum manager
OSPF: Open shortest path first
P-/I-/S-CSCF: Proxy-/interrogating-/serving-call session control
functions

P2MP: Point-to-multipoint
P2P: Peer-to-peer, point-to-point
PBA: Proxy binding acknowledgement
PBU: Proxy binding update
PDN-GW: Packet data network gateway
PDR: Packet delivery ratio
PGSA: Predefined gateway set algorithm
PHY: Physical
PIR: Passive infrared
PIT: Pending interest table
PLC: Power line communication
PLL: Phase locked loop
PLME: Physical layer management entity
PMI: Preferred matrix index/precoding matrix indicator
PMIPv6: Proxy MIPv6
PNC: Piconet coordinator
PoA: Point of attachment
PoI: Point of interest
PRN: Primary radio network
PROC: Progressive route calculation
PrRtAdv: Proxy router advertisement
PS: Presence service
PSK: Phase shift keying
PSNR: Peak signal-to-noise ratio
PSU: Power supply unit
PU: Primary user
QAM: Quadrature amplitude modulation
QCCP-PS: Queue-based congestion control protocol with priority
          support
QoE: Quality of experience
QoS: Quality of service
QPSK: Quadrature PSK
RAN: Radio access network
RAT: Radio access technology
RC: Relative closeness
REAP: Reachability protocol
REAR: Real-time and energy aware routing
RED: Random early detection
REST: Representational state transfer
RF: Radiofrequency
RFD: Reduced function device
RFID: Radiofrequency identification
RH: Resolution handler

RIP: Routing information protocol
ROLL: Routing over low-power and lossy networks
ROMER: Resilient opportunistic mesh routing
RS: Relay station
RSS: Received signal strength
RSSI: Received signal strength indicator
RSTP: Reliable synchronous transport protocol
RTLD: Real time routing protocol with load distribution
RTP: Real-time pricing
RtrAdv: Router advertisements
RTS: Request-to-send
RtSolPr: Router solicitation for proxy advertisement
RTT: Round-trip time
RTUs: Remote terminal units
SA: Spectrum automaton
SAE: Service architecture evolution
SAP: Service access point
SAR: Sequential assignment routing
SAW: Simple additive weighting
SCADA: Supervisory control and data acquisition
SC-FDM: Single-carrier frequency-division multiplexing
SCGA: Self-constituted gateway algorithm
SCMA: Stream-controlled medium access
SCTP: Stream control transmission protocol
SDMA: Space-division multiple access
SD-MAC: Spatial diversity for MAC
SDR: Software-defined radio
SecGW: Security gateway
SG: Smart grid
SGIP: Smart grid interoperability panel
SGP: Stratum gateway point
SGSN: Serving GPRS support node
S-GW: Serving gateway
SH: Storage hub
SIMO: Single-input multiple-output
SINR: Signal to interference and noise ratio
SIP: Session initiation protocol
SIPTO: Selected IP traffic off-load
SIR: Signal to interference ratio
SLA: Service level agreement
SM: Spectrum manager
S-MAC: Sensor MAC
SMDs: Smart metering devices

SME: Station management entity
SMG: Smart microgrid
SN: Sink node
SNAIL: Sensor networks for an all-IP world
SNIR: Signal-to-noise-plus-interference ratio
SNR: Signal-to-noise ratio
SNTP: Simple NTP
SOAR: Simple opportunistic adaptive routing
SOHO: Small office/home office
SON: Self-organizing network
SPP: Success probability product
SS: Subscriber station, scalar sensor
SSF: Spectrum sensing function
SSP: Stratum service point
STA: Station
SU: Secondary user
SU-MIMO: Single user MIMO
SVC: Scalable video coding
SYN-MAC: Synchronized MAC
TCP: Transmission control protocol
TDD: Time–division duplex
TDMA: Time–division multiple access
TD-SCDMA: Time–division synchronous code–division multiple access
TG: Task group
TH-UWB: Time hopping UWB
T-MAC: Time-out MAC
TOPSIS: Technique for order preference by similarity to an ideal solution
TPC: Transmission power control
TVWS: Television white spaces
UAHO: User-assisted handover
UCG: Unified channel graph
UDP: User datagram protocol
UE: User equipment
UHF: Ultrahigh frequency
ULID: Upper layer identifier
UMTS: Universal mobile telecommunications system
URI: Uniform resource identifier
USIM: Universal subscriber identity module
USN: Ubiquitous sensor network
UWB: Ultra-wideband
VANET: Vehicular ad hoc network
VAS: Video and audio sensor

VCA: Video content analysis
VCN: Vehicular communication network
VCO: Voltage-controlled oscillator
VCS: Virtual carrier sensing
VHD: Vertical handover decision
VHF: Very high frequency
VHT: Very high throughput
VNet: Virtual network
VNO: Virtual network operator
VNP: Virtual network provider
VoIP: Voice over Internet protocol
WCDMA: Wideband CDMA
WCETT: Weighted cumulative ETT
WDC: Wireless distributed computing
WDS: Wireless distribution system
WFAP: WiMAX femtocell access point
WG: Working group
Wi-Fi: Wireless fidelity
WiMAX: Worldwide interoperability for microwave access
WInnF: Wireless Innovation Forum
WLAN: Wireless local area network
WMAN: Wireless metropolitan area network
WMCS: Wireless multimedia communication system
WMN: Wireless mesh network
WMSN: Wireless multimedia sensor network
WNC: Wireless network control
WPAN: Wireless personal area network
WPKI: Wireless public key infrastructure
WPSN: Wireless passive sensor network
WRAN: Wireless regional area network
WRR: Weighted round-robin
WSN: Wireless sensor networks
WSRN: Wireless sensor and robot network
W-SVC: Wavelet-based SVC
WWAN: Wireless wide area network
XCP: eXplicit control protocol
XML: Extensible markup language
μIP: Micro IP

1 Next Generation Wireless Technologies

Wireless technologies have penetrated every aspect of our lives and affects research fields in several disciplines. At the same time, wireless research has also benefited from research ideas developed in other fields. How to effectively utilize interdisciplinary research ideas to develop more efficient wireless technologies opens a new thread of research and design. After many years of research activity related to the optimization of the physical layer of radio networks, the focus is shifting in part toward system and applications level engineering and optimization for an improved performance of wireless multimedia communications. New topics are introduced, most of them not only for incrementally improving current systems but also for providing superior capacity, spectral efficiency, and flexible operation to next generation wireless systems (NGWS). This can be achieved by extending the system bandwidth and pushing up the transmission rate by coordinating radio resources across adjacent cells for reduced interference in dense reuse cellular networks, by introducing ad hoc networking components to cellular networks, and by introducing smart antenna technology for spatial multiplex-based transmission. Rapid progress in wireless communication technologies has created different types of systems that are envisioned to coordinate with one another to provide ubiquitous high data rate services for mobile users. On the other hand, mobile users are demanding anywhere-and-anytime access to high-speed real-time and non-real-time multimedia services. These services have different requirements in terms of latency, bandwidth, and error rate. However, none of the existing wireless systems can simultaneously satisfy the needs of mobile users at a low cost. This necessitates a new direction in the design of NGWS.

1.1 INTRODUCTION

Nowadays, the most exciting advances in the access networks field concerns wireless networking. Generally, they are identified as data packet switched networks that can be categorized according to the size of their coverage area. There are two approaches in designing NGWS. The first is to develop a new wireless system with radio interfaces and technologies that satisfy the requirements of future mobile users. This approach is not practical because it is expensive and uses more time for development and deployment. The second one is a more feasible option, which integrates the existing wireless systems so that users may receive their services via the best available wireless network [1]. Following this approach, heterogeneous wireless systems, each optimized for some specific service demands and coverage area, will cooperate with one another to provide the best connection to mobile users. In the integrated overlay heterogeneous networks architecture, each user is always connected in the best available network(s) [2]. The integrated NGWS must have some characteristics, such as support for the best network selection based on the user's service needs, mechanisms to ensure high-quality security and privacy, and protocols to guarantee seamless intersystem mobility. Also, the architecture should be able to integrate any number of wireless systems of different service providers who may not have direct service level agreements (SLAs) among them.

In NGWS, users move between different networks. They want to maintain their ongoing communications while moving from one network to another. The heterogeneous networks may or may not belong to the same service provider. Thus, support for intersystem movement between networks of different service providers is required. The architecture of NGWS should have the following characteristics: economical, scalable, transparency to heterogeneous access technologies, secure, and seamless mobility support [3]. In an integrated wireless system, managing resources among multiple wireless networks in an economical and computationally practical way is crucial to preserving system robustness. When each network manages its resources and reacts to congestion independently, it can lead to instability. Quality of service (QoS) provisioning in heterogeneous wireless networks introduces new mobility management problems such as timely service delivery and QoS negotiations.

Future networked media infrastructures will be based on different technologies, while at the same time have to support the entire range of content types, formats, and delivery modes. The goal is to design an integrated content services infrastructure comprising content delivery and content services within one framework. This infrastructure should provide seamless, content-aware, end-to-end (E2E) content delivery in a heterogeneous environment in a community context. A number of challenges are identified: dynamic adaptability and managing complexity of infrastructures, overlay challenges (e.g., coding, cashing, and organization), protection, multiple domain multicasting, distribution modes, peer-to-peer (P2P) distribution storage systems, bandwidth, symmetry, etc.

The higher data rate (up to 100 Mb/s) and strict QoS requirements of multimedia applications may not be fully supported over the current cellular networks. Because the transmit power of a data link increases with the data rate when a specific link quality is maintained, providing very high data rate services will require either the expenditure of high amounts of power or limiting the link to a short distance. Therefore, there is interest in integrating next-generation cellular network structures for higher data rates and coverage as well as scalability in addition to the continuing research on advanced wireless communication technologies [4].

The practical success of service platform technology for next generation wireless multimedia relies largely on its flexibility in providing adaptive and cost-effective services [5]. The demand for high bandwidth is a key issue for these multimedia services (video broadcasting, video conferencing, combined voice–video applications, etc.), but is not sufficient. Among other major requirements that should also be considered, such as seamless mobility, security, and flexible changing, is a unified QoS support. Future service platforms are expected to integrate paradigms known from traditional telecommunications environments and the Internet [6].

The first part of this chapter summarizes the specifications for wireless networking standards that have emerged to support a broad range of applications: wireless personal area networks (WPANs), wireless local area networks (WLANs), wireless metropolitan area networks (WMANs), wireless wide area networks (WWANs), and wireless regional area networks (WRANs). The standards have been mainly developed under the auspices of the Institute of Electrical and Electronics Engineers (IEEE) and Third Generation Partnership Project (3GPP) bodies. Technology underlies all developments in communication networks. However, wide-scale networks deployment is not based directly on technology but on standards that embody technology, along with the economic realities. Because standards mediate between the technology and the application, they provide an observation point from which to understand current and future technological opportunities.

This chapter continues with interworking in heterogeneous wireless environments. Interworking mechanisms are of prime importance to achieve access and seamless mobility in heterogeneous wireless networks. Heterogeneity is an important characteristic in traditional and future wireless communication scenarios. It refers to the coexistence of multiple and diverse wireless networks with their corresponding radio access technologies (RATs). Heterogeneity is directly associated with the fact that no single RAT is able to optimally cover all the wireless communications scenarios. Despite RAT heterogeneity, the service model under NGWS is intended to facilitate the deployment of applications and services independent of the RAT. Hence, it is expected that mobile users could enjoy seamless mobility and service access in an always best connected mode, employing the most efficient combination of available RANs anytime and anywhere. From this point of view, an appropriate interworking of different wireless access systems is crucial to meet the mobile user's expectations, while making the coexistence of diverse RATs possible.

After that, cooperative communications, such as multiple-input, multiple-output (MIMO) technologies, which have been taken into consideration for future wireless systems development, are presented. There are many challenges making them practical to use because viable collaborations among communication entities have to be addressed more effectively to gather the needed information [7]. Various MIMO technologies have been introduced to improve the performance of NGWS [8]. For example, in single-cell environments, closed-loop MIMO schemes increase the system capacity as well as cell coverage. In mobile systems, in which interference coming from other cells is usually a dominant factor, several multiple base stations cooperation schemes have been introduced to mitigate the intercell interference. Closed-loop MIMO techniques can enhance throughput for both average and cell edge users by exploiting channel state information feedback from the receiver [9]. Cell edge area users are still vulnerable to intercell interference from adjacent cells. Thus, handling the interference property is one of the main issues in cellular systems design.

Next, we deal with cooperation techniques from a networking perspective. The diffusion of mobile network devices has significantly increased the need for reliable and high-performance wireless communications. Two main aspects characterize the physical layer of wireless networks. On one hand, the use of wireless terminals is desirable for mobility and ease of deployment. On the other hand, the radio channel is known to be strongly influenced by multipath fading and interference problems. To improve this situation and to achieve reliability at the upper layers when relaying on a wireless physical layer, many researchers have proposed the adoption of cooperative paradigms [10,11]. This means that one or more intermediate nodes intervene in the communication between transmitter and a receiver so that either the communication is rerouted over a buffer path, or the original line is kept in use but its quality is strengthened thanks to diversity provided by these cooperators.

In the final part of this chapter, wireless loss-tolerant congestion control protocol will be invoked. In response to the broadband isolation of the medium access control (MAC) layer and the wide use of Internet protocols (IP), the role of transmission control protocols (TCP) in controlling the transmission rate over IP networks has been expanded. Although TCP is the standard control protocol in the Internet, its shortcomings in the heterogeneous wireless environment are widely known [12]. Most existing congestion control protocols can be classified into two categories: additive-increase multiplicative-decrease (AIMD) and feedback control system (FCS) [13]. Recent advances in radio technology provide great flexibility and enhanced capability in existing wireless services. One of these capabilities that can provide significant advantages over traditional approaches is the collaborative computing concept. With collaborative radio nodes, multiple independent nodes operate together to form a wireless distributed computing (WDC) network with significantly increased performance operating efficiency and ability over a single node [14]. WDC exploits wireless connectivity to share processing-intensive tasks among multipath devices. The goals are to reduce per node and network resource requirements and enable complex applications not otherwise possible.

1.2   WIRELESS NETWORKING STANDARDIZATION PROCESS

Standardization on a global basis is of fundamental importance and even more so for wireless networks because of the need for worldwide interoperability. Wireless networks are most frequently divided, based on coverage area, into four specific groups (WPAN, WLAN, WMAN, and WWAN). In addition to these well-known wireless network environments, recently WRAN standards have become more and more actual. A review of wireless technologies together with corresponding network environment, standards, frequency band, peak data rates as well as coverage is presented in Table 1.1.

In what follows, special attention is given to emerging and perspective standards for NGWS, whereas older but actual standards are analyzed by Rao et al. [5].

1.2.1   WIRELESS PERSONAL AREA NETWORKS

With the rapid evolution of wireless technologies, ubiquitous and always-on wireless systems in homes and enterprises are expected to soon emerge. Facilitating these ubiquitous wireless systems is one of the ultimate goals of fourth-generation (4G) wireless technologies being discussed worldwide today [15]. WPANs are an emerging technology for future short-range indoor and outdoor multimedia and data centric applications. They enable ad hoc connectivity among portable consumer electronics and communications devices. The coverage area for a WPAN is generally within a 10 m radius. The term ad hoc connectivity refers to both the ability for a device to assume either master or slave functionality, and the ease in which devices may join or leave an existing network [16]. The IEEE 802.15 standards suite aims at providing wireless connectivity solutions for such networks without having any significant effect on their form factor, weight, power, requirements, cost, ease of use, or other traits [17].

The first step in introducing WPAN to the worldwide market is the standardization of Bluetooth technology's physical and MAC layer over

TABLE 1.1

Actual and Emerging Wireless Technologies Basic Characteristics

Network Environment	Standard	Frequency Bands	Peak Data Rates	Coverage
WPAN	IEEE 802.15.1 Bluetooth	2.4 GHz	3 Mb/s	10 m
	IEEE 802.15.3a (disbanded) UWB	3.1–10.6 GHz	480 Mb/s	
	IEEE 802.15.3c mmWave	60 GHz	5.78 Gb/s	
	IEEE 802.15.4 ZigBee	2.4 GHz (915/868 MHz)	250 kb/s	20 m
WLAN	IEEE 802.11n Wi-Fi	2.4/5 GHz	600 Mb/s	200 m
	IEEE 802.11ac Gigabit Wi-Fi	5 GHz	6.95 Gb/s	100 m
	IEEE 802.11ad Gigabit Wi-Fi	60 GHz	6.8 Gb/s	10 m
WMAN	IEEE 802.16	10–66 GHz	120 Mb/s	50 km
	IEEE 802.16e Mobile WiMAX	2.3/2.5–2.7/3.5 GHz	63 Mb/s	10 km
	IEEE 802.16m Advanced WiMAX	<6 GHz	100 Mb/s (mobile) 1 Gb/s (fixed)	100 km
WWAN	GSM (2G)	450–1900 MHz	19.2 kb/s	35 km
	GPRS (2.5G)		171.2 kb/s	
	EDGE (2.75G)		384 kb/s	
	UMTS (3G)	700–2600 MHz	2 Mb/s	20 km
	HSPA (3.5G)		14 Mb/s	
	HSPA+ (3.75G)		42.2 Mb/s	
	LTE		300 Mb/s	
	IEEE 802.20 MobileFi	<3.5 GHz	16 Mb/s	15 km
WRAN	IEEE 802.22 MBWA	54–862 MHz	31 Mb/s	100 km

IEEE 802.15.1 [18]. Two types of WPAN standards have been developed for advanced short-range wireless communications: IEEE 802.15.3 for high-rate WPAN (HR-WPAN) [19] and IEEE 802.15.4 for low-rate WPAN (LR-WPAN) [20]. The first one defines the protocols and their primitives for supporting high-rate multimedia applications with better QoS. On the other hand, LR-WPAN standards also define the protocols and their primitives for supporting low data rate communication over a short-range

transmission channel. In general, WPAN environment is featured with low-cost and very low power consumption nodes, ease of installation, simple protocol structure, and reliable transfer. Table 1.2 shows a general comparison of the main characteristics of HR-WPANs and LR-WPANs.

An HR-WPAN topology consists of several nodes that implement a full set or a subset of the standard. The HR-WPAN is also known as piconet. The formation of a piconet requires one node to assume the role of a piconet coordinator (PNC), which provides synchronization for the piconet nodes to the periodically transmitted PNC beacon frames, support QoS, and manage nodal power control and channel access control mechanisms. The piconet topology is created in an ad hoc manner, which requires no pre-planning, so nodes can join and leave the network unconditionally. There are three types of piconet topologies which are defined in the standard [21]:

- Independent piconet—standalone topology that consists of a single network coordinator and one or more network nodes. The network coordinator manages the network operation through the periodi-cally transmitted beacon frames in which other network nodes use to synchronize to be able to communicate with the network coor-dinator as well as performing P2P communication.
- Parent piconet—topology that controls the functionality of one or more other piconets. In addition to managing communication of its nodes, the PNC of the parent piconet controls the operation of one or more dependent network coordinators.
- Dependent piconet—classified according to the purpose of their creation into "child" piconet and "neighbor" piconet. A child piconet is created by a node from a parent piconet to extend net-work coverage or to provide computational and memory resources to the parent piconet. The resources allocation such as channel time allocation for network nodes in such a piconet is controlled by the parent network coordinator node.

TABLE 1.2
HR-WPAN and LR-WPAN Main Characteristics

	HR-WPAN	LR-WPAN
Data rate	11–55 Mb/s	20–250 kb/s
Transmission range	10–100 m	
Battery life	Moderate	Long
Topology	Mesh	Star and mesh
Traffic type	Multimedia	Data centric
Frequency band	Unlicensed (2.4 GHz)	Unlicensed (2.4 GHz, 868/915 MHz)

The IEEE 802.15.4 Task Group (TG) has defined star and P2P network topologies to suit different application requirements. These network topologies consist of different devices which are classified by the standard into full function device (FFD) and reduced function device (RFD). The FFD implements all features of the IEEE 802.15.4 standard and can serve as a regular device or as a network controller known as PAN coordinator. Also, the FFD can serve as a relaying node in a LR-WPAN mesh network topology or as an end node performing only specific functions. On the other hand, RFD implements the minimum required functionalities by the standard and can only function as an end node such as home automation sensor, which communicates its sensed information only to the PAN coordinator of the corresponding network. The main characteristics and the related applications of each network topology are summarized by Ali and Mouftah [21]:

- The star network topology is composed of a PAN coordinator and a number of FFD or RFD nodes. When a beacon-enabled MAC frame structure is used, network devices synchronize their communication with the PAN coordinator through the periodic transmission of the coordinator's beacon frames and they use slotted carrier sense multiple access with collision avoidance (CSMA/CA) for accessing the transmission channels. Also, the PAN coordinator allocates, upon nodal request, collision-free time slots for supporting mission-critical services. However, when a nonbeacon-enabled frame structure is used, there is no beacon frame transmission by the coordinator and the unslotted CSMA/CA is used instead for accessing the transmission channel. The PAN coordinator may have a specific application, but also can be the initiation or termination communication point as well as routing communication in the network. A variety of applications can benefit from the star network topology, such as health care, personal computer peripherals, building automation, logistics, etc.
- A P2P LR-WPAN topology is an ad hoc, self-organizing, and self-healing network topology. It is also formed around a PAN coordinator but differs from the star network topology in which any device may have a P2P communication with any other device in the network through a single hop or multiple hops. Complex network topologies can be created that facilitate the formation of mesh networking topology. Only nonbeacon-enabled MAC frames with unslotted CSMA/CA mechanisms are used in this topology. The targeted applications for P2P network topology are wireless sensor networks, industrial control, monitoring, etc.

As presented in Table 1.3, the development of short-range next generation wireless networks is focused on three high-perspective technologies [15]: ultra-wideband (UWB), 60 GHz millimeter-wave radio, and ZigBee.

UWB has its origin in the spark-gap transmission design in the late 1890s. In fact, the first wireless communication system was based on this technology. UWB refers to bandwidths in excess of 2 GHz, whose utilization by radar systems dates back to the late 1960s. Recent advances in semiconductor technology have made UWB technology ready for commercial applications [22]. The renewed and rapidly growing interest for UWB was triggered by the first report and order on UWB issued by the Federal Communications Commission (FCC) [23], and is well motivated by attractive features. UWB brings to commercial communications: low-power carrier-free transmissions, ample multipath diversity, enhanced penetration capability, low-complexity transceivers, ability to overlay existing systems, and a potential for increase in capacity. The allowed commercialized UWB devices in the United Sates provided momentum to worldwide regulation and standardization.

TABLE 1.3
Basic Characteristics of Perspective WPAN Technologies

System	UWB	60 GHz WPAN	ZigBee
IEEE standard	802.15.3a	802.15.3c	802.15.4
Frequency band	3.1–10.6 GHz	57–64 GHz (US)	2.4–2.4835 GHz
		59–66 GHz (Japan)	901–928 MHz
		57–66 GHz (Europe)	868–868.6 MHz
Channel bandwidth	≥500 MHz	2.16 GHz	2 MHz
			0.6 MHz
			0.3 MHz
Number of channels	2/14	4	16/10/1
Maximum data rate	100 Mb/s (10 m)	2 Gb/s (at least)	250 kb/s
	200 Mb/s (4 m)	≥3 Gb/s (optional)	40 kb/s
	480 Mb/s (optional)		20 kb/s
Modulation	DS-UWB	QPSK	BPSK
	MB-OFDM	16-QAM	OQPSK
Channel access	Hybrid multiple access (random and guaranteed access)		CSMA/CA (optional) Guaranteed time slot

According to the requirements of the FCC, the fractional bandwidth and the transmission bandwidth of UWB signals should be greater than 0.2 and 500 MHz, respectively. There are two competing modulation techniques:

- Direct spectrum UWB (DS-UWB), based on the transmission of very short pulses with relatively low energy, and
- Multiband orthogonal frequency division multiplexing (MB-OFDM) scheme, which divides the frequency spectrum to multiple nonoverlapping bands with OFDM transmission.

These two technologies are compared from different aspects such as channel, interference, performance, and complexity points of view by Nikookar and Prasad [24]. The further standardization of UWB–air interface is disbanded from IEEE 802.15.3a TG in 2006.

The *60 GHz millimeter-wave radio* can provide medium-range and short-range wireless communications with a variety of advantages. Huge and readily available spectrum allocation, dense deployment or high-frequency reuse, and small form factor can pave the way to multigigabit wireless networks [25]. A dedicated unlicensed frequency spectrum for this purpose was insufficient until the FCC declared that the 57 to 64 GHz band can be used. Japan, in turn, allocated the 59 to 66 GHz band. With the latest adoption of the European Telecommunications Standards Institute's (ETSI) 57 to 66 GHz band, there is now a common continuous 5 GHz band available around 60 GHz in most of the major markets. Another important breakthrough was the introduction of relatively cheap and power-efficient complementary metal oxide semiconductor (CMOS) processing for semiconductor production. As a result, the price and power requirements for consumer 60 GHz devices were met. For successful commercialization, the final need for developers was a standard that would support almost all usage models.

Within the IEEE, interest in developing a mmWave physical layer began in July 2003, with the formation of an interest group under the 802.15 Working Group (WG). In March 2004, a Study Group for mmWave PHY was formed, with the goal of developing a new PHY layer to transmit 1 Gb/s or higher data rates. It was decided to reuse an existing medium MAC layer (IEEE 802.15.3b), with the necessary modifications and extensions. After the approval of the project authorization request, a TG was created. The TG first focused on creating usage models, a 60 GHz indoor channel model, and evaluation criteria. After 2 years, three PHY modes and multiple MAC improvements were selected to support different usage models. In September 2009, after various ballots and resulting improvements, the IEEE 802.15.3c standard was approved [26].

During the beginning of the standardization process, the 802.15.3c TG conducted a detailed analysis of the possible consumer applications in the 60 GHz band, and five usage models were accepted [27]:

1. Uncompressed video streaming: The particularly large bandwidth available in the 60 GHz band enables sending HDTV signals, thus eliminating the need for video cables from HD video players to display devices. The 802.15.3c assumed 1920 × 1080 resolution and 24 b/pixel for video signals. Assuming a rate of 60 frames/s, the required data rate is found to be more than 3.5 Gb/s.

2. Uncompressed multivideo streaming: In some applications, multiple video signals can be supplied by a single transmitter (e.g., a TV and a DVD recorder receive signal from a single set-top box). The 802.15.3c system should be able to provide video signals for at least two 0.62 Gb/s streams. This data rate corresponds to a video signal with 720 × 480 pixels/frame.

3. Office desktop: It is assumed that a personal computer communicates with external peripherals (printers, display and storage devices, etc.), unidirectionally or bidirectionally. The model mainly assumes data communications, where retransmissions are possible.

4. Conference ad hoc: This model considers a scenario in which many computers are communicating with one another using the 802.15.3 network. Most communications are bidirectional, asynchronous, and packet-based. The conference ad hoc usage model requires longer ranges than the office desktop model for improved QoS.

5. Kiosk file downloading: In the last model, the TG assumed electronic kiosks that enable wireless data uploads and downloads with their fixed antennas. Users will operate handheld devices, such as cell phones and cameras with low-complexity, low-power transceivers. This model requires 1.5 Gb/s data rate at 1 m range.

Due to conflicting requirements of different usage models, three different PHY modes have been developed: single carrier mode, high-speed interface mode, and audio/visual mode (Table 1.4). Channel model, PHY layer design, and MAC enhancements were analyzed in detail by Baykas et al. [28].

Although wireless control and sensor networking are at the other end of high-speed networks with QoS support, these low-power consumption and low-data rate WPANs are driven by applications in the building automation, medical, and logistics fields. This low-throughput and low-cost wireless technology, named *ZigBee*, has adopted the IEEE 802.15.4 standard.

TABLE 1.4
Comparison of mmWave PHY Modes

PHY Mode	Single Carrier	High-Speed Interface	Audio/Visual
Maximum data rates	5.28 Gb/s	5.78 Gb/s	3.8 Gb/s
Modulation scheme	Single carrier	OFDM	OFDM
Forward error control coding options	Reed Solomon code, low-density parity check codes	Low-density parity check codes	Reed Solomon code, convolutional coding
Block size/fast Fourier transform size		512	
Main usage	File downloading	Conference ad hoc	Multimedia streaming

The ZigBee specification, defined by ZigBee Alliance [29], provides upper layer stacks and application profiles that are compatible with the IEEE 802.15.4 PHY and MAC layers [20]. The main ZigBee characteristics are presented in Table 1.3, whereas elaboration of complete protocol stack is provided by Baronti et al. [30].

1.2.2 Wireless Local Area Networks

WLANs have become an ubiquitous networking technology. The IEEE 802.11 WLAN is being deployed widely and rapidly for many different environments including enterprise, home, and public access networking. Low infrastructure cost, ease of deployment, and support for nomadic communication are among the strengths of WLANs.

In a broadcast network such as a WLAN, the MAC sublayer is responsible for arbitrating multiple stations accessing a shared transmission medium [31]. The IEEE 802.11 PHY layer specification concentrates on radio transmission and performs secondary functions such as accessing the state of the wireless medium and repeating it back to the MAC sublayer. The legacy standard includes a primitive MAC architecture and three basic over-the-air communication technologies with maximum raw data rates of a few Mb/s megabits per second. Because of their fairly low data bandwidth, further amendments have been proposed for many years: IEEE 802.11a [32], IEEE 802.11b [33], and IEEE 802.11g [34]. Both IEEE 802.11a and IEEE 802.11b were finalized in 1999 and support data rates of up to 54 Mb/s and 11 Mb/s, respectively. In June 2003, the third PHY specification IEEE 802.11g was introduced with the same maximum data rate as IEEE 802.11a, but operating in separate frequency bands. For this period, there were many amendments and research work done for improved PHY specifications that mostly aimed to provide reliable connections and higher data rates.

Furthermore, the IEEE 802.11 WG is pursuing IEEE 802.11n [35] as an amendment for higher throughput and higher speed enhancements. Different from the goal of IEEE 802.11b/a/g, that is, to provide higher speed data rates with different PHY specifications, IEEE 802.11n aims at higher throughput instead of higher data rates with PHY and MAC enhancements. This standard seeks to improve the peak throughput to at least 100 Mb/s, measured at the MAC data service access point (SAP). This represents an improvement of at least four times the throughput obtainable using existing WLAN systems. The highest throughput mode shall achieve a spectral efficiency of at least 3 b/s/Hz. Primary steps undertaken by IEEE 802.11n TG include

- Identification and definition of usage models, channel models, and related MAC and application assumptions, and
- Identification and definition of evaluation metrics that characterize the important aspects of a particular usage model.

Initial usage models include hotspot, enterprise, and residential. Evaluation metrics include throughput, range, aggregate network capacity, power consumption (peak and average), spectral flexibility, cost/complexity flexibility, backward compatibility, and coexistence [36].

Two basic concepts are employed in IEEE 802.11n to increase the PHY data rates: MIMO and 40 MHz bandwidth channels. Increasing from a single spatial stream and one transmit antenna, to four spatial streams and four antennas, increases the data rate by a factor of four. The term "spatial stream" is defined by the IEEE 802.11n standard [35] as one of several bitstreams that are transmitted over multiple spatial dimensions created by the use of multiple antennas at both ends of a communications link. However, due to the inherent increased cost associated with increasing the number of antennas, modes that use three and four spatial streams are optional, and to allow for handheld devices, the two spatial stream mode is only mandatory in a SAP. As shown in Figure 1.1, the 40 MHz bandwidth channel operation is optional in the standard regarding interoperability between 20 and 40 MHz bandwidth devices, the permissibility of the use of 40 MHz bandwidth channels in the various regulatory domains, and spectral efficiency. However, the 40 MHz bandwidth channel mode has become a core feature due to the low cost of doubling the data rate compared with doubling the bandwidth.

Other minor modifications were also made to the 802.11a/g waveform to increase the data rate. The highest encoder rate in previous WLAN

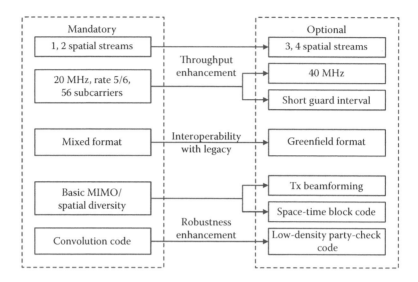

FIGURE 1.1   Mandatory and optional IEEE 802.11n PHY features.

standards was 3/4. This was increased to 5/6 in IEEE 802.11n for an 11% increase in data rate. With the improvement in radiofrequency technology, it was demonstrated that two extra frequency subcarriers could be squeezed into the guard band on each side of the spectral waveform and still meet the transmit spectral mask. This increased the data rate by 8% compared with IEEE 802.11a/g. The waveform from old standards and from mandatory operations in IEEE 802.11n contains an 800 ns guard interval between each OFDM symbol. An optional mode was defined with a 400 ns guard interval between each OFDM symbol to increase the data rates by another 11% [37].

Another functional requirement for new WLAN technology was interoperability between IEEE 802.11 a/g and n standards. The TG decided to meet this requirement in the PHY layer by defining a waveform that was backward-compatible with IEEE 802.11a and OFDM modes of IEEE 802.11g. The preamble of the 802.11n mixed format waveform begins with the preamble of the 802.11a/g waveform. This allows 802.11a/g devices to detect the 802.11n mixed format packet and decode the signal field. Even though the 802.11a/g devices are not able to decode the remainder of the 802.11n packet, they are able to properly defer their own transmission based on the length specified in the signal field.

MIMO training and backward compatibility increases the overhead, which reduces the efficiency. In an environment free from legacy devices (greenfield) backward compatibility is not required; therefore, IEEE 802.11n includes an optional greenfield format. By eliminating the components of the preamble that supports backward compatibility, the greenfield format preamble is shorter than the mixed format preamble. This difference in efficiency becomes more pronounced when the packet length is short, as in the case of voice over Internet protocol (VoIP) traffic. Therefore, the use of the greenfield format is permitted even in the presence of legacy devices with proper MAC protection, although the overhead of the MAC protection may reduce the efficiency gained from the PHY [37]. The comprehensive overview that describes the underlying principles, implementation details, and key enhancing features of IEEE 802.11n can be found in the study by Perahia and Stacey [38].

To overcome the limitations of single hop communication, data packets need to traverse over multiple wireless hops, and wireless mesh networks (WMNs) are called for. WMNs, as low-cost and reliable systems for rapid network deployment, have attracted considerable attention from academia and industry. The IEEE 802.11s [39] standard defines a WLAN mesh using the IEEE 802.11 MAC and PHY layers, and is one of the most active standards with increasing commercial opportunities.

IEEE 802.11s defines an IEEE 802.11–based WMN that supports broadcast and unicast delivery over a self-configured multihop link-layer topology. Figure 1.2 shows an IEEE 802.11s–based mesh network, which typically contains three types of nodes: the mesh point (MP), mesh access point (MAP), and mesh portal point (MPP). The MP is the basic mesh unit that provides topology construction, routing, and traffic forwarding. This type of node can also be designed as a terminal device for end users

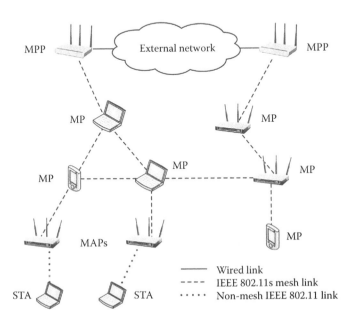

FIGURE 1.2   IEEE 802.11s mesh network.

to directly connect with peer MPs and access the mesh. Non-mesh IEEE 802.11 stations (STAs) must first associate with a MAP, which is an MP capable of IEEE 802.11 access point (AP) functions, before accessing a mesh. An MPP is an MP with the additional functionality of acting as a gateway between the mesh cloud and external networks.

The IEEE 802.11s standard specifies the boot sequence procedure for an MP joining a mesh network based on the procedure for an STA associating with an AP in a conventional non-mesh WLAN. MP performs an active or passive scan to obtain a list of existing MPs in each channel. After that, the MP uses the mesh peer link management protocol to associate with an MP matching its own preferences. A mesh peer link is univocally identified by the MAC addresses of both participants and a pair of link identifiers, generated by each of the MPs to minimize reuse in short time intervals [40].

IEEE 802.11s introduces the unified channel graph (UCG), which presents one mesh sharing the same preferences in the same channel to handle several meshes spanning different channels. When forming a new mesh, the initial MP randomly decides a channel precedence value and embeds that value in the management frames. After the channel scan, other MPs select the channel with the highest precedence as their operating channels. This procedure forms a UCG called the simple channel unification protocol. To resolve multiple UCGs in different channels due to spatial division and the needs of channel switching, IEEE 802.11s proposes a channel graph switch protocol. In this protocol, an MP sets a waiting timer and broadcasts a mesh channel switch announcement [41].

Also, IEEE 802.11s introduces a mesh header subfield in the beginning of the frame body to address multihop transmissions. When conveying packets whose source and destination are both inside the mesh, the subfield indicates that the four-address format in the frame header is used. The frame header includes the MP addresses of the next-hop receiver, transmitter, destination, and source, and is processed by MPs as it would be in a wireless distribution system (WDS).

Although current IEEE 802.11n devices have reached data rates of up to 600 Mb/s, the IEEE started working on an amendment to achieve aggregate throughputs beyond 1 Gb/s in the 5 GHz band. This is the first time that an IEEE 802.11 amendment is targeting to improve the overall network throughput rather than only improving the throughput of a single link. In the IEEE 802.11ac draft standard [42], throughput is increased by the following mechanisms:

- Multiuser MIMO (up to eight spatial streams divided across up to four different clients)
- Larger channel bandwidths of 80 MHz (two adjacent 40 MHz channels with some extra subcarriers) and 160 MHz (two separate

80 MHz channels without any tone filling in the middle of these two subchannels)
- 256-quadrature amplitude modulation (QAM)

Although IEEE 802.11ac will be largely an evolution standard, it is based on IEEE 802.11n in some parts such as in channel coding and MIMO mode. The cumulative benefit of IEEE 802.11ac features will enable Wi-Fi solutions to meet today's demand for high-capacity and high-quality mobile real-time applications like video and voice.

On the other hand, the IEEE 802.11ad standard [43] defines modifications to both the PHY layer and the MAC layer, to enable operation in frequencies of approximately 60 GHz and to be capable of very high throughput. As already mentioned, the wide harmonized spectrum in the unlicensed millimeter-wave band is considered as the most prominent candidate to support the evolution toward multi-Gb/s data rates. IEEE 802.11ad enhances the PHY layer by using the modulation coding scheme, which combines the benefits of both OFDM and single-carrier techniques. Also, some enhancements are proposed for the MAC layer with the aim of achieving very high throughput delivery and to support directionality of antenna layer (e.g., bidirectional aggregation frame with aggregated acknowledgement, directional associations, and beamforming). It is significant to note that IEEE 802.11ad may provide enhanced coexistence among homogeneous (legacy 802.11 and 802.11ad) systems, as well as between heterogeneous systems (e.g., IEEE 802.15.3c) that is working at the 60 GHz band.

The data rates offered by IEEE 802.11ac and IEEE 802.11ad can meet the needs of many applications, with the replacement of wired digital interface cables arguably being the most prominent new use of these technologies. Several companies have already announced products implementing these standards, with a few of those products already available, or will soon be available to consumers [44].

1.2.3 WIRELESS METROPOLITAN AREA NETWORKS

Despite the challenges faced when transmitting data through varying wireless channels, broadband WMAN systems are becoming a reality, partly thanks to the increasingly sophisticated designs that are being employed. Although not as widely deployed as WLANs, WMANs are expected to be deployed with increasing numbers in the next few years. Companies developing products for WMAN networks have formed a forum named Worldwide Interoperability for Microwave Access (WiMAX). Similar to the Wireless Fidelity (Wi-Fi) Alliance related to WLANs, the WiMAX forum aims at overcoming interoperability problems between WMAN devices from different companies.

First published in 2001, the IEEE 802.16 standard [45], as a solution for broadband wireless access (BWA), specified a frequency range of 10 to 66 GHz, with a theoretical maximum data rate of 120 Mb/s and maximum transmission range of 50 km. The standard specifies the air interface, including the MAC and PHY layers, of BWA. The key development in the PHY layer includes OFDM, in which multiple access is achieved by assigning a subset of subcarriers to each individual user. This resembles code-division multiple access (CDMA) spread spectrum in that it can provide different QoS for each user. Users achieve different data rates by assigning different code spreading factors or different numbers of spreading codes. In an OFDM system, the data is divided into multiple parallel substreams at a reduced data rate, and each is modulated and transmitted on a separate orthogonal subcarrier. This increases symbol duration and improves system robustness. OFDM is achieved by providing multiplexing on users' data streams on both uplink and downlink transmissions [46].

However, the initial standard only supports line-of-sight (LoS) transmission and thus does not seem to favor deployment in urban areas. A variant of the standard, IEEE 802.16a-2003, approved in 2003, can support non-LoS transmission and adopts OFDM at the PHY layer. It also adds support for the 2 to 11 GHz band. From the initial variants, the IEEE 802.16 standard has undergone several amendments and evolved to the IEEE 802.16-2004 standard (also known as IEEE 802.16d). Lack of mobility support seems to be one of the major hindrances to its deployment compared with other standards, such as IEEE 802.11, because mobility support is widely considered as one of the key features in wireless networks. It was natural that the IEEE 802.16e, legacy standard for Mobile WiMAX released in 2007, has added mobility support. Mobile WiMAX adds significant enhancements [46]:

- It improves non-LoS coverage by utilizing advanced antenna diversity schemes and hybrid automatic repeat request (HARQ)
- It adopts dense subchannelization, thus increasing system gain and improving indoor penetration. It uses adaptive antenna system and MIMO technologies to improve coverage
- It introduces a downlink subchannelization scheme, enabling better coverage and capacity trade-off.

This brings potential benefits in terms of coverage, power consumption, self-installation, frequency reuse, and bandwidth efficiency.

QoS provisioning is one of the most important issues given the inherent QoS specification in the WiMAX MAC layer definition. It is anticipated that IEEE 802.16 is fully capable of supporting multimedia transmissions with differentiated QoS requirements through the use of scheduling mechanisms. With the rising popularity of multimedia applications in the Internet, IEEE 802.16 provides the capability to offer new services such as VoIP, multimedia streaming and conferencing, real-time surveillance, and others. Due to its long-range and high-data rates transmission, WiMAX has also been considered in areas where it can serve as the backbone network with long separation among the infrastructure nodes. Another promising area is replacing cellular technology using VoIP over WiMAX if QoS requirements can be satisfied [47].

Advanced WiMAX, also known as next generation mobile WiMAX, which will provide up to 1 Gb/s peak throughput with the IEEE 802.16m [48] update in 2011, is one of the technologies for the International Mobile Telecommunications (IMT)–Advanced Program [49] for the fourth generation (4G) mobile systems. IEEE 802.16m defines the WMAN advanced air interface as an amendment to the ratified IEEE 802.16-2009 specification with the purpose of enhancing performance such that IMT-Advanced requirements are fulfilled. A new amendment provides enhancements including higher throughput, mobility, and user capacity, as well as lower latency while maintaining full backward compatibility with the existing mobile WiMAX systems (IEEE 802.16e). A detailed overview of the prominent features and structure of the IEEE 802.16m PHY layer and simulation results of spectral efficiency performance are provided by Cho et al. [50], whereas most important features and system requirements of mobile WiMAX are given in Table 1.5.

The PHY layer structure refers to the design of multiplexing user and system data with control signaling to ensure proper use of radio resources. A frame structure first partitions the time domain resources into frames within which the time–frequency–space resources are further organized according to a more refined resource structure. OFDM access and MIMO, as two keystones of 4G PHY layer technologies, allow the minimal resource unit to be one subcarrier for a duration of one OFDM symbol in each spatial layer. PHY layer design specifies how to group, map, and allocate time–frequency–space resources as either reference signals or to form various physical channels.

In contrast with contention-based WLAN networks, where each station needs to compete for access to the medium, the IEEE 802.16 standard is related to the connection-oriented wireless networks. A basic understanding of connection-oriented and WiMAX network architectures is assumed in MAC, which is divided into three sublayers: convergence sublayer, radio resource control and management sublayer, and MAC sublayer. The general architecture of the IEEE 802.16m MAC layer is shown in Figure 1.3. The purpose of the sublayers is to encapsulate wireline technologies on the air interface, and to introduce the state-of-the-art connection-oriented features [51].

TABLE 1.5

Most Important Features and System Requirements of Mobile WiMAX Standards

Requirement	IEEE 802.16e	IEEE 802.16m
Data rates	63 Mb/s	100 Mb/s (mobile)
		1 Gb/s (fixed)
Frequency band	2.3 GHz	<6 GHz
	2.5–2.7 GHz	
	3.5 GHz	
MIMO support	Up to four streams	Four or eight streams
	No limit on antennas	No limit on antennas
Coverage	10 km	<100 km
Handover interfrequency interruption time	35–50 ms	30 ms
Handover intrafrequency interruption time	Not specified	100 ms
Handover between 802.16 standards	From 802.16e serving BS to 802.16e target BS	Overall
Handover with other technologies	Not specified	IEEE 802.11, 3GPP2, GSM/EDGE, LTE using IEEE 802.21
Mobility speed	120 km/h	350 km/h
IDLE to ACTIVE state transition	390 ms	50 ms

The explosive increase in demand for wireless data traffic has created opportunities for new network architectures incorporating multitier base stations (BSs) with diverse sizes. Support for small-sized low-power BSs such as femtocell BSs is gaining momentum in mobile systems. Because of their potential advantages such as low cost deployments, traffic offloading from macrocells, and the capability to deliver services to mobile stations which require large amounts of data, femtocells are presenting perspective concept for NGWS, such as advanced WiMAX [52,53].

In advanced WiMAX, a femtocell BS or WiMAX femtocell access point (WFAP), is a low-power BS intended to provide in-house or small office/home office (SOHO) coverage. With traditional macrocells or microcells, indoor coverage is challenging and expensive for deployment due to the high penetration losses of most buildings. WFAPs are usually self-deployed by end-users inside their premises, and connected to the radio access network via available wired broadband connections.

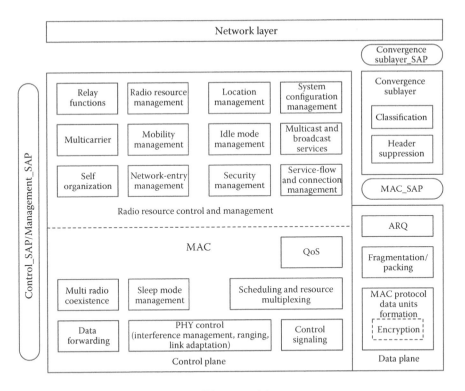

FIGURE 1.3   IEEE 802.16m MAC layer architecture.

Figure 1.4 represents the general WiMAX network architecture with additional support for femtocells. Here, the control/management planes' interfaces are presented as solid lines, whereas dashed lines are used for data plane interfaces. A WiMAX operator may act as a network service provider (NSP) and a network access provider (NAP). NAP implements the infrastructure using one or more access service nodes (ASNs). An ASN is composed of one or more gateways and one or more BSs to provide mobile Internet services to subscribers. The ASN gateway serves as the portal to an ASN by aggregating BS control plane and data plane traffic to be transferred to a connectivity service network (CSN). An ASN may be shared by more than one CSN. A CSN may be deployed as part of a WiMAX NSP. A CSN may comprise the authentication, authorization, and accounting (AAA) server and the home agent (HA) to provide a set of network functions (e.g., roaming, mobility, and user's profiles).

Femto NAP, femto NSP, as well as self-organizing network functionalities are included in the general WiMAX architecture for femtocell support [53]. A femto NAP implements the infrastructure using one or more femto ASNs to provide short-range radio access services. A femto ASN is mainly

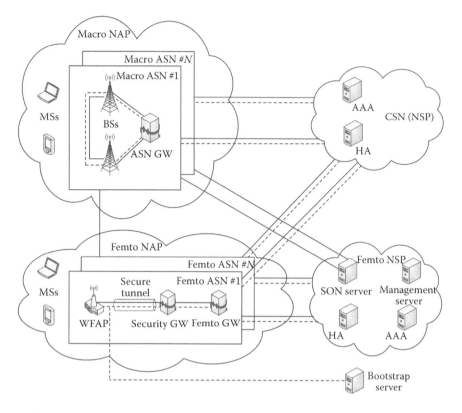

FIGURE 1.4   General WiMAX network architecture.

differentiated from a macro ASN in that the WFAP backhaul is transported over the public Internet. Therefore, a security gateway (GW) is needed for WFAP authentication. When a WFAP is booted, it first communicates with the bootstrap server to download the initial configuration information, including the address of the security GW. The WFAP and security GW authenticate each other, and create an Internet protocol security (IPSec) tunnel. The security GW then communicates with the AAA server in the femto NSP to verify whether the WFAP is authorized. The femto GW acts as the portal to a femto ASN that transfers both control and bearer data between MS and CSN, and control data between WFAP and femto NSP.

The femto NSP manages and controls entities in the femto ASN. The management server implements management plane protocols and procedures to provide operation, administration, maintenance, and provisioning functions to entities in the femto ASN. WFAP management includes fault, configuration, performance, and security management.

For proper integration into the operator's RAN, the WFAP enters the initialization state before becoming operational. In this state, it performs procedures such as attachment to the operators' network, configuration of radio interface parameters, time/frequency synchronization, and network topology acquisition. In the operational state, normal and low duty operational modes are supported. In low duty mode, the WFAP reduces radio interface activity to reduce energy consumption and interference to neighboring cells.

1.2.4 WIRELESS WIDE AREA NETWORKS

The success of second generation (2G) mobile systems (e.g., global system for mobile communications; GSM), along with the IP support provided by 2.5G technologies, such as the general packet radio service (GPRS), paved the way toward evolved WWANs. In this sense, technologies like enhanced data rates for GSM evolution (EDGE) provided higher data rates using inherited 2G network infrastructure and frequency spectrum. Confirming the rule that every 10 years, a new advanced generation is developing, in parallel with third generation (3G) systems implementation, the fourth generation (4G) systems development started, too. Some technology solutions seek to be their representatives while there exist some expectations in concepts concerning the fifth generation (5G) [54,55]. Nowadays, 5G is not a term officially used for any particular specification, or in any official document yet made public by telecommunication companies or standardization bodies such as 3GPP, IEEE, etc.

With the target of creating a collaboration entity among different telecommunications associations, the 3GPP was established in 1998. It started working on the radio, core network, and service architecture of a globally applicable 3G technology specification. Even though 3G data rates were already real in theory, initial systems like universal mobile telecommunications system (UMTS) did not immediately meet the IMT-2000 requirements in their practical deployments. Hence, the standards needed to be improved to meet or even exceed them. The combination of high-speed downlink packet access (HSDPA) and the subsequent addition of an enhanced dedicated channel, also known as high-speed uplink packet access (HSUPA), led to the development of the technology referred to as high-speed packet access (HSPA) or, more informally, 3.5G [56].

Motivated by the increasing demand for network services, such as web browsing, IP telephony, and video streaming, with constraints on QoS requirements, 3GPP [57] started working on long-term evolution (LTE). LTE allows cellular operators to use new and wider spectrum and complements 3G networks with higher user data rates, lower latency, and flat IP-based network architecture. The very first commercial LTE networks were deployed on a limited scale in Scandinavia at the end of 2009, and currently, large-scale deployments are taking place in several regions, including North America, Europe, and Asia [58]. According to the latest

Global mobile Suppliers Association (GSA) Evolution to LTE Report, 338 operators in 101 countries have committed to commercial LTE network deployments or are engaged in trials, technology testing, or studies [59].

3GPP Release 8 is the first LTE release, which was finalized in 2008. Release 8 provides downlink and uplink peak rates up to 300 and 75 Mb/s, respectively, a one-way radio network delay of less than 5 ms, and a significant increase in spectrum efficiency. LTE provides extensive support for spectrum flexibility, supports both frequency-division duplex (FDD) and time-division duplex (TDD), and targets a smooth evolution from earlier 3GPP technologies such as time-division synchronous code-division multiple access (TD-SCDMA) and wideband CDMA (WCDMA)/HSPA as well as 3GPP2 technologies such as cdma2000. LTE technology is based on conventional OFDM waveform for downlink and single-carrier frequency-division multiplexing (SC-FDM) waveform for uplink communications mainly to improve the user experience for broadband data communications. SC-FDM is a modified form of OFDM, in which time-domain data symbols are transformed to frequency-domain by a discrete Fourier transform (DFT) before going through the standard OFDM modulation. The use of OFDM on the downlink combined with SC-FDM on the uplink thus minimizes terminal complexity on the receiver side as well as on the transmitter side, leading to an overall reduction in terminal complexity and power consumption.

The transmitted signal is organized into subframes of 1 ms duration with 10 subframes forming a LTE radio frame (Figure 1.5). Each downlink subframe consists of a control region of one to three OFDM symbols, used for control signaling from the BS to the terminals, and a data region comprising the remaining part and used for data transmission to the terminals. The transmissions in each subframe are dynamically scheduled by the BS. Cell-specific reference signals are also transmitted in each downlink subframe. These reference signals are used for data demodulation at the terminal, and for measurement purposes (e.g., for channel status reports sent from the terminals to the BS) [60].

Guard periods are created by splitting one or two subframes (special subframes), in each radio frame into three fields [61]:

- *Downlink part*—ordinary, albeit shorter, downlink subframe used for downlink data transmission
- *Uplink part*—one or two OFDM symbols duration, used for transmission of uplink sounding-reference signals and random access, and
- *Guard period*—the remaining symbols in the special subframe, which have not been allocated to downlink or uplink part

Support for multiantenna transmission is an integral part of LTE from the first release. Downlink multiantenna schemes supported by LTE include

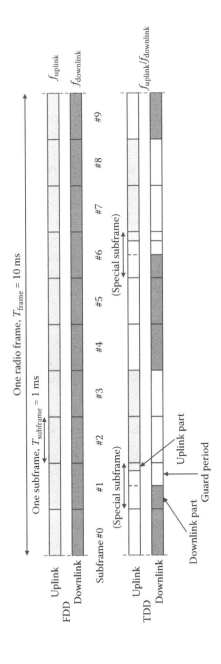

FIGURE 1.5 LTE frame structure.

transmit diversity, spatial multiplexing (including both single-user MIMO as well as multiuser MIMO), with up to four antennas, and beamforming.

Anticipating the invitation from the International Telecommunication Union (ITU), by March 2008, 3GPP had already initiated a study item on LTE-Advanced, with the task of defining requirements and investigating potential technology components for the LTE evolution. This study item, completed in March 2010 and forming the basis for the Release 10 work, aimed beyond IMT-Advanced. In 2010, 3GPP submitted LTE Release 10 to the ITU and, based on this submission, ITU approved LTE as one of two IMT-Advanced technologies. As it can be seen from Table 1.6, LTE-Advanced will not only fulfill the IMT-Advanced requirements but in many cases even surpass them.

3GPP Release 10, also known as LTE-Advanced, is not a new radio access technology but is the evolution of LTE for further improved performance. Release 10 includes all the features of Release 8/9 and adds several new features, such as carrier aggregation, enhanced multiantenna support, improved support for heterogeneous deployments, and relaying [60]. Evolving LTE rather than designing a new radio access technology is important from an operator's perspective as it allows for the smooth introduction of new technologies without jeopardizing existing investments. A Release 10 terminal is back-compatible with networks from earlier releases, and a Release 8/9 terminal can connect to a network supporting the new enhancements. Hence, an operator can deploy a Release 8 network, and later, when the need arises, upgrade to Release 10 functionality where needed. In fact, most of the Release 10 features can be implemented into the network as simple software upgrades.

The LTE system is based on a flat architecture, known as the service architecture evolution (SAE), with guarantees for seamless mobility support and

TABLE 1.6
ITU Requirements and LTE Fulfillment

	IMT-Advanced	LTE Release 8	LTE Release 10
Peak data rate	1 Gb/s	300 Mb/s	1 Gb/s
Transmission bandwidth	>40 MHz	<20 MHz	<100 MHz
Peak Spectral Efficiency			
Uplink	15 b/s/Hz	16 b/s/Hz	30 b/s/Hz
Downlink	6.75 b/s/Hz	4 b/s/Hz	16.1 b/s/Hz
Latency			
Control plane	<100 ms	50 ms	50 ms
User plane	<10 ms	4.9 ms	4.9 ms

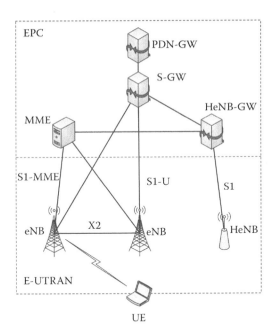

UE

FIGURE 1.6 LTE network architecture.

high-speed delivery for data and signaling. As presented in Figure 1.6, SAE consists of an all-IP–based core network, known as the evolved packet core (EPC) and the evolved-universal terrestrial radio access network (E-UTRAN).

The main part in the E-UTRAN segment is the enhanced Node B (eNB), which provides an air interface with the user plane and control plane protocol terminations toward the user equipment (UE). Each of the eNBs is a logical component that serves one or several E-UTRAN cells, and the interface interconnecting the eNBs over X2 interface. Additionally, home eNBs (HeNBs, also called femtocells), which are eNBs of lower cost for indoor coverage improvement, can be connected to the EPC directly or via a gateway that provides additional support for a large number of HeNBs. The main components of the EPC are

- Mobility management entity (MME) is in charge of managing security functions, handling idle state mobility, roaming, and handovers. Also, selecting the serving gateway (S-GW) and packet data network gateway (PDN-GW) nodes is part of its tasks. The S1-MME interface connects the EPC with the eNBs.
- S-GW is the edge node of EPC, and it is connected to the E-UTRAN via the S1-U interface. Each UE is associated to a unique S-GW, while hosting several functions. It is the mobility anchor point for both local inter-eNB handover and inter-3GPP

mobility, and it performs interoperator charging as well as packet routing and forwarding.
- PDN-GW provides interconnectivity among LTE UEs and external all-IP–based networks.

In LTE-Advanced systems, the transmission bandwidth can be further extended up to 100 MHz, with the potential of achieving more than 1 Gb/s throughput for downlink and 500 Mb/s for uplink, through the support of a so-called *carrier aggregation* concept [62]. According to this solution, multiple component carriers are aggregated and jointly used for transmission to/from a single mobile terminal, as illustrated in Figure 1.7. In addition, it can be used to effectively support different component carrier types that may be deployed in heterogeneous environments. Carrier aggregation is attractive because it allows operators to deploy a system with extended bandwidth by aggregating several smaller component carriers while providing backward compatibility to legacy users.

Up to five component carriers, possibly each of different bandwidth, can be aggregated, whereas backward compatibility is provided, as each component carrier uses the Release 8 structure. Hence, to a Release 8/9 terminal, each component carrier will appear as a Release 8 carrier, whereas a carrier aggregation–capable terminal can exploit the total's aggregated bandwidth enabling higher data rates. In general, different numbers of

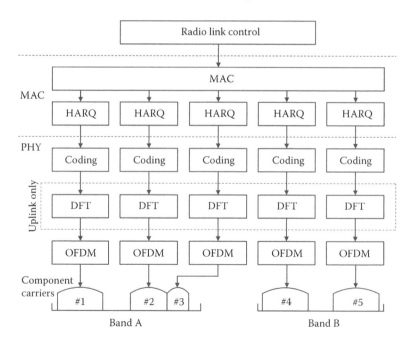

FIGURE 1.7   Carrier aggregation in LTE-Advanced.

component carriers can be aggregated for the downlink and uplink. Three different cases can be identified with respect to the frequency location of the different component carriers:

- Intraband aggregation with noncontiguous carriers, e.g., #1 and #2 (#4 and #5)
- Intraband aggregation with contiguous carriers, e.g., aggregation of component carriers #2 and #3, and
- Interband aggregation, e.g., aggregation of component carriers #1 and #4

The possibility of aggregating nonadjacent component carriers enables the exploitation of fragmented spectrums. Operators with a fragmented spectrum can provide high–data rate services based on the availability of wide overall bandwidth even though they do not possess a single wideband spectrum allocation. Although the exploitation of fragmented spectrum and expansion of the total bandwidth beyond 20 MHz are two important uses of carrier aggregation, there are also scenarios in which carrier aggregation within 20 MHz of the contiguous spectrum is useful, for example, in heterogeneous environment [60].

It can be seen that scheduling HARQ retransmissions are handled independently for each component carrier. As a baseline, control signaling is transmitted on the same component carrier as the corresponding data. As a complement, it is possible to use so-called cross-carrier scheduling, where the scheduling decision is transmitted to the terminal on another component carrier. To reduce the power consumption, a carrier aggregation–capable terminal receives on one component carrier only, the primary component carrier. Reception of additional secondary component carriers can be rapidly turned on/off in the terminal by the BS through MAC signaling, which is in close relation with the radio link control and HARQ. Similarly, in the uplink, all the feedback signaling is transmitted on the primary component carrier, and secondary component carriers are only enabled when necessary for data transmission.

The IEEE 802.20 mobile broadband wireless access (MBWA) standard is the new high-mobility standard developed by the 802.20 WG [63] to enable high-speed, reliable, and cost-effective mobile wireless connectivity. Although this standard, considering the cell size, could be perhaps represented in a WMAN environment, it undoubtedly belongs to the WWAN group of standards, taking into account his application and technical features. This standard, also known as MobileFi, is optimized to provide IP-based BWA in a mobile environment and it can operate in a wide range of deployments, thereby affording network operators with superior flexibility in optimizing their networks. It is targeted for use in a wide variety

TABLE 1.7
General Characteristics of IEEE 802.20

Frequency bands	<3.5 GHz	
Bandwidth	1.25 MHz	5 MHz
Spectral efficiency	>1 b/s/Hz/cell	
Frequency arrangements	FDD and TDD	
Peak aggregate data rate (downlink)	>4 Mb/s	>16 Mb/s
Peak aggregate data rate (uplink)	>0.8 Mb/s	>3.2 Mb/s
Peak user data rate (downlink)	>1 Mb/s	>4 Mb/s
Peak user data rate (uplink)	>0.3 Mb/s	>1.2 Mb/s
Number of active users per sector/cell	>100	
Mobility	Full mobility up to 250 km/h	
Intersector/cell handover time	<200 ms	

of licensed frequency bands and regulatory environments. General characteristics of IEEE 802.20 are presented in Table 1.7, whereas detailed overviews of the draft and final standard specifications can be found in articles by Bolton et al. [64], Bakmaz et al. [65], and Greenspan et al. [66].

IEEE 802.20 specifies two modes of operation, a wideband mode and a 625k-MC mode, utilizing distinct and optimized MAC and PHY layers [66]. TDD is supported by both the 625k-MC mode and the OFDM wideband mode, whereas FDD is supported by the OFDM wideband mode. Both modes are designed to support a full range of QoS attributes, making this technology suitable to support real-time streaming service that has low delay and jitter requirements, as well as near-real-time data services, where low error rate can be traded off for delay. According to the proposed characteristics, the IEEE 802.20 standard is suitable to meet user requirements for mobile access and is competitive to similar technologies (e.g., 3G and Mobile WiMAX), but currently, corresponding WG is in hibernation status due to lack of activity.

1.2.5 Wireless Regional Area Networks

In 2004, the FCC indicated that the unutilized TV channels in both very high frequency (VHF) and ultrahigh frequency (UHF) bands can be used for fixed broadband access [67]. From then, there has been overwhelming interest from the research community to develop a standard for WRAN systems operating on TV white spaces. IEEE 802.22 [68] is the first international standard based on cognitive technology, which aims at providing broadband access in a large coverage area by effectively utilizing the unused TV channels.

WRAN cells consists of a BS and the associated customer premise equipment (CPE), which communicates with the BS via a fixed point-to-multipoint radio air interface as shown in Figure 1.8. Apart from coexisting with digital TV services, WRAN cells also must be aware of FCC Part 74 devices (e.g., wireless microphones) and other licensed devices in the TV bands. It is envisioned that frequency availability for data transmission of a WRAN cell is determined by referring to an up-to-date incumbent database, augmented by distributed spectrum sensing, performed continuously both by the BS and the CPE [69].

The IEEE 802.22 standard will provide wireless broadband access to a rural area of typically 30 km or more (up to 100 km) in radius from a BS and serving up to 255 fixed units of CPE with outdoor directional antennas located at nominally 10 m above ground level, similar to a typical VHF/ UHF TV receiving installation.

Due to the extended coverage afforded by the use of these lower frequencies, the PHY layer must be optimized to absorb longer multipath excess delays rather than being accommodated by other IEEE wireless standards. An excess delay of up to 37 μs can be absorbed by the OFDM modulation [70]. Beyond the 30 km for which the PHY layer has been specified, the MAC layer will absorb additional propagation delays for covering distances of up to 100 km through intelligent scheduling to cover cases in which advantageous topography allows such coverage.

The reference architecture for IEEE 802.22 systems, shown in Figure 1.9, addresses the PHY and MAC layers, and the interfaces to a station management entity (SME) through PHY and MAC layer management

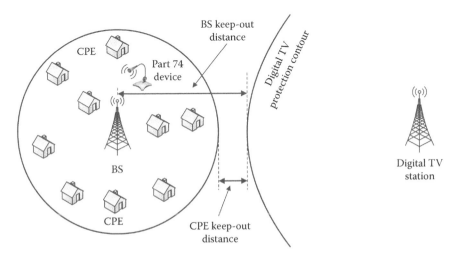

FIGURE 1.8   WRAN cell coexisting with digital TV and Part 74 devices.

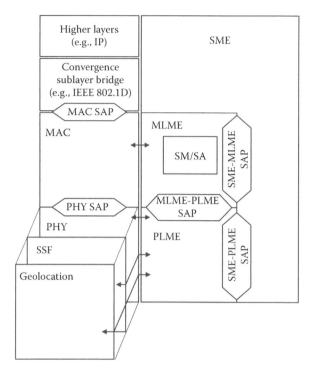

FIGURE 1.9   The IEEE 802.22 reference architecture.

entities (MLMEs), as well as to higher layers such as IP, through an IEEE 802.1D [71] compliant convergence sublayer.

At the PHY layer, there are three primary functions: the main data communications, the spectrum sensing function (SSF), and the geolocation function, with the latter two providing necessary functionality to support the cognitive abilities of the system. The PHY interfaces with the MAC through the PHY SAP, as well as to the MLME and the SME through the PHY layer management entity (PLME) and its SAPs. A functional entity known as the spectrum manager (SM) exists in the MLME at the BS and a "light" version of the SM, known as a spectrum automaton (SA), exists in the MLME at the CPE. The SM at the BS controls the use of and access to spectral resources for the entire cell and all associated CPEs served by the BS. The SA at each CPE provides the autonomous behaviors necessary to ensure proper noninterfering operation of CPE in all cases, including during startup/initialization, during channel changes, and in case of temporary loss of connection with the BS. IEEE 802.22 PHY and MAC layers' main features, related to capacity and coverage, are presented in Table 1.8.

It is obvious that deployment of the IEEE 802.22 standard in sparsely populated areas that cannot be economically served by wireline technologies, or other wireless solutions at higher frequencies, will increase the

TABLE 1.8
IEEE 802.22 General Characteristics

Frequency	54–868 MHz
Bandwidth	6, 7, or 8 MHz
Average spectrum efficiency	3 bit/s/Hz
Maximum data rate	31 Mb/s
System capacity per user	1.5 Mb/s (downlink)
	384 kb/s (uplink)
Coverage	30 km (up to 100 km without power limitation)
Effective isotropic radiated power	98.3 W

efficiency of spectrum utilization, and provide large economic and societal benefits.

1.3 INTERWORKING IN HETEROGENEOUS WIRELESS ENVIRONMENT

Rapid progress in the research and development of wireless networking technologies has created different types of wireless systems. Heterogeneity, in terms of wireless networking, is directly associated with the fact that no single RAT is able to optimally cover all the different wireless communications scenarios. Heterogeneous wireless networks are based on multi-network environments that provide multiservices with multimode mobile terminals (so-called 3M concept) [72]. The development of interworking solutions for heterogeneous wireless networks has spurred a considerable amount of research in this topic, especially in the context of WLANs and mobile networks' integration [73]. Interworking is linked to many technical challenges such as the development of enhanced network architectures, new techniques for seamless handover, and advanced management functionality for the joint exploitation of heterogeneous wireless networks.

1.3.1 INTERWORKING ARCHITECTURE

Attending to current architectural trends in next generation networks, wireless systems are mainly devoted to providing network connectivity services that may be characterized by a given QoS profile [74]. End user service provisioning is supported by specialized service platforms that become accessible to the users via those bearer services.

Figure 1.10 illustrates generic wireless network architecture [75] in terms of its main network nodes and protocol layer allocation. The wireless network provides network layer connectivity to external networks and

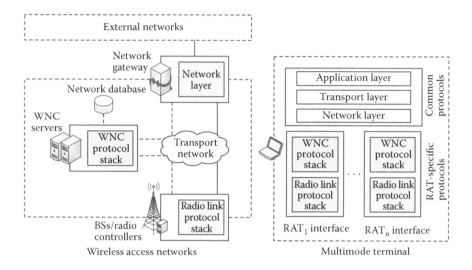

FIGURE 1.10 Generic wireless access network architecture and multimode terminal protocol stack.

service platforms over some type of network gateway. This gateway can allocate mechanisms to dynamically acquire operator policies related to QoS and accounting, and enforce them on a packet-by-packet basis for each mobile user. On the other hand, a RAT-specific radio link protocol stack would be used in the air interface. This protocol stack can be entirely allocated in BSs or distributed in a hierarchical manner between BSs and some type of radio controllers. The radio link protocol stack is composed of PHY, MAC, and radio link control layers. BSs and network gateways constitute the two key elements within the data plane functions (i.e., functions that are executed directly on the traffic flows). The management of the overall connectivity service is achieved by a network control plane. The purpose of the network control plane is to handle mechanisms such as network access control (e.g., AAA, mobility management, security management, and session management). This set of control plane mechanisms is referred to as wireless network control (WNC) mechanisms [75].

The protocol stack of a multimode terminal consists of RAT-specific protocols for the lower layers and a common set of protocols for the higher layers [72]. RAT-specific protocols comprise the corresponding radio link protocol stack to handle data transfer in the air interface along with the protocols used for WNC-related functionality in each wireless network. As to the common protocol layers, the network layer has a fundamental role in the interworking model because it provides a uniform substrate over which transport and application protocols can efficiently run independent of the access technologies used.

1.3.2 INTERWORKING MECHANISMS

Several interworking levels can be envisioned with a different range of interworking requirements. The definition of interworking levels is conducted attending to network architecture aspects or the level of support for specific service and operational capabilities [75]. In a general case, four interworking levels are distinguished. They are presented in Figure 1.11, together with corresponding interworking mechanisms.

Visited network service access (Level A) would allow users to gain access to a set of services available in visited networks while relying on home network credentials. The users can be charged for service usage in the visited network through its own home network billing system. Mechanisms included here aim at extending AAA functions among wireless networks, allowing users to perform authentication and authorization processes in a visited network attending to security suites and subscription profiles provided by their home networks. Intersystem AAA functionalities are basically achieved by

- Adoption of flexible AAA frameworks able to support multiple authentication methods, e.g., the extensible authentication protocol (EAP) [76]
- Deployment of additional functionality such as AAA proxy/relay functions and related signaling interfaces between networks, e.g., the Diameter protocol [77]
- Enhanced network discovery mechanisms for identity selection so that mobile terminals can know in advance whether their home network's credentials are valid for AAA control in a visited network

Intersystem service access (Level B) provide users the availability of specific services located in the home network while they are connected via a visited RAN. Intersystem user data transfer is a mechanism that enables the transfer of user data between interworking networks. A common approach to enforce user data transfer between networks relies on tunneling protocols. Tunnels may be established either directly between mobile terminals and remote gateways or may require additional dedicated nodes.

Intersystem service continuity (Level C) extends level C so that the user is not required to re-establish active sessions when moving between networks. However, a temporary QoS degradation can be tolerated during this period. Network layer handover is a mandatory mechanism when service continuity between wireless networks relies on the maintenance of a permanent mobile terminal IP address. Several versions of IP mobility

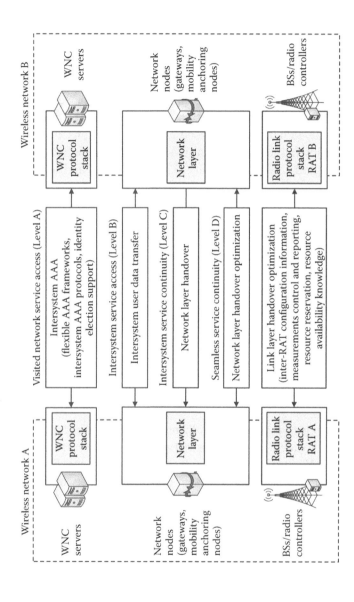

FIGURE 1.11 Interworking levels and mechanisms. (From R. Ferrus, O. Sallent, and R. Agusti. Interworking in heterogeneous wireless networks: Comprehensive framework and future trends. *IEEE Wireless Communications* 17, 2 (April 2010): 22–31.)

protocols have been proposed over the past several years to complement or enhance network layer intersystem service continuity [78] and they will be analyzed in Chapter 3.

Intersystem seamless service continuity (Level D) is aimed at providing a seamless mobility experience. Seamless service continuity can be achieved by enabling mobile terminals to conduct seamless handover [79] across heterogeneous RANs. A seamless handover is commonly related to the achievement of low handover latencies. Because of that, this level imposes the hardest requirements on the interworking mechanisms. Although simple network or application layer handover solutions may suffice for intersystem service continuity, they may not be able to satisfy the requirements for seamless mobility. In particular, during a handover, latencies related to radio link layers (e.g., new radio link establishment) and network layer operation (e.g., movement detection, new IP address configuration, and binding updates) can turn into a period during which the terminal is unable to send or receive packets [75]. Radio link layer operations (e.g., scanning, authentication, and association) can introduce additional delays in the handover process.

1.4   MULTIPLE-INPUT, MULTIPLE-OUTPUT SYSTEMS

MIMO is a communication system in which the transmitter and receiver are equipped with multiple antennas. Unlike traditional phased array or diversity techniques that improve the sensitivity to one signal of interest, MIMO systems employ antenna arrays jointly upon transmition and receive on spatially multiplex signals over multipath or near-field channels. Measuring a system performance in terms of channel capacity, MIMO systems offer the exciting possibility of linear capacity increase with additional antennas compared with the logarithmic growth (Shannon's channel capacity formula) of traditional diversity systems. The large potential of MIMO techniques in combination with OFDM is evidenced by the rapid adoption of recent wireless standards, such as Wi-Fi, LTE, and WiMAX.

A general MIMO configuration is shown in Figure 1.12. Here, N_t and N_r refer to the numbers of transmit and receive antennas, respectively. Special cases of MIMO are single-input, multiple-output (SIMO), where $N_t = 1$ and $N_r > 1$, and multi-input, single-output (MISO), where $N_t > 1$ and $N_r = 1$.

During the last decade, MIMO technology has received a lot of attention because of the potential improvement in throughput and reduction in energy consumption that this technology provides. Generally speaking, MIMO technology increases spectral efficiency by exploiting the spatial diversity that is attained through the physical separation of multiple

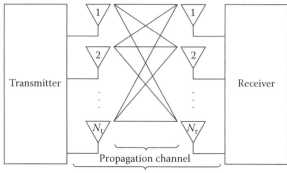

FIGURE 1.12   MIMO system configuration.

antennas in space. The advantages of MIMO systems can be divided into three main categories [80]:

- Spatial multiplexing for enhancing the data transmission rate
- Transmit diversity using space–time coding for enhancing the transmission robustness, and
- Beamforming for improving the received signal and reducing interference to other users

1.4.1 CLASSIFICATION AND FEATURES OF MIMO TECHNIQUES

Spatial multiplexing [81] is a MIMO technique in which independent and separately encoded data signals, called streams, are transmitted from multiple antennas. This technique requires multiple antennas at both ends of the radio link ($N_t \geq 2$ and $N_r \geq 2$), and also there is no need for channel information at the transmitter. Under the spatial multiplexing mode, the achievable capacity (maximum spatial multiplexing order) is $N_s = \min(N_t, N_r)$. For linear receivers, this means that N_s streams can be transmitted in parallel, leading to an N_s increase in the spectral efficiency. If the transmitted streams arrive at the receiver with sufficiently different spatial signatures, the receiver can separate them. As a result, an increment in the channel capacity is achieved.

Antenna diversity is a transmission technique in which similar data (replicas) signals are transmitted from multiple antennas to improve the signal-to-noise ratio (SNR). Such a gain is equal to $N_t N_r$ and can be achieved by using multiple receiver antennas (diversity reception) or by using multiple transmitting antennas (transmit diversity). A diversity-combining circuit combines or selects the signals from the receiver antennas to provide an improved signal quality.

Beamforming is a signal-processing technique that is used to control the directionality of the emission pattern of an antenna system. When receiving a signal, beamforming can increase the receiver sensitivity in the direction of the desired signals and decrease the sensitivity in the direction of the interference and noise. On the other hand, when transmitting a signal, a beamforming antenna system can increase the radiated power in the intended direction.

1.4.2 MIMO-Based Protocols

The channel access in existing WLANs based on IEEE 802.11 standard is performed according to a CSMA/CA scheme with an optional virtual carrier sensing (VCS) mechanism. Because packets in CSMA/CA are transmitted at a fixed power level, this strategy has been proven to be inefficient in terms of energy consumption and channel spatial reuse. The fixed power strategy affects channel utilization by not allowing multiple transmissions to take place concurrently within the same neighborhood [80]. This phenomenon is presented as an example in Figure 1.13, where nodes A and B communicate using the maximum transmission power.

According to the classic CSMA/CA scheme, nodes C and D cannot communicate while the A–B communication is taking place because they are within B's transmission range. However, those two transmissions actually may be able to occur concurrently if transmission powers can be appropriately adjusted. Several transmission power control (TPC) protocols have been proposed to overcome this problem, some of which are aimed at energy conservation, whereas others are throughput oriented [82].

A simple MIMO extension to CSMA/CA is CSMA/CA(k). It performs like CSMA/CA except that all transmissions are performed using k streams

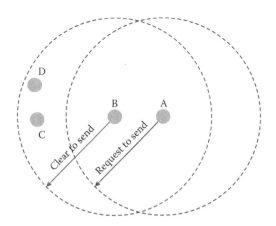

FIGURE 1.13   Inefficiency of the fixed transmission power strategy.

to provide spatial multiplexing gain. Such a protocol, when compared with CSMA/CA operating in the same network topology, but with omnidirectional antennas, achieves k times the throughput performance as the latter. In an article by Sundaresan et al. [83], it is shown that the unique characteristics of MIMO links (e.g., transmitter range versus capacity trade off, and robustness to multipath fading) require an entirely new MAC design. They discussed several optimization considerations that can help in obtaining an effective MAC protocol from such an environment, such as control gains, partial interference suppression, and receiver overloading. Also, a centralized algorithm called stream-controlled medium access (SCMA) incorporates those optimization considerations into its design. The objective of the SCMA algorithm is to maximize the network utilization subject to a given fairness model. The operation of the SCMA algorithm depends on the receiver overloading problem.

The mobile IP reservation (MIR) protocol [84] is a routing protocol proposed for MIMO links, which adapts between the different strategies based on the network conditions. MIR controls the various characteristics of MIMO links (e.g., network density, mobility, and link quality) to improve the network performance. The goal is to obtain routes that support high rates and allow for maximum spatial reuse in the network, which means that the number of range links in the route should be minimized. On the other hand, the goal is to detect link failures due to mobility and channel degradation proactively and switch from multiplexing to a diversity operation. In that way, the increased communication range or the increased reliability is exploited to increase the lifetime of the link during mobility and channel degradation, respectively.

The application of MIMO techniques in mobile ad hoc networks (MANETs) was explored by Hu and Zhang [85]. The focus was on the application of spatial diversity to reduce the effects of fading and achieve robustness in the presence of user mobility. Also, the effect of spatial diversity on MAC design was examined and accordingly, spatial diversity for MAC (SD-MAC) protocol was proposed. SD-MAC has the following features:

- Space–time codes are used for four-way handshaking to achieve full-order spatial diversity
- For carrier sensing, if the average interference across the antenna elements is higher than the threshold, the channel is marked as busy
- The transmitting node adapts the data rate for the data packet according to the channel conditions

As mentioned earlier, MIMO is used mostly in combination with OFDM because MIMO is most compatible with flat-fading channels, which ensure low complexity and receiver power consumption. The problem of designing a MAC protocol for MIMO-OFDM ad hoc networks was considered by Hoang and Iltis [86]. The proposed protocol reserves some carriers for the control channel and others for data communications. The control carriers are used to transport the control packets omnidirectionally. After the hand-shaking procedures have been completed on the signaling channel, the data exchange and additional P2P signaling follow. The proposed protocol uses appropriate power control and allocation strategies to allow multiple transmissions on the data channel. The advantage of this protocol over the protocols that use the same channel for both control and data transmissions is that the data channel is collision-free. In that way, data channels are used more efficiently.

Another representative MIMO-adaptive protocol for WLANs is MIMO power-controlled MAC (MIMO-POWMAC) [87]. It is a classic throughput-oriented protocol. According to this protocol, every node is equipped with two antennas and has the ability to use MIMO-based multiplexing gain to maximize the perceived throughput of the network. MIMO-POWMAC allows multiple concurrent transmissions to occur simultaneously because it uses collision avoidance information in the control packets to bound the transmission power of potentially interfering terminals in the vicinity of a receiving terminal.

1.4.3   MIMO SYSTEM CONFIGURATION FOR IEEE 802.16M AND 3GPP RELEASE 10

The application of MIMO technologies is one of the most crucial distinctions between 3G and 4G systems. It not only enhances the conventional point-to-point link but also enables new types of links such as downlink multiuser MIMO. A large family of MIMO techniques have been developed for various links and with various amounts of available channel state information in both IEEE 802.16m and 3GPP Release 10. The MIMO systems can be configured as single-user MIMO (SU-MIMO), multiuser MIMO (MU-MIMO), and multicell MIMO [88]. Their downlink configurations are illustrated in Figure 1.14.

SU-MIMO transmissions occur in time–frequency resources dedicated to a single-terminal LTE advanced mobile station (AMS) or WiMAX user equipment (UE), and allow achieving the peak user spectral efficiency. They encompass techniques ranging from transmit diversity to spatial multiplexing and beamforming. These techniques are supported in both standards, with a most noticeable difference in the approach taken for spatial multiplexing. Both open-loop SU-MIMO and closed-loop SU-MIMO are supported for the antenna configurations.

For open-loop SU-MIMO, both spatial multiplexing and transmit diversity schemes are supported. For closed-loop SU-MIMO, codebook-based

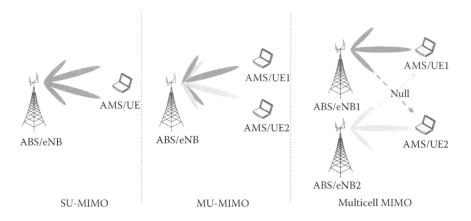

FIGURE 1.14 MIMO downlink configurations.

precoding is supported for both TDD and FDD systems. Channel quality indicator (CQI), preferred matrix index/indicator (PMI), and rank feedback can be transmitted by the AMS/UE to assist the advanced base station's (ABS) or eNB's scheduling, resource allocation, and rate adaptation decisions. CQI, PMI, and rank feedback may or may not be frequency dependent. For closed-loop SU-MIMO, sounding-based precoding is supported for TDD systems.

MU-MIMO allocates multiple users in one time–frequency resource to exploit multiuser diversity in the spatial domain, which results in significant gains over SU-MIMO, especially in spatially correlated channels. In configurations such as downlink 4 × 2 (four transmit antennas and two receive antennas) and uplink 2 × 4, single-user transmission only allows spatially multiplexing a maximum of two streams. On the other hand, linear schemes allow sending as many as four spatial streams from four transmit antennas, or receiving as many as four spatial streams with four receive antennas by multiplexing four spatial streams to or from multiple users. MU-MIMO techniques provide large sector throughputs in heavy traffic conditions.

For open-loop MU-MIMO, CQI and preferred stream index feedback may be transmitted to assist the base station's scheduling, transmission mode switching, and rate adaptation. The CQI is frequency dependent. For closed-loop MU-MIMO, codebook-based precoding is also supported for both TDD and FDD systems. CQI and PMI feedback can be transmitted by the AMS/UE to assist the ABS/eNB scheduling, resource allocation, and rate adaptation decisions. For closed-loop MU-MIMO, sounding-based precoding is also supported for TDD systems.

Multicell MIMO and uplink MIMO techniques are defined in IEEE 802.16m and under discussion in 3GPP Release 10. Key techniques in

TABLE 1.9
Key Techniques in MIMO Downlink

Downlink MIMO Techniques	IEEE 802.16	3GPP Release 10
Open-loop transmit diversity	Space-frequency block coding with precoded cycling	Space-frequency block coding with precoded cycling Frequency-switched transmit diversity
Open-loop spatial multiplexing	Single codeword with precoder cycling	Multiple codewords with large delay
Closed-loop spatial multiplexing	Advanced beamforming and precoding	
MU-MIMO	Closed-loop and open-loop	Closed-loop

TABLE 1.10
MIMO Capabilities

Downlink	SU-MIMO	Up to eight streams
	MU-MIMO	Up to four users
Uplink	SU-MIMO	Up to four streams
	MU-MIMO	Up to four users

MIMO downlink and the MIMO system capabilities are summarized in Tables 1.9 and 1.10, respectively. An overview of the ongoing research related to these techniques can be found in articles by Li et al. [88] and Liu et al. [89].

1.4.4 MULTIPLE-BASE STATION MIMO COOPERATION

Multiple-base station (multi-BS) MIMO cooperation schemes are expected to play an important role in terms of intercell interference mitigation. However, practical challenges also act as a big obstacle to gaining the benefits of multi-BS MIMO techniques [9]. The multi-BS MIMO cooperation schemes can be categorized into two types: single-BS precoding with multi-BS coordination, also known as coordinate beamforming, and multi-BS MIMO joint processing.

In the coordinate beamforming scheme, a BS transmits the precoded data to its serving mobile stations (MSs) to reduce the intercell interference. On the other hand, in joint processing, multiple BSs transmit data not only to serving MSs, but also to other coordinated BSs with jointly optimized precoding matrices. Also, coordinate beamforming does not require data forwarding among different BSs, whereas joint processing does. As a result, coordinate beamforming can be implemented without an increase in backbone capacity.

In a cellular system, based on the level of available information on the channel of a neighboring mobile station, three scenarios for downlink multi-BS MIMO cooperation can be considered [9]:

- No channel knowledge—each base station only knows the channel information of its subordinate mobile station
- Partial channel knowledge—each base station has partial channel state information (CSI) between the base station and its neighboring mobile stations
- Complete channel knowledge—each base station knows all CSI on its own and neighboring mobile stations

Generally, the more CSI is available, the higher the performance improvement of MIMO techniques is obtained, at the cost of increased overhead. Full CSI acquisition is difficult to achieve in FDD systems, although it can be supported in TDD systems by using sounding signals from a mobile station. However, even in TDD systems, an unacceptably high bandwidth sounding channel may be required to provide accurate estimation performance in the case of high mobility or poor link quality.

To reduce the operational overhead, partial CSI schemes using channel quantization methods are proposed as alternative solutions. Among them, a codebook-based technique [90] is widely used and adopted in many standards (e.g., IEEE 802.16m and 3GPP Release 10). Here, the codebook implies a set of codewords known to both network elements. The mobile station conveys the best codeword within the codebook using a predetermined strategy.

Precoding matrix indicator (PMI), also known as *preferred matrix index* in IEEE 802.16m terminology, is one of the codebook-based coordinate beamforming schemes [9]. In closed-loop MIMO systems, the use of a certain precoder or beamformer can incur much higher interference than others. Especially in codebook-based MIMO techniques, a precoding matrix based on the reported PMI for improving user throughput may in turn decrease the neighboring user throughput. Conversely, a certain precoding matrix can be less damaging to the neighboring user throughput. Intercell interference level can be managed by controlling the transmit precoding matrix with two different schemes: PMI restriction and PMI recommendation.

Example 1.1

A diagram of PMI coordination is shown in Figure 1.15. This is an example of an operational scenario of the PMI restriction and the PMI recommendation. Suppose that there are three cells in the network. The performance of MS_{edge} is highly affected by the signals from BS_1, BS_2, and BS_3. The other MSs (MS_1, MS_2, and MS_3), located at the cell center, receive high signal power from each serving BS, and low interference power from interfering BSs.

For PMI restriction, an MS reports the PMI(s) for interfering BS(s) that incur the strongest interference so that the neighboring BS(s) can restrict the use of those PMI(s) for its own cell operation. On the other hand, for PMI recommendation, the MS issues the PMI(s) for neighboring BS(s) that cause the weakest interference so that the neighboring BS(s) can use one of those PMIs for its own cell operation. The restricted PMIs are bad precoders for the serving cell operation, whereas the recommended PMIs are beneficial for the serving cell operation. The BS can choose one of the PMI coordination methods depending on an interference mitigation strategy.

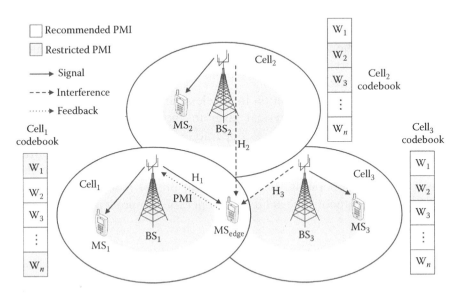

FIGURE 1.15 PMI coordination diagram. (From W. Lee et al. Multi-BS MIMO cooperation: Challenges and practical solutions in 4G systems. *IEEE Wireless Communications* 19, 1 (February 2012): 89–96.)

Each BS uses different resources to send reference signals, whereas the other resources that belong to other BSs are set to zero so that there is no interference from the other two BSs when MS_{edge} measures the channel response from each BS, H_1, H_2, and H_3.

1.5   COOPERATION TECHNIQUES FROM A NETWORKING PERSPECTIVE

The concept of cooperation in wireless networks has gained much more attention from researchers because it can be effective in addressing performance limitations due to user mobility and the scarcity of resources. Cooperation techniques are increasingly important to enhance the performance of wireless communications, with their ability to decrease power consumption and packet loss rate and increase system capacity, computation, and network resilience [91]. The idea of employing cooperation in wireless communication networks has emerged in response to user mobility support and limited energy and radio spectrum resources, which pose challenges in the development of wireless networks and services in terms of capacity and performance [92]. As for user cooperation, it takes many forms, including the physical layer cooperative communications, the link layer cooperative and cognitive medium access, the network layer cooperative routing and load balancing, the collaborative E2E congestion control in the transport layer, and the cooperative P2P services [93]. The development of cooperation techniques requires interdisciplinary efforts, from advanced signal processing to network protocol design and optimization.

The research on cooperation in wireless networks focuses on developing strategies at the physical layer to support such a cooperative transmission. However, such a cooperative operation introduces challenging issues at different layers of the protocol stack. Some modifications to the networking protocol stack are required to achieve the objectives of cooperation. In fact, without proper modification of the higher layer protocols, the achieved cooperation improvements may not be significant.

1.5.1   BENEFITS OF COOPERATION

The potential benefits of employing cooperation in wireless networks [92] include improved channel reliability, improved system throughput, seamless service provision, and operation cost reduction (as shown in Table 1.11).

TABLE 1.11

Potential Benefits and Corresponding Techniques of Employing Cooperation in Wireless Networks

Benefits	Cooperative Technique	Description
Improved channel reliability	Mitigating channel impairments using spatial diversity [94]	When the channel between the original source and destination is unreliable, other network entities can cooperate with the source to create a virtual antenna array and forward the data toward the destination
	Interference reduction [95]	Using the cooperative relays, the transmitted power from the original source can be significantly reduced due to a better channel condition of the relaying links
Improved system throughput	Resource aggregation [96]	Cooperation can increase the achieved throughput through aggregating the offered resources from different cooperating entities
Seamless service provision	Service continuity over substitute paths [97]	When the service is interrupted along one path, it still can be continued using another cooperative path
Operation cost reduction	Energy savings [98]	Green communications can improve the energy efficiency, which reduces the energy costs

Example 1.2

Spatial diversity and interference reduction as cooperative techniques for channel reliability improvement are illustrated in Figure 1.16. Downlink transmission from a base station to a mobile station is observed, where the source node transmits its data packets toward the destination over cooperating entities. In this context, a cooperating entity is a relay node with an improved channel condition over the direct transmission channel from the source to the destination. This relay node can be a mobile station or a dedicated relay station.

In the sense of interference reduction, using the cooperative relays, the transmitted power from the original source can be significantly reduced due to the better channel condition of the

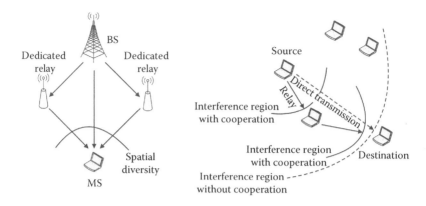

FIGURE 1.16   Cooperation techniques for improving channel reliability.

relaying links, which greatly reduces the interference region. This also helps improve the energy efficiency of the communication system. In addition to reducing the interference region, cooperation can solve the hidden terminal problem and, hence, results in interference reduction.

Example 1.3

Mobile users are more sensitive to call dropping than call blocking. Call dropping interrupts service continuity for different reasons depending on the networking scenario. Cooperation for seamless service provision is shown in Figure 1.17.

When the service is interrupted along one path (Ch_1), it still can be continued using another cooperative path (Ch_2, Ch_3). In this context, a cooperating entity can be a MT, BS, or AP, which can create a substitute path between the source and destination nodes.

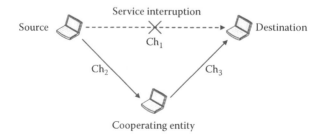

FIGURE 1.17   Cooperation for seamless service provision.

1.5.2 COOPERATIVE ROUTING

To achieve reliability at the upper layers when relying on a wireless physical layer, many researchers have proposed the adoption of cooperative paradigms. This means that one or more intermediate nodes intervene in the connection so that either the communication is rerouted over a better path, or the original link is kept in use but its quality is strengthened thanks to diversity provided by these cooperators. This concept is known as cooperative (opportunistic) routing [11].

In the simplest version, one intermediate node simply acts as a relay between the source and the destination node [99]. However, if the communication between these two nodes is part of a multihop transmission, it may be rerouted over an entirely new path that no longer involves this receiver. To this end, a distance metric is needed to verify that the selection of a given intermediate node as the next hop still sends the message toward the destination node and not further away from it. Additionally, a negotiation phase is also required in which intermediate nodes can volunteer as the next hops whenever the link from the transmitter to the intended receiver does not guarantee sufficient quality. Figure 1.18 represents a network in which the elements of cooperative routing (direct links and opportunistic hops) have been highlighted.

To participate as a cooperative relay, node C must be able to listen to both the clear-to-send (CTS) and request-to-send (RTS) control packets. From the RTS, it can estimate the signal-to-noise-plus-interference ratio (SNIR) between itself and the source, SNIR(AC). Also, it is possible to derive the distance $l(A, D)$ of node A from the final destination D because the RTS contains the identifier of these nodes. From the CTS, which carries the SNIR(AB) value, it may learn whether the quality of the source–destination direct link is poor. If this is the case, node C checks the following two conditions to determine whether or not to contribute:

1. The value of SNIR(AC) must be above $SNIR_{th}$
2. Node C should represent an advancement toward the final destination D

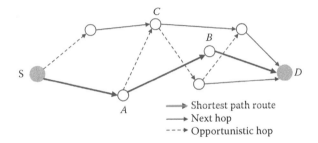

> Shortest path route
> Next hop
---> Opportunistic hop

FIGURE 1.18   Cooperative routing.

It should be noted that C is not required to know the complete route exactly, just its own distance to the destination. The first condition means that the channel AC is good enough, and it is also better than channel AB, which is below the threshold. The second condition expresses that, even though the route is changed, the packet is still advancing toward the destination node. According to the second condition, the hop count $l(C, D)$ must be less than $l(A, D)$. That is, C has to be chosen among the neighbors of A for which $l(C, D) = l(A, D) - 1$. To simplify, the assumption that l must be decreased at every step has been chosen. It is possible to consider extensions where even nodes with the same distance can be accepted, provided that some form of hysteresis is introduced to avoid the packet continually moving within the same set of nodes without advancing toward D [99]. In the same way, it is even possible to consider cooperative paths where the l is actually increased, provided that a better overall quality of the route is envisioned.

Several intermediate nodes can be a cooperative next hop from A if they satisfy the predefined conditions. Such nodes declare their availability to take charge of the forwarding of the packet by sending a cooperative CTS message, which can be received by both A and B. The transmission of such a message requires an additional time slot to be left unused after the CTS.

The cooperative routing approach implies additional overhead due to the cooperative CTS packets. The transmission of such control packets may also cause interference peaks for other neighboring nodes because many potential cooperators might send a CTS simultaneously. However, this problem does not occur very frequently because the transmission of cooperative CTS happens only when the direct link AB fails. Because this link is chosen as belonging to the best path to D, its quality is frequently bad only in the case of a scattered network. On the other hand, the phenomenon of simultaneous transmission of cooperative CTS messages causes high interference levels only if the network is densely structured.

1.5.3 WIRELESS RELAYING PROTOCOLS

Over the last 10 years, cooperative relay-aided (multihop) techniques have been intensively researched as means for potentially improving link performance of future wireless multimedia systems. In the relay station (RS) aided concept, the MS communicates with the BS assisted by a single or multiple relays, which may be expected to provide better link quality than direct single link–based communication. RS-aided systems are capable of increasing the attainable data rate, especially in the cell edge region, where MSs typically suffer from both low-power reception and intense intercell interference [100]. Shadow fading can also be attenuated by relays. RSs also offer high flexibility in terms of their geographic position. For example, when the traffic intensity rapidly increases in a certain area, RSs can be immediately engaged to provide high-speed communications. The MSs currently not engaged in active communication with the BS may also act as relays without any additional deployment cost.

Besides all benefits invoked by the RSs application in terms of link quality, they require additional radio resources. More explicitly, the traditional relaying concept requires four phases to transmit a pair of downlink and uplink packets, which is twice as many as direct communications operating without a relay node. This additional radio resource allocation halves the effective throughput and, hence, mitigates the advantages of relaying. As shown in Figure 1.19a, during the first phase, the BS transmits the downlink signal to the RS. During the second phase, the RS forwards the received signal to the MS. The uplink signal is also sent from the MS to the BS during the remaining two phases in a two-phase reverse-direction manner. As for the relaying operation, it can be classified into two types [100]: the amplify-and-forward (AF) and decode-and-forward (DF) relaying protocols.

In AF relaying, the relay simply retransmits a scaled version of the received signal without performing any detection or decoding. The simple operation of AF relaying leads to a low-cost, low-complexity implementation. The desired signal and additive noise are jointly amplified at the relay, and thus, the AF protocol fails to improve the SNR. By contrast, in DF relaying, the RS entirely decodes the received signal and, hence, may succeed in regenerating the transmitted signal before it forwards the

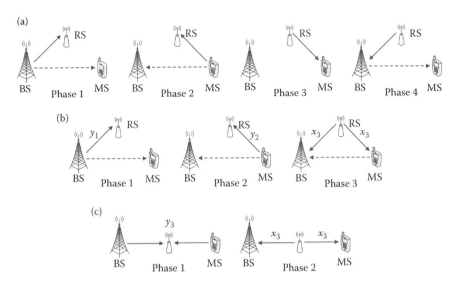

FIGURE 1.19 Relaying protocols: (a) traditional four-phase relaying, (b) three-phase relaying, and (c) two-phase relaying.

re-encoded packet to the destination. In this case, SNR is sufficiently high, although complexity is high.

Recently, *network coding* (NC) has emerged as a new cooperative technique for improving network throughputs over traditional routing techniques [101,102]. Figure 1.19b depicts the basic philosophy of the NC-aided three-phase relaying system, whereas the corresponding relay's operation is presented in Figure 1.20. The BS and MS transmit the codewords c_1 and c_2 during the first and second communication phases to the RS, respectively. Then, the RS independently decodes the corresponding received signals y_1 as well as y_2 and combines the decoded codewords into a single stream.

In the third phase (Figure 1.19b), the composite network-coded packet is modulated and broadcast to both the BS and the MS. The signal received at each destination can be regarded as the combination of the uplink and downlink packets. To recover the desired packet at each destination node, each node utilizes a priori knowledge of its own transmitted packet that was transmitted in the previous communication phase. The three-phase DF relaying requires only three time slots to complete a full cycle transmission. Therefore, a 33% throughput increase can be expected, compared with the traditional relaying concept, which requires four communication phases [100].

To further reduce the required communication resources, three different types of two-phase relaying protocols may be invoked: two-phase AF, denoise-and-forward (DNF), and DF protocols. The two-phase relaying protocols of Figure 1.19c are often referred to as two-way relaying or bidirectional relaying, as opposed to the traditional four-phase relaying protocol, which is also often referred to as one-way relaying. The relay's actions in various two-phase relaying protocols are illustrated in Figure 1.21.

In two-phase AF relaying, the MS and BS simultaneously transmit their signals to the RS during the first phase, as shown in Figure 1.19c. In the second phase, the RS amplifies the composite received signal while obeying a specific total power constraint, and forwards it to both the MS and BS.

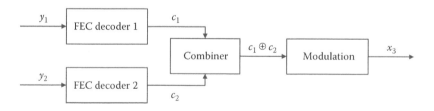

FIGURE 1.20   Three-phase DF relaying protocol.

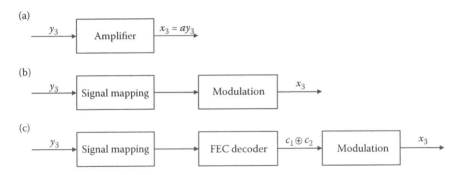

FIGURE 1.21   Block diagrams of two-phase relaying protocols: (a) two-phase AF, (b) two-phase DNF, and (c) two-phase DF.

To reduce the effect of noise amplification in the two-phase AF regime, the DNF relaying protocol, which is based on the NC coding scheme, was proposed by Zhang and Liew [103]. As in the AF two-phase relaying regime (Figure 1.21a), the two-phase DNF (Figure 1.21b) receives both signals in the first phase. The denoising operation may also be referred to as detect and forward because it is constituted by the noise elimination process of the detector's slicing or decision operation. In contrast to the two-phase AF relaying, the noise amplification problem does not occur in the DNF relaying protocol because the denoised symbol stream is transmitted from the RS with the aid of the specific mapping function. However, the RS of the DNF scheme does not take advantage of channel coding and, hence, may have a high error probability for the relayed signal.

In contrast to the DNF regime (Figure 1.21b), which dispenses with FEC coding, in the two-phase DF relaying of Figure 1.21c, the RS performs FEC decoding to mitigate the effects of error propagation. The RS receives the superposition of the uplink and downlink signals in the first phase and attempts to decode both of them. During the second phase (Figure 1.19c), the RS generates the composite packet, which is broadcast to both the MS and BS, as in the three-phase DF relaying scheme. In the second phase, each destination node decodes its desired signal in the same way as the three-phase relaying regime. Nonetheless, its decoding performance at the RS is typically worse than that of the three-phase DF relaying protocol, where only one of the uplink and downlink packets arrives and is decoded at the RS in each phase.

A comparison of relaying protocol characteristics is summarized in Table 1.12, whereas more detailed analyses and numerical results, which justify the application of network coding aided protocols as the most promising solution, can be found in an article by Lee and Hanzo [100].

In practical RS-aided systems, the destination node should estimate both the relay channel as well as the direct channel. In AF, relaying the

TABLE 1.12

Comparison of Relaying Protocols

Relaying Protocol	4-AF	4-DF	3-DF	2-AF	2-DNF	2-DF
Spectral efficiency	Low	Low	Medium	High	High	High
Effective SINR	Low	High	High	Very low	Medium	High
System constraint	Low	Low	Medium	Low	Medium	Very high
Complexity	Low	High	High	Low	Medium	High

channel information between the source and the relay is also required at the destination, which implies a high loading on the overall system, especially when the signal is relayed over more than two hops. To resolve this problem, it is worth studying the efficient estimation, quantization, and transmission of the channel impulse response information in relaying systems. Furthermore, noncoherent relaying algorithms, which do not require channel information, have to be studied.

1.6 HIGH-PERFORMANCE CONGESTION CONTROL PROTOCOL

Most current Internet applications rely on TCPs to deliver data reliably across the networks [104]. The performance characteristics of a particular TCP version are defined by the congestion control algorithm it employs. Although TCP is the de facto standard congestion control protocol in the Internet, its shortcomings in the wireless heterogeneous environment are widely known and have been actively addressed by many researchers. Congestion control protocols can be classified into two groups, AIMD and FCS, according to the theory used for rate control [13]. AIMD approaches are traditional schemes that adopt corresponding theory for window control. Because static AIMD control limits its utility, the challenge is to design a dynamic AIMD control scheme that is able to adapt to a variety of network environments.

Several FCS approaches such as explicit control protocol (XCP) [105] and adaptive congestion protocol (ACP) [106] have been proposed. In these protocols, a source node and network nodes exchange useful information for congestion control, and the source regulates its own window size according to the feedback from network nodes. As the window size of each flow is controlled based on the FCS theory, aggregate traffic can be stabilized and the throughput of each flow becomes steady after a reasonable time lapse. In general, FCS approaches tend to achieve better performance than AIMD approaches because they can operate according to the actual traffic conditions observed at a bottleneck node. However, in FCS approaches, there is a significant drawback involving the network nodes, which are required to handle the packet headers of the upper layers. It is not easy to read and write the corresponding TCP headers in network nodes when an IP payload is encrypted or encapsulated. Although the IP option field may be applied, such usage is generally avoided as it would substantially slow down switching.

1.6.1 TCP ENHANCEMENTS FOR HETEROGENEOUS WIRELESS ENVIRONMENT

The poor TCP performance in wireless networks stems from the unique characteristics of wireless links as compared with wired links, and the current TCP's design assumption of the packet loss model. The problems manifest in various applications as degradation of throughput, inefficiency in network resource utilization, and excessive interruption of data transmissions.

In TCP, all packet losses are assumed to be caused by network congestion, thus leading to the reduction of window size by the MD algorithm. Therefore, the throughput of TCP can be improved if it is possible to distinguish link error–related losses from congestive ones. TCP Veno and Jitter-based TCP (JTCP) are typical examples of this type of solution. These schemes estimate the cause of packet losses from the changes in round-trip time (RTT), which reflect network congestion. TCP Westwood and TCP Jersey are examples that estimate the available bandwidth and accordingly update the window size. In addition, TCP Jersey has a mechanism for specifying the cause of packet losses.

TCP Veno [107] supports the estimation of the backlogged packets in the network. It further suggests a way to differentiate the cause of packet loss. If the number of backlogged packets is below a threshold, the loss is considered to be random. Otherwise, the loss is said to be congestive. The number of backlogged packets can be estimated from the latest congestion window size and RTT values monitored at the source node. RTTs are measured by using a millisecond resolution time stamp mechanism.

JTCP [108] is a congestion control protocol that distinguishes the cause of packet losses by observing the interarrival jitter. Because an increase in interarrival jitter implies an increase in queuing delay caused by network congestion, it can be used as an indicator of congestion. In JTCP, jitter ratio is calculated from the measured interarrival jitters. Upon detecting packet loss, the jitter ratio is used to determine whether or not to halve the window size.

TCP Westwood [109] is a rate-based E2E approach in which the sender estimates the available network bandwidth dynamically by measuring and averaging the rate of returning ACKs. Instead of the MD parameter, the value of the available bandwidth is used for setting a new window size in every MD procedure. By setting transmission rate to be the available bandwidth, TCP Westwood is able to achieve efficient link utilization and robustness to link error–related losses. However, its performance largely depends on the estimation accuracy.

TCP Jersey [110] is another proactive approach that adapts the sending rate proactively according to the network condition. It is similar to TCP Westwood, but TCP Jersey employs a more complex mechanism by using the TCP time stamp option to compute the available bandwidth more accurately. It consists of two key components, the available bandwidth estimation algorithm and the congestion warning router configuration. Using these components, TCP Jersey calculates the optimum congestion window size at the sender.

1.6.2 EXPLICITLY SYNCHRONIZED TCP

High-performance congestion control protocol, referred to as explicitly synchronized TCP (ESTCP) is based on dynamic AIMD theory [13]. The motivation of this protocol is to dynamically control AIMD parameters by adjusting them to network congestion. ESTCP consists of two key components: the window controller, which executes the AIMD window adjustment at the source node; and the traffic controller, equipped at the bottleneck node for scaling the rate of the window size to appropriately stabilize the aggregate traffic.

In the equilibrium state, windows of all flows sharing the same bottleneck are synchronized [13], as shown in Figure 1.22. Although the rate

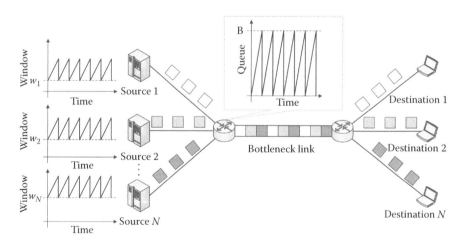

FIGURE 1.22 Equilibrium state in ESTCP.

of increase of the window size in TCP depends on the flow's RTT, it is equal for all flows in ESTCP and is simultaneously scaled by following the feedback from the bottleneck node. On the other hand, the MD parameter is independently adjusted by each source to keep full link utilization. By doing so, the amount of network traffic can be matched to bandwidth capacity, thus leading to almost zero buffer occupancy with synchronized MDs. The timing of running the MD algorithm in each flow is concentrically controlled by the bottleneck node to exactly synchronize all flows. Because the aggregate traffic passing through the bottleneck node is the superposition of window sizes of all flows, the bottleneck queue occupancy demonstrates the change as a sawtooth wave. By combining these mechanisms, ESTCP is able to overcome the TCP problems exhibited in static AIMD.

To constantly maintain full utilization of the link capacity, the fluctuation range of the aggregate traffic must lie within the capacity of the bottleneck buffer. In the equilibrium state, all flows simultaneously and periodically experience packet drops due to buffer overflow, thus leading to invocation of MD, whereas throughputs of all flows do not change much and remain the same. Equilibrium state can be presented by

$$\frac{w_1}{d_1} = \frac{w_2}{d_2} = \cdots = \frac{w_i}{d_i} = \cdots = \frac{w_N}{d_N} = \frac{BW}{N}, i \in [1,\ldots,N], \qquad (1.1)$$

where N is number of flows, BW is bottleneck link bandwidth, w_i is window size of flow i, and d_i is equal to the minimum value of RTT for corresponding flow.

The throughput at the moment before invoking MD can be expressed as the congestion window size, $cwnd_{cur}$, divided by the RTT value, RTT_{cur}. RTT_{cur} is greater than the minimum value, RTT_{min}, due to the queuing delay. On the other hand, the throughput at the moment just after MD is equal to ($\beta_{ESTCP} = cwnd_{cur}/RTT_{min}$) because the queuing delay is near zero. According to the fact that the throughputs before and after invoking MD are the same in each flow, the ideal value of β_{ESTCP} can be defined as

$$\beta_{ESTCP} = \frac{RTT_{min}}{RTT_{cur}}, \qquad (1.2)$$

where RTT_{min} denotes the minimum RTT measured since the beginning of the flow, and RTT_{cur} shows the RTT value before invoking the MD mechanism.

In protocols that adopt the ACK clocking mechanism such as TCP, the rate of increase of the window size depends on the flow's RTT. A larger

RTT results in slower growth of the congestion window because the congestion window can increase only upon receiving a new ACK packet. The simplest solution to remove the effect of RTT on the rate of increase of the window size is to design the AI parameter proportional to the flow's RTT. ESTCP uses the following AI parameter

$$\alpha_{ESTCP} = g^{-1} \cdot RTT_{min} \qquad (1.3)$$

where g is a scaling parameter. In ESTCP, all flows are synchronized by feedback from the same bottleneck node. In addition, the aggregate traffic fluctuates within the buffer space. Hence, it is indeed important to control and stabilize the traffic changes. Fortunately, traffic can easily be handled by controlling g. A small value of g should be avoided because it makes traffic increase rapidly, leading to significant packet drops and breaking synchronization. On the other hand, large values of g prolong the invocation of MDs by the buffer overflow; thus, the system takes a long time to reach the equilibrium state because bandwidth fair sharing can be achieved by repeating AI and MD alternately.

The results of extensive simulations provided by Nishiyama et al. [13] confirm the robust performance of ESTCP when dynamic AIMD control is applied. Also, it is suggested that assistance from the network side dramatically enhances performance, whereas additional packet headers (increased complexity) in the protocol stack should be avoided.

1.7 WIRELESS DISTRIBUTED COMPUTING

Traditional wired distributed computing research has been around for many years. However, the main challenge for WDC over traditional distributed computing is the presence of the wireless channel. Wireless systems consume higher power, and energy and may induce delays due to the retransmissions required at the link layers [111]. The radio environment influences computational workload balancing and limits scalability in terms of network coverage, node density, and overall computational load.

WDC can potentially offer several benefits over local computing, including reduced energy and power consumption per node, reduced resources consumption of individual nodes, enhanced computational capability of wireless networks, and robustness. WDC networks can offer high-performance computational services with the ability to perform complex tasks by utilizing the leveraged computational power of several radio nodes [112].

WDC exploits wireless connectivity to share processing-intensive tasks among multiple devices [14]. The goals are to reduce per node and network power, energy, and processing resource requirements. WDC aims to extend traditional distributed computing approaches to allow operation in dynamic network environments, as well as meet challenges unique to this concept, with the help of recently available enabling technologies. Today, WDC is possible due to the availability of key technologies such as

- Fault-tolerant computing
- Distributed computing
- Software-defined radio (SDR) and
- Cognitive radio (CR)

These technologies can offer capabilities such as collaboration among wirelessly connected computational devices, flexible link design, and adaptation and autonomous operation in dynamic environments. These capabilities enable flexible and optimized use of resources, reliable communications, joint optimization of computation and communication processes, and customization for the QoS requirements.

In WDC, however, a node's processing and communication capacity may be limited. In addition, mobility and wireless channel variations introduce uncertainty into the otherwise traditional distributed computing environment, posing a challenge to the design and execution of distributed algorithms and ensuring their correctness, efficiency, and promptness.

1.7.1 WDC APPLICATIONS

Several classes of applications can potentially benefit from WDC [14]:

- Real-time data capture, processing and dissemination (time-critical data applications, such as portable mobile scientific computing, in which the captured data is processed to provide information to a remote node)
- Complex communication waveforms (robust communication signal processing approaches often involve complex computations that can be executed in a distributed manner)
- Information sharing (real-time sharing of multimedia data may require complex information coding and ad hoc networking capabilities)
- Robust process control (control applications require time-critical coordination among wireless sensors and actuators)

Applications include: image processing and pattern recognition, distributed data storage and database search to avoid communication with a remote server over unreliable and unsecure links. Also, there is an application to enable the web indexing of user-generated content that is stored at the mobile device and not on a remote server. The next is synthetic aperture radar processing, jamming, signal detection and classification, as well as position location.

Example 1.4

Example scenario with WDC nodes and remote supernode is shown in Figure 1.23. Consider a collaborative wireless ad hoc network that may be composed of tactical handheld radios, radio nodes attached to unmanned aerial vehicles (UAVs), or sensor nodes. The master node in the network captures data (such as images) that holds the information required at a remote base.

Generally speaking, the data can be processed using two options [14]:

1. Remote data processing, where unprocessed data is transmitted over a long backhaul to a node abundant in resources located in the remote base
2. In-network processing, where data is processed within the network before transmitting the processed information to the remote base

The second option, when performed in a distributed manner, also allows the use of a large number of relatively small, simple,

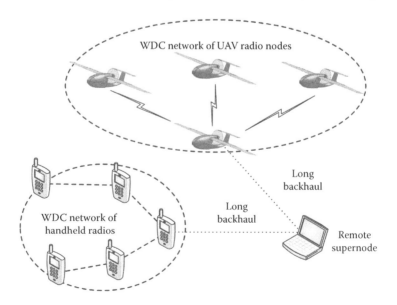

FIGURE 1.23   Scenario with WDC nodes and remote supernode.

and inexpensive radio nodes in place of a single complex and expensive one. In addition, WDC minimizes the dependency of the handheld radios on remote supernodes for mission-critical computation services. For WDC network configuration, the master node broadcasts data to the slave nodes, which return processed data to the master node. Then, the master node transmits processed data to the remote base. Processed data accounts for a relatively smaller communication payload compared with raw data. The reduction in energy consumption of the wireless nodes is achieved in several ways. First, WDC enables in-network processing, which significantly reduces the number of data bits transmitted over the long backhaul at the cost of some computational energy consumption. Second, power-hungry processors activate cooling mechanisms only when the processor operates at a certain high clock frequency. In WDC, the required computational latency can be achieved by concurrently processing data in all the nodes at a fraction of the required rate.

1.7.2 WDC DESIGN CHALLENGES

Limited and variable data rates over wireless channels can pose a bottleneck for applications involving high-speed multimedia computations. In such cases, buffering between communication and computational subsystems can control message losses and delays. However, the limited buffering capacity of portable WDC nodes makes it challenging to address this problem.

Some distributed applications require the computational processes on different nodes in a WDC network to be synchronized with one another. Synchronization is particularly important when the processes have to interact with one another while executing the application. Synchronization involves the establishment of a temporal relationship between these processes. In WDC networks, synchronization is a challenge when the computational processes that have to be synchronized are heterogeneous in terms of their execution times. The execution time is uncertain when there are other computational processes contending for limited computational resources within each node.

Control of distributed application execution in the WDC network includes enabling cooperation between the processes, segmentation of computational task into subtasks, and optimal resource allocation. The different aspects of control involve additional challenges in terms of flexible network protocol design. Control can be either centralized, distributed, or a combination of both. In the case of distributed control, the consensus among nodes is made through agreement protocols.

1.8   CONCLUDING REMARKS

With the rapid growth in the number of wireless applications, services, and devices, using a single wireless technology would not be efficient enough to deliver high-speed data rate and QoS support to mobile users in a seamless way. Next generation wireless networks will integrate (in a seamless manner) emerging radio technologies, ranked from personal to regional level, to create a heterogeneous system with improved capacity and complementary coverage. Because of the obvious heterogeneous nature of NGWS, the first and most important step in the presentation of emerging and perspective radio technologies is the standardization process overview. Although some standards enhance older ones (e.g., Wi-Fi and WiMAX), others have introduced new wireless access concepts (LTE and cognitive radio). Taking into account the future of existing wireless technologies and the possibility of invoking a heterogeneous environment, this chapter seeks to provide a brief overview of different challenges for next generation wireless multimedia system deployment.

Some of these techniques, applied in a wireless environment to improve the existing and new standard developments, have drawn important attention to this area of research. Among the most significant concepts for wireless technologies' interworking architectures, MIMO systems, cooperation techniques, relaying protocols, congestion control, as well as distributed computing, are imposed. The second part of this chapter is devoted to introducing these techniques.

From the interworking perspective, generic architecture for heterogeneous wireless networks together with access networks and multimodal terminal protocol stack are considered. A definition based on the level of service integration among networks is considered because of its independence from underlying network technologies and architectures. Four interworking levels: video network service access, intersystem service access, intersystem service continuity, and seamless intersystem service continuity are distinguished.

MIMO systems have great potential to improve the throughput performance of next generation wireless networks and simultaneously reduce energy consumption by exploiting multiplexing and diversity gains. Multifunctional MIMO schemes are capable of combining the benefits of several MIMO schemes attaining diversity, multiplexing, and beamforming gains. Several schemes have been introduced or discussed to mitigate, eliminate, or reduce the intercell interference. Among them, multi-BS MIMO cooperation schemes are expected to play a significant role in terms of interference mitigation.

The concept of cooperation is adopted in wireless networks in response to seamless user mobility support. Today, three cooperation techniques are often used: cooperation to improve channel reliability through spatial diversity, cooperation to improve throughput by resource aggregation, and cooperation to achieve seamless service provisioning. The potential benefits, such as improved reliability and throughput, reduced service cost and energy consumption, and support for seamless service provision, come with various challenges at different layers of the protocol stack. Also, fundamental cooperative approaches can be implemented in a network context with special emphasis given to cooperative routing.

The traditional relaying protocol is capable of improving the achievable E2E channel quality, but it requires four communication phases (i.e., time slots), which is twice as many as in classic direct communication. The successive relaying scheme requires nearly the same number of time slots as direct communication, but it needs an additional relay. As another approach to save valuable communication resources, network coding may be adopted, and here it is envisioned as a promising relaying protocol solution for NGWS.

In TCP, all packet losses are assumed to be caused by network congestion, thus leading to the reduction of window size by multiplicative-decrease algorithm. Therefore, the throughput of TCP can be improved if it is possible to distinguish link error–related losses from congestive ones. Drastic performance improvement cannot be expected without network node support, which can directly observe traffic conditions.

Finally, WDC enables powerful computing and extended service execution capabilities with the help of a collaborative network of small form factor radio nodes that operate with reasonably low-capacity batteries. The WDC network can potentially operate as an overlay on diverse wireless platforms. A comprehensive framework for executing distributed applications in a generic WDC network is still in the nascent stage.

2 Cognitive Radio Networks

Because cognitive radio technology can significantly help spectrum utilization by exploiting some of the parts unused by licensed users, it is rapidly gaining popularity and inspiring numerous applications. It should be mentioned that cognitive radio is the driver technology that enables next generation wireless networks to use spectrum in a dynamic manner. It can be defined as a radio that can change its transmitter parameters based on interaction with the environment in which it operates. The current concept shows the big picture of international standardization processes related to cognitive radio systems. Understanding these standardization activities is very important for both academia and industry to select important research topics and promising business directions. Most of the research on cognitive radio technology to date has focused on single-hop scenarios, tackling physical or medium access control layer issues, including the definition of effective spectrum sensing, decision, and sharing techniques. The research community started to realize the potential of multihop cognitive radio networks (CRNs), which can open up new and unexplored service possibilities, enabling a wide range of pervasive communication applications. Routing in multihop CRNs is a promising research area in well-explored wireless environments. A growing number of solutions targeting routing and channel assignments have been proposed by the research community. Designing routing algorithms and protocols for dynamic CRNs raises several issues related to route stability, control information exchange, and channel synchronization. The medium access control (MAC) protocols should consider the key features of cognitive networks such as the lack of a central unit to coordinate the communication, dynamic topology, requirements to keep interference to primary users minimal, and variation of spectrum availability with time and location.

2.1 INTRODUCTION

Cognitive radio (CR), in its original meaning, is a wireless communication paradigm utilizing all available resources more efficiently with its ability to self-organize, self-plan, and self-regulate [1]. CR-based technology aims to combat scarcity in the radio spectrum using dynamic spectrum access

(DSA). DSA technologies are based on the principle of opportunistically using available spectrum segments in a somewhat intelligent manner [2]. Because of the complexities involved in designing and developing CR systems (CRS) [3], more emphasis has been placed on the development of hardware platforms for full experimentation and CR features testing [4]. Since 1999, numerous different platforms and experimental deployments have been presented. These CR test beds differ significantly in their design and scope. Thus, the questions that often arise are how mature these platforms are, what has been learned from them, and if any trends from the analysis of functionalities provided by these platforms can be identified.

CR transceivers have the capability of completely changing their transmitter parameters (operating spectrum, modulation, transmission power, and communication technology) based on interactions with the surrounding spectral environment. They can sense a wide spectrum range, dynamically identify currently unused spectrum blocks for data communications, and intelligently access the unoccupied spectrum called spectrum opportunities [5]. Devices with cognitive capabilities can be networked to create CRNs, which are recently gaining momentum as viable architectural solutions to address the limited spectrum availability and the inefficiency in spectrum usage [6]. The most general scenario of CRNs distinguishes two types of users sharing a common spectrum portion with different rules:

- Primary (licensed) users (PUs) who have priority in spectrum utilization within the band and
- Secondary users (SUs) who must access the spectrum in a nonintrusive manner

PUs use traditional wireless communication systems with static spectrum allocation, whereas SUs are equipped with CRs and exploit spectrum opportunities to sustain their communication activities without interfering with PU's transmissions.

CR was first identified as a preferred technology for high-end applications in the military and public safety domains when the general concept had emerged [1]. Then CR research was also oriented toward the needs of civil wide area (cellular) and short-range communication systems. Early civil CR research was mainly motivated by ensuring an efficient operation of equipment in the 5 GHz band, whereas recent studies further investigate operation in lower frequency bands [7]. The involved industrial, regulatory, and academic partners were attracted to CR-based technology by the prospect of a hugely increased level of spectral efficiency and improved overall system capacity exploitation, among others, thanks to the dual exploitation of spectrum by applying opportunistic spectrum usage, as well as a mobile terminal being aware of its (heterogeneous) context and dynamically adapting its parameters such that its operational objectives are reached in an optimum way. For example, mobile terminals aware of surrounding radio access technologies (RATs) and select those that guarantee to fulfill its quality of service (QoS) requirements at the lowest cost in terms of subscription cost, power consumption, etc. This concept (aka network selection) will be analyzed in Chapter 4 in detail.

Standards are crucial for the development of new CR technologies as they encourage innovation in the industry and shorten the time to market of products and technologies [8]. Standardization efforts are taking place within the Institute of Electrical and Electronics Engineers (IEEE), European Computer Manufacturers Association (ECMA), European Telecommunications Standards Institute (ETSI), and International Telecommunication Union (ITU). One of the first IEEE Working Groups to consider CR technology was IEEE 802.22, created in 2004 and developing a standard for WRANs using white spaces in the TV frequency spectrum (see Chapter 1 part 1.2). Growing interest in CR was demonstrated starting in 2005 of the IEEE Communications Society Technical Committee on Cognitive Networks. Moreover, due to the importance of CR, the IEEE initiated a set of standardization projects called IEEE P1900 in 2004, which evolved into the IEEE Standards Coordinating Committee 41 (IEEE SCC41) in 2006. The activities of the IEEE SCC41 aim at facilitating the development of research ideas into standards to expedite the results of research for public use. Recently, the IEEE SCC41 has become the premier forum for CR standardizing concepts, and was renamed as IEEE Dynamic Spectrum Access Networks Standards Committee (DySPAN-SC) [9]. Another relevant standard is ECMA-392 [10], initially published in 2009 and revised in 2012, which specifies a physical (PHY) layer and MAC sublayer for personal/portable CRNs operating in TV bands. The work in ETSI Reconfigurable Radio Systems Technical Committee is complementary to the activities of the IEEE DySPAN and IEEE 802, with the focus on the software-defined radio (SDR) standards beyond the IEEE scope, CR/SDR standards addressing the specific needs of the European Regulation Framework, and TV white spaces (TVWS) standards adapted to the digital TV signal characteristics in Europe [11].

Most of the research related to CRNs to date has focused on single-hop scenarios tackling PHY and MAC layers issues, including the definition of effective spectrum sensing, decision, and sharing techniques [12]. Recently, the research community has started realizing the potentials of multihop CRNs, which can open up new and unexplored dimension enabling a wide range of pervasive communications applications. The cognitive paradigm can be applied to different scenarios of multihop wireless systems including mesh networks featuring a semistatic network infrastructure [13] and ad hoc networks characterized by completely self-configuring architecture [14]. Effective routing solutions must be integrated into the work carried out on the lower layers while accounting for the unique properties of the cognitive environment.

Traditional wireless networks (such as cellular systems) have been designed from a centralized perspective with a predefined infrastructure, but this rigid approach lacks flexibility and adaptability, which are advantages of next generation CRSs. Thus, it is of paramount importance to design self-configurable CR-based networks that are aware of and adaptable to the changing environment to coexist harmoniously with other wireless systems that use a variety of protocols in the same frequency bands [8]. The design of such networks requires the development of a number of new access and transmission technologies to ensure successful coexistence and to avoid interference. CR technologies such as DSA create huge opportunities for research and development in a wide range of applications including spectrum sensing, navigation, biomedicine, etc.

This chapter starts with a CRS concept. Next, dealing with CR-related applications is performed. Spectrum sensing is also included. After that, the importance of multihop CRNs is presented. Access control in distributed CRNs concludes the chapter.

2.2   COGNITIVE RADIO SYSTEM CONCEPT

Cognition is "the process involved in knowing, or the act of knowing, which, in its completeness, includes awareness and judgment" [15]. Clearly, and as often pointed out, perceiving, recognizing, and reasoning are also closely related to the cognitive process. Cognition is not a completely new concept in wireless multimedia communications. As a matter of fact, system states like channel condition and usage of available resources can usually be obtained by sensing the surrounding wireless environment from a received signal.

The term cognitive radio, coined by Mitola and Maguire [1], "identifies the point at which wireless personal digital assistants and the related networks are sufficiently computationally intelligent about radio resources and related computer-to-computer communications to detect communications needs as a function of use context, and to provide radio resources and services most appropriate to those needs." CR and SDR are two relatively new concepts in wireless multimedia communications that will change the way that radio systems are designed and operated. Also, these changes will have significant effects on antenna requirements in applications from mobile to satellite communications [16].

2.2.1   COGNITIVE RADIO

The existence of many wireless communications applications, especially in the region of 0.8 to 3 GHz, has increased the spectrum use, causing

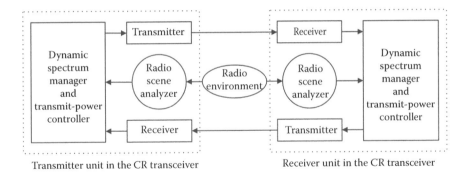

FIGURE 2.1 Information flow in cognitive radio.

significant spectrum congestion [17]. As presented in Figure 2.1, there are three fundamental cognitive tasks in the perception–action cycle of CR [18]:

a. Radio scene analysis of the environment, which is performed in the receiver
b. Dynamic spectrum management and transmit power control, both of which are performed in the transmitter, and
c. Global feedback, enabling the transmitter to act in light of information about the radio environment feedback to it by the receiver

There are a number of DSA models possible, such as shared use by underlay methods, as in the ultra-wideband (UWB) systems, overlay as in CR used in industrial, scientific, and medical (ISM) bands.

2.2.2 SOFTWARE-DEFINED RADIO

The current trust in SDR was first described by Mitola [19]. Since that first description, interest has risen, driven to some extent by the great promise of low cost and available processing. SDR is seen as an enabling technology for CR. It offers much promise to increase spectrum usage efficiencies to users in a wide variety of applications, covering commercial, military, and space communications [16]. SDR does not represent a single concept, but has several definitions. For example, one is given by Wireless Innovation Forum (WinnF) [20], working in collaboration with the IEEE P1900.1 group, as "radio in which some or all of the physical layer functions are software-defined." Others include software-based radio, reconfigurable radio, and flexible architecture radio. The early concept of an amplifier followed by analog to digital converter is used at very low frequencies [21].

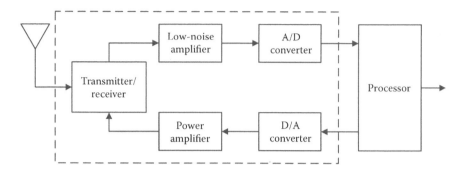

FIGURE 2.2 An example of SDR architecture.

An example of SDR architecture is shown in Figure 2.2 with the addition of low-noise and high-power amplifiers. This reflects the travel in SDR transceiver design, toward integrating radiofrequency (RF) front ends and signal processing circuits onto one chip, which is driven by requirements for small and low-cost equipment.

2.2.3 CAPABILITIES OF COGNITIVE RADIO SYSTEM

ITU-Radio Communication Sector (ITU-R) Working Party 1B defines the CRS as a radio system employing technology that provides the system the capability to obtain knowledge of its environment, to adjust operational parameters and protocols, and finally, to learn from the obtained results [22]. The knowledge used by the CRS includes operational radio and geographic environment, internal state, established policies, usage patterns, and users' needs. The methods for obtaining knowledge include [23]: getting information from component radio systems, geolocation, spectrum sensing, and access to a cognitive pilot channel (CPC) and white spaces database. Component radio systems of the CRS include received signal power, signal to interference and noise ratio (SINR), and load. Frequency bands and RATs used by base stations and terminals, as well as transmission power values, contribute a lot to the knowledge of the CRS. The positions of base stations, terminals, and other radio systems can be obtained using geolocation. It can be performed using localization systems such as global positioning system (GPS) or wireless positioning system. White spaces database access and spectrum sensing are very important in some deployment scenarios of the CRS. These two approaches are used to identify white spaces and detect PUs, although they may also be used to detect SUs. As for CPC, it serves as a means to exchange information between components of the CRS, and in such cases the CPC is typically considered a part of the CRS.

The next characteristic of the CRS is its capability to dynamically and autonomously adjust its operational parameters and protocols according to the obtained knowledge to achieve some predefined objectives. Adjustment consists of two stages: decision making and reconfiguration. CRS includes an intelligent management system responsible for making decisions regarding parameters and protocols that need to be adjusted. The reconfiguration stage may include a change of the following parameters: output power, frequency band, and RAT.

The third key characteristic of the CRS is its capability to learn from the results of its actions to further improve its performance. Figure 2.3 shows CRS concept summary. The main components of the CRS are the intelligent management system and reconfigurable radios. The four main actions of the CRS are obtaining knowledge, making decisions, reconfiguration, and learning [23].

Using the obtained knowledge, the CRS makes reconfiguration decisions according to predefined objectives to improve the efficiency of spectrum usage. Based on the decisions made, the CRS adjusts operational

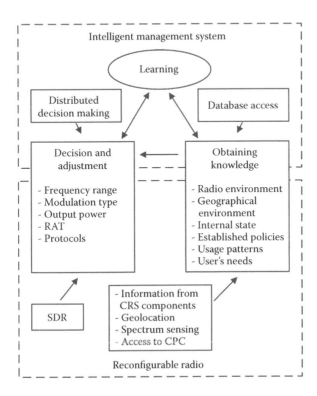

FIGURE 2.3 Cognitive radio system summary.

parameters and protocols, whereas the learning results contribute to both obtaining knowledge and decision making.

2.2.4 CENTRALIZED AND DECENTRALIZED COGNITIVE RADIO SYSTEM

The centralized and decentralized solutions for CRS are shown in Figure 2.4. The centralized, operator-driven solution is designed for wide area utilization. On the other hand, the decentralized solution is used for local area ad hoc/mesh networking [7].

The centralized CRS concept is represented by the composite wireless network including cognitive network management systems. Here, the key components are the operator spectrum manager (OSM) and joint radio resource management (JRRM). The decentralized CRS concept is represented by the cognitive mesh network controlled by the cognitive control network. As for the key enabling technologies, they include SDR and multi-radio user equipment (UE), reconfigurable base stations (BSs) management, spectrum sensing, CPC, cognitive control radio and networking, geolocation, primary protection database, and distributed decision making [7].

2.2.5 COGNITIVE PILOT CHANNEL

The CPC is defined in ETSI TR 102 683 V1.1.1 [24] as a channel that conveys the elements of necessary information facilitating the operations of the CRS. The CPC provides information on which radio accesses can be expected in a certain geographical area. This information includes operator

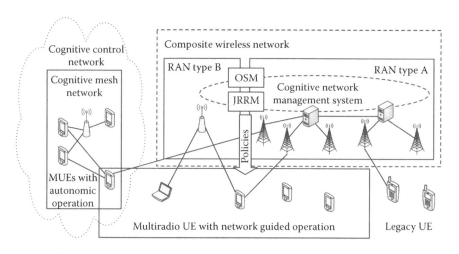

FIGURE 2.4 Centralized and decentralized cognitive radio system.

information, RAT type as well as used frequencies. Besides this, the CPC can potentially also communicate other data such as pricing information and (potentially time variant) usage policies, and can even be used to transmit missing protocols [25]. In this way, a CPC can eliminate the need for continuous scanning of the entire spectrum, while allowing services and RATs to be changed without limits. Moreover, if applied on a regional or global scale, a harmonized CPC frequency can greatly improve the cross-border functionality of devices.

The CPC is used to support a terminal during the startup phase in an environment where the terminal does not yet know the available RATs and corresponding used frequencies [7]. In the context of a secondary system, the CPC is used to exchange sensing information between MTs and BSs to perform collaborative/cooperative sensing, facilitating the searching of white spaces to start communication. Also, the CPC is used for an efficient level of collaboration between a network and the terminals by supporting radio resource management optimization procedures.

The basic principle of the CPC is shown in Figure 2.5 as an example of its usage in a heterogeneous RAT environment. Different CPC deployment approaches are possible.

In the out-band CPC solution, the CPC is conceived as a radio channel outside the component RAT. The CPC uses either a new radio interface, or alternatively, an adaptation of legacy technology with appropriate characteristics. In the in-band CPC solution, the CPC is conceived as a logical channel within the technologies of the heterogeneous radio environment. CPC can be observed as a promising solution for the distribution of required information for network selection initiation and will be analyzed in Chapter 4.

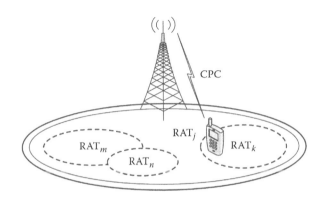

FIGURE 2.5 Basic principle of CPC.

2.2.6   Key Types of Cognitive Radio Systems

Two key types of the CRS can be identified [22]: heterogeneous CRS and spectrum-sharing CRS.

In the *heterogeneous CRS*, one or several operators are exploiting more RANs using the same or different RATs. Frequency bands allocated to these RANs are fixed. Operators provide services to users having different terminals. One type of terminal is legacy, designed to use a particular RAT. Such a terminal can connect to one particular operator or other operators having roaming agreements with the home operator. Another type of terminal has the capability to reconfigure itself to use different RATs operated by different operators. Optionally, this type of terminal can support multiple simultaneous links with RANs. In one scenario, the CRS has only legacy BSs, although some of the terminals are reconfigurable. Such terminals can make decisions to reconfigure themselves to connect to different component RANs inside the CRS. The heterogeneous CRS is considered in the IEEE 802.21 (see Chapter 4) and IEEE 1900.4 [26] standards.

In the *spectrum-sharing CRS*, several RANs using the same or different RATs can share the same frequency band. One deployment scenario of this type of CRS is when several RANs operate in unlicensed or lightly licensed spectrums, where the CRS capabilities can enable the coexistence of such systems. Another deployment scenario is when a secondary system operates in the white spaces of a TV broadcast operator frequency band. In such a scenario, the CRS capabilities should provide protection for the primary service (TV broadcast) and coexistence between secondary systems. Spectrum-sharing CRS is considered in the following standards: IEEE 1900.4, IEEE 1900.6 [27], IEEE 802.11y [28], IEEE draft standard P802.11af [29], IEEE draft standard P802.19.1 [30], IEEE 802.22 (see Chapter 1), and ECMA-392 [10].

2.2.7   Cognitive Cycle

The cognitive capability of a CR enables real-time interaction with its environment to determine appropriate communication parameters. Also, adoption into the dynamic radio environment is of great importance. The cognitive cycle is shown in Figure 2.6. There are three main steps in the cognitive cycle [6]: spectrum sensing, spectrum analysis, and spectrum decision.

In spectrum sensing, a CR monitors the available spectrum bands, records their information, and then detects the spectrum holes. Spectrum analysis takes into account the spectrum hole's characteristics that are detected and estimated through spectrum sensing. Spectrum decision

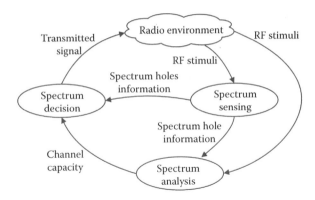

FIGURE 2.6 Cognitive cycle.

understands that a CR determines the data rate, the transmission mode, and the bandwidth of the transmission. Then, the appropriate band is chosen according to the spectrum's characteristics and the user's requirements.

When the operating spectrum band is determined, the communication can be performed over this spectrum band. However, because the radio environment dynamically changes over time and space, the CR should keep track of the changes in the radio environment. If the current spectrum band becomes unavailable, the spectrum mobility function is performed to provide a seamless transmission. Any environmental change during the transmission such as PU appearance, user movement, or traffic variation can trigger this adjustment.

2.2.8 Reconfigurable Radio Systems Management and Control

Reconfigurability is the capability to adjust operating parameters for transmission without any modifications on the hardware's components. This capability enables the CR to adapt easily to the dynamic radio environment [6]. The reconfigurable parameters that can be incorporated into the CR include operating frequency, modulation, transmission power, and communication technology.

A CR is capable of changing the operating frequency. Based on the information about the radio environment, the most suitable operating frequency can be determined and the communication can be dynamically performed on this appropriate operating frequency.

A CR should reconfigure the modulation scheme adaptive to the users' requirements and channel conditions. For example, in some delay-sensitive applications, the data rate is more important than the error rate. Thus, the modulation scheme that enables the higher spectral efficiency should be selected. On the other side, loss-sensitive applications focus on the error rate, which necessitate modulation schemes with low bit error rates.

Transmission power can be reconfigured within the power constraints. Power control enables dynamic transmission power configuration within the admissible power limit. If higher power operation is not necessary, the CR reduces the transmitter power to a lower level to allow more users to share the spectrum and to decrease the interference.

Communication technology understands that a CR can be used to provide interoperability among different systems. The CR transmission parameters can be reconfigured at the beginning of a transmission, as well as during the transmission.

ETSI RRS WG3 has collected and reported the following set of requirements for defining a functional architecture that is able to provide optimized radio and spectrum resources management [31]:

- Personalization, to support various classes of users
- Support of pervasive computing, enabled by the existence of sensors, actuators, and WLANs in all application areas
- Context awareness, for efficiently handling multiple, dynamically changing, and unexpected situations
- Always best connectivity, for optimally serving equipment and users in terms of QoS and cost
- Ubiquitous application provision for the applications above
- Seamless mobility, for rendering the user's agnostic of the heterogeneity of the underlying infrastructure
- Collaboration with alternate RATs for contributing to the achievement of always best connectivity
- Scalability, for responding to frequent context changes

To address these requirements, a proper functional decomposition was proposed in ETSI TR 102 682 V1.1.1 [31]. The functional blocks, together with the interfaces among them and their distribution between networks and terminals, are depicted in Figure 2.7.

The dynamic spectrum management (DSM) block is responsible for the medium-term and long-term (both technical and economical) management of spectra and, as such, it incorporates functionalities like provisioning of information for spectrum assignments and occupancy evaluation and decision making on spectrum sharing. Dynamic self-organizing network planning and management (DSONPM) caters to medium-term and long-term management at the level of a reconfigurable network segment (e.g., incorporating several BSs). It provides decision-making functionality for QoS assignments, traffic distribution, network performance optimization, RATs activation, configuration of radio parameters, and so on.

JRRM block functionalities mainly include radio access selection, neighborhood information provision, and QoS/bandwidth allocation/

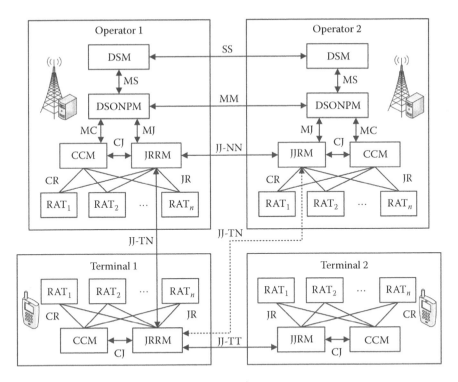

FIGURE 2.7 ETSI reconfigurable radio system functional architecture.

admission control. Finally, configuration control module (CCM) is responsible for the enforcement of the reconfiguration decisions typically made by the DSONPM and JRRM. The interfaces between the identified functional blocks used in this architecture are summarized in Table 2.1.

Example 2.1

The characteristic of CR transceiver is a wideband sensing capability of the RF front-end [6]. This function is mainly related to RF hardware technologies such as wideband antenna, power amplifier, and adaptive filter. The wideband RF signal presented at the antenna includes signals from close and widely separated transmitters and from transmitters operating at widely different power levels and channel bandwidths. As a result, detection of weak signals must frequently be performed in the presence of very strong signals. Thus, there will be extremely stringent requirements placed on the linearity of the RF analog circuits as well as their ability to operate over wide bandwidths. RF hardware for the cognitive radio should be capable of tuning to any part of a large range of frequency spectrum. Such spectrum sensing

TABLE 2.1

Interfaces in ETSI Reconfigurable Radio System Architecture

Interface	Location	Description
MS	Between DSM and DSONPM	Requests and information exchange about available spectrum for the different RATs, spectrum opportunities, as well as the cost of service provision
MC	Between DSONPM and CCM	Exchange of configuration information
MJ	Between DSONPM and JRRM	Information exchange on the current context, i.e., the amount of resources used in each RAT and cell as well as other relevant context and status information
CJ	Between CCM and JRRM	Reconfiguration related synchronization
CR	Between CCM and RAT	Underlying RATs configuration/reconfiguration execution
JR	Between JRRM and RAT	Exchange resource status information, like cell load or measurements, of the current active and candidate links
JJ-TN	Between the JRRMs on terminal and network side	Send neighborhood information from the network to the terminal, provide access selection information (e.g., policies or handover decisions), exchange measurement information (link performances, spectrum usage, etc.)
SS	Between DSM instances	Exchange information about spectrum usage policies, negotiate on the spectrum usage between operators
MM	Between DSONPM instances	Information exchange on the network configuration to avoid or reduce interference
JJ-NN	Between two JRRM instances on network side	Support the handovers negotiation and execution between different operators

enables real-time measurements of spectrum information from the radio environment. The wideband CR front-end architecture [32] is shown in Figure 2.8.

The components of a front-end architecture are as follows. The RF filter selects the desired band by bandpass filtering the received RF signal. A low-noise amplifier (LNA) amplifies the desired signal while simultaneously minimizing the noise component. In the mixer, the received signal is mixed with a locally

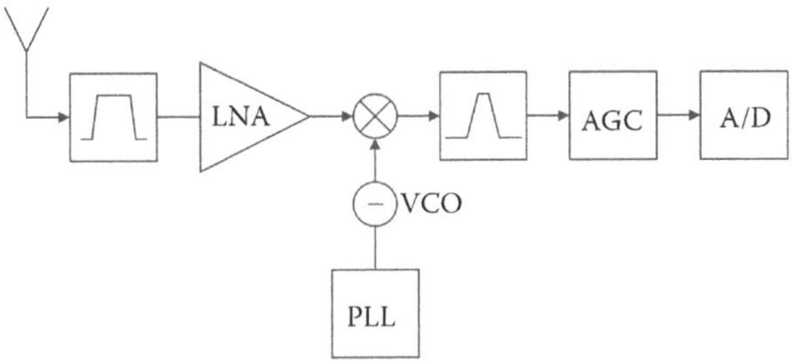

FIGURE 2.8 Wideband front-end architecture for cognitive radio.

generated RF frequency and converted to the baseband or the intermediate frequency. A voltage-controlled oscillator (VCO) generates a signal at a specific frequency for a given voltage to mix with the incoming signal. A phase-locked loop (PLL) ensures that a signal is fixed on a specific frequency and can also be used to generate precise frequencies with fine resolution. The channel selection filter is used to select the desired channel and to reject the adjacent channels. The direct conversion receiver uses a low-pass filter for channel selection. On the other hand, the superheterodyne receiver adopts a bandpass filter. Automatic gain control (AGC) maintains the gain or output power level of an amplifier constant over a wide range of input signal levels. Wideband signal is received through the RF front-end, sampled by the high-speed A/D converter with high resolution, and measurements are performed for the detection of the licensed user signal.

Example 2.2

An example of a prototype spectrum-sharing CRS can be designed based on the system architecture and functions described in the previous part of this chapter. The CRS is operated in frequency bands from 400 MHz to 6 GHz for user communication. The CRS also utilizes 720 MHz band for out-of-band pilot channel (OPC) transmitter. Because the frequency band for the OPC is not standardized, the 720 MHz band is adopted for empirical evaluation purposes [33].

In this prototype, the OPC transmitter is implemented inside the CR base station (CRBS). Therefore, the CRBS is composed of a user data communication part and an OPC part, as shown in Figure 2.9. Each part is composed of CPU, field-programmable gate array (FPGA), and RF boards. The CPU board executes processing of layer 3 and above, whereas the FPGA board executes

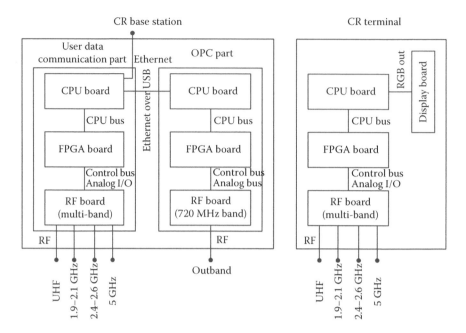

FIGURE 2.9 Hardware prototype architecture of cognitive radio base station and terminal. (From K. Ishizu et al., Feasibility study on spectrum sharing type cognitive radio system with outband pilot channel. *Proc. 6th CROWNCOM 2011. Osaka, Japan,* June 2011: 286–290.)

PHY baseband processing and MAC. The RF board executes PHY modulation and demodulation. The two CPU boards are connected via Ethernet and radio configuration information is provided from the user data communication part to the OPC part over IP. The CPU and FPGA boards are connected via CPU bus. The FPGA and RF boards are connected via analog I/O and control bus. The RF board has four RF interfaces according to frequency bands. The hardware architecture of the CR terminal is similar as the user data communication part of CRBS. Additionally, the CR terminal has a display connected to the CPU board. Specifications for the user data communications and the OPC radio terminal are shown in Table 2.2.

The user data communications is capable of selecting operational frequency from a wider range of frequencies and a higher data rate. On the other hand, the OPC is capable of using only fixed frequencies and low data rate. This is because a narrower frequency band of 4.15 MHz is assigned to the OPC compared with the 20 MHz assigned to the user data communication.

TABLE 2.2
Spectrum-Sharing Prototype Terminals' Specification

Characteristic	User Data Communications Terminal	OPC Terminal
Frequency	UHF, 1.9–2.1, 2.4–2.6, 5GHz	720/722 MHz
Bandwidth	20 MHz	4.15 MHz
PHY modulation	OFDM (total 52 carriers: 48 for data + 4 for control)	
Subcarrier modulation	16QAM, FEC (coding rate 1/2, constraint length 7)	BPSK, FEC (coding rate 1/2, constraint length 7)
MAC	IEEE 802.11 compatible	
Output power	+10 dBm	+20.66 dBm

Because the OPC should be capable of more reliable communications, binary phase shift keying (BPSK) for OFDM subcarrier and higher output power are used for the OPC.

2.3 COGNITIVE RADIO DEPLOYMENT ISSUES

There are new opportunities for CR to enable a variety of emerging applications. CR has emerged as a promising technology to enhance spectrum utilization through opportunistic on-demand spectrum access [34]. The cognitive capability provides spectrum awareness, whereas reconfigurability enables a CR user to dynamically adapt its operating parameters to the surrounding wireless environment. More specifically, the CR can be programmed to transmit and receive over widely separate frequency bands, adapt its transmit power, and determine its optimal transmission strategy. At the same time, the licensed users of the spectrum are not affected. This necessitates adapting to the dynamically changing spectrum resource, learning about spectrum occupancy, and making decisions on the quality of the available spectrum resource including its expected duration of use and disruption probability. Thus, CRNs help make efficient use of the available spectrum by using bands, such as TV broadcast frequencies below 700 MHz.

CR is being intensively researched as the enabling technology for secondary access to TVWS [35]. The TVWS comprises large portions of the UHF spectrum (and VHF in the United States) that is becoming available on a geographical basis for sharing as a result of the switchover from analog to digital TV. This is where SUs can, using unlicensed equipment, share the spectrum with the digital TV transmitters and other primary (licensed) users such as wireless microphones.

In both the United States [36] and the United Kingdom [37], the regulators have given conditional endorsement to this new sharing mode of access. Also, there is significant industry effort under way toward standardization, trials, and test beds. These include geolocation databases and sensing for PU protection, agile transmission techniques, and the so-called etiquette protocols for intrasystem and intersystem coexistence in TVWS [35]. The provision of commercial services based on the technology (e.g., unlicensed mobile or wireless home networks) will involve situations with systems of multiple cognitive equipment types that may belong to either the same or different service providers.

2.3.1 TVWS Services

There is a growing demand for high data-rate wireless services such as data, video, and multimedia to users in homes, in offices, and on the move. On the other side, telecom operators are under pressure to cost-effectively provide universal broadband service to rural communities, and are investigating wireless options as an alternative to wired technologies [38].

Network operating costs cause traditional macro network designs to be uneconomical in providing equivalent coverage and capacity using licensed 3G and LTE bands because of the need for smaller cells. This need arises both from demand for higher bit rates and to support more users in the system. Therefore, alternative solutions based on CR technology operating on an exempt–exempt basis in the TVWS spectrum are becoming commercially interesting, in particular for operators with significant network infrastructure. Also, TVWS can provide a viable and highly scalable alternative to conventional solutions based on cellular or Wi-Fi technologies (or both) [35].

Implementations of TVWS services are likely to start with point-to-multipoint deployments (zero mobility), such as rural broadband access and backhaul to small 3G/4G cells, and later progress to more mobile and QoS-aware systems. Access to TVWS will enable more powerful Internet connections with extended coverage and improved download speeds. Time-division duplex (TDD) systems are preferable to frequency division duplex (FDD) when using TVWS because FDD requires a fixed separation of BS transmit and terminal transmit frequencies. This condition restricts the number of available TVWS channels. TDD is free from this restriction and is also better suited to asymmetrical links. These factors point toward Wi-Fi, WiMAX, and LTE being suitable candidates that have mature standards.

The commercial case for using TVWS will depend on the amount of spectrum that becomes available for sharing, on how the availability of this spectrum varies with location, and on transmit power allowed by cognitive devices [39].

User cases in TVWS can be classified into three scenarios [35]:

1. Indoor services, which generally require small coverage, and hence, power levels that are either significantly lower, due to better propagation characteristics in the VHF/UHF bands, or comparable to that used in current ISM bands
2. Outdoor coverage from indoor equipment, which requires penetration through barriers with medium range coverage (a few hundred meters), and hence, power levels that are generally higher than or comparable to that in the ISM bands
3. Outdoor services, which may require significantly higher transmit power levels that are currently permitted in the ISM bands, comparable to those used by cellular systems

Most indoor applications of TVWS can already be realized using Wi-Fi and ZigBee technology operating in the 2.4 and 5 GHz ISM bands. The main advantage of using TVWS is that the additional capacity offered will help relieve congestion, in particular in the 2.4 GHz band. This use can also result in better indoor propagation of signals through the home. Furthermore, the lower frequencies of TVWS bands can result in lower energy consumption compared with Wi-Fi and ZigBee. This is an interesting advantage for use case scenarios that involve battery-powered devices.

2.3.2 Secondary Access to White Spaces Using Cognitive Radio

Secondary operation of CR in TV bands is conditioned by regulators on the ability of these devices to avoid harmful interference to incumbents, which in addition to TV stations, include program producers and special event users. Furthermore, successful operation in these bands relies on the ability of CRs to reliably detect and use TVWS. To achieve these objectives, the main methods have been considered and evaluated by a number of regulators such as geolocation databases, beacons, and spectrum sensing [40].

In *geolocation databases* approach, a CR device queries a central database with its location and other device specifications to find out which TVWS frequencies are available for its operation at a given location and time, height, and required service area. The geolocation database then uses this information along with a database of location, transmit power, frequencies, and antenna radiation patterns of all TV transmitters to perform a set of propagation modeling calculations [41]. The outcome of these calculations is a list of available TVWS channels that can be used by the requesting device without causing harmful interference to TV services. Limits on allowed transmit powers, and possibly time validity parameters for each channel are also included. Protection via a geolocation database is mainly applicable to systems that have usage patterns that are either fixed in time or vary slowly. In that way, information stored in a database does not require frequent updating. Furthermore, devices need to know their locations with a level of accuracy prescribed by regulators (50–100 m for TVWS access). First, to access the database, a device needs to either be connected to the Internet over a wired link or be able establish a wireless link that does not require secondary spectrum. Some of these issues can be addressed in master–slave communication architectures in which a master device, such as AP or BS, is already connected to the Internet via a wireless or fixed link and can also geolocate itself. Then, the master node uses its location to query the geolocation database about available secondary spectrum within a predefined service range.

With the *beacons* method, CRs only transmit if they receive an enabling beacon granting them use of vacant channels. Alternatively, a CR may transmit as long as it has not received a disabling beacon denying it the use of these channels. One issue with this approach is that it requires a beacon infrastructure. Furthermore, beacon signals can be lost due to mechanisms similar to the hidden node problem, which occurs when there is blockage between the secondary device and a primary transmitter [39].

Spectrum sensing is a crucial technique in CRNs to accurately and efficiently detect PUs for avoiding interference. In this method, devices autonomously detect the presence (or absence) of primary system signals using a detection algorithm. Many unpredictable problems (for example, channel instability and noise uncertainty) can significantly degrade the performance of spectrum sensing, which is characterized by both sensing accuracy and sensing efficiency [42].

Sensing accuracy refers to the precision in detecting PU signals such that the primary transmissions are not interfered. It is represented by a false alarm probability and a detection probability. On the other hand, sensing efficiency refers to the number of spectrum opportunities discovered by consuming a unit of sensing cost in terms of sensing overhead and throughput. Sensing accuracy and efficiency are two opposite aspects that reflect the performance of spectrum sensing. It is observed that more sensing SUs will lead to higher sensing accuracy but more sensing overhead, that is, less sensing efficiency. Because the overall system performance of CRNs potentially depends on both aspects, the trade-off between them should be optimally addressed.

To solve the hidden node problem, cooperative sensing algorithms have been proposed [42], in which multiple users of the secondary system cooperate to combat the unpredictable dynamics in wireless environments and improve sensing accuracy and efficiency. The procedure for cooperative spectrum sensing is shown in Figure 2.10.

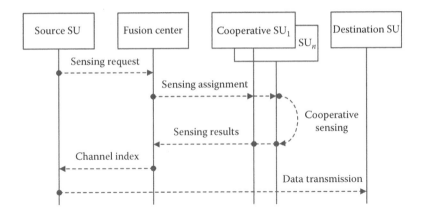

FIGURE 2.10 Cooperative spectrum sensing procedure.

The source SU that intends to transmit packets will send a request to the fusion center, claiming for the cooperative spectrum sensing. The fusion center is responsible for receiving and combining the sensing results to make a final decision. It selectively assigns several cooperative SUs for cooperation. The sensing results are sent back to the fusion center for collaborative decision. Once the spectrum opportunities are discovered, the packets can be transmitted from the source to the destination SU on the detected channel specified by the fusion center.

Cooperation mechanisms in spectrum sensing are different because of the following [42]:

a. The cooperative SUs are selected and scheduled for cooperation.
b. The sensing results are transmitted to a fusion center.
c. The sensing results are combined.

According to the number of sensed channels in one sensing period, cooperative spectrum sensing can be broadly categorized into sequential and parallel cooperative sensing. In sequential cooperative sensing, all the SUs are scheduled to sense an identical channel in each period. Channels are sensed one by one sequentially. This sequential cooperative sensing is the so-called traditional cooperative spectrum sensing scheme for the sake of improving sensing accuracy by inherently exploiting the spatial diversity of the cooperative SUs. On the other hand, in parallel cooperative sensing, more than one channel is sensed in each period. The cooperative SUs are divided into multiple groups while each group senses one channel. The motivation of parallel cooperative sensing is to enhance sensing efficiency by allowing the cooperative SUs to sense distinct channels in one sensing period. Because multiple channels are detected in one sensing period, the period for finding all available channels is much shorter than that in

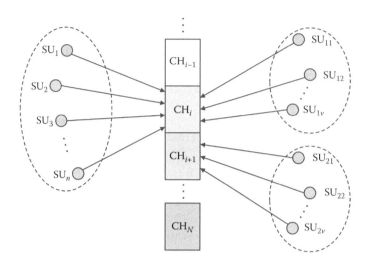

FIGURE 2.11   Sequential (left) and parallel (right) cooperative sensing.

sequential cooperative sensing. The sensing efficiency is then significantly enhanced. Figure 2.11 shows sequential cooperative sensing, together with parallel cooperative sensing.

The SU's cooperation can also be categorized into synchronous and asynchronous mechanisms based on the moments when sensing operations are carried out. In synchronous cooperative sensing, all cooperative SUs have the same sensing period, and perform sensing at the same time. In asynchronous cooperative sensing, each SU performs spectrum sensing according to its own sensing period. Hence, the moments when SUs perform sensing may be different. It should be noted that one of the key problems with cooperative sensing is that the gains, compared with a single sensor, depend on location and number of SUs, which will typically be random.

Example 2.3

In the full-parallel cooperative scheme, each cooperative SU senses a distinct channel in a centralized and synchronized mode. Upon receiving the request of parallel cooperative sensing from the source SU, the fusion center will deliberately select a subset of cooperative SUs to perform sensing. Each of these selected SUs is assigned to sense a different channel at the same time during the sensing period. Thereby, these SUs perform spectrum sensing in a parallel manner. After each round of sensing, the cooperative SUs will send back the sensing results to the fusion center, indicating the channel availability (busy or idle). When an available channel with a satisfactory data rate has been found, the fusion center will broadcast the stopping command to terminate the parallel cooperative sensing. The fusion center selects the discovered available channel with the highest achievable rate for the source SU. The fusion center then delivers the channel index to the source SU, which will transmit over this allocated channel.

Let n and x denote the number of the cooperative SUs and the channel rate threshold, respectively. The achievable throughput of the full-parallel cooperative sensing is defined as

$$T_{FPCS} = \max_{n,x}\{G(n,x) - O(n,x)\}, \qquad (2.1)$$

where $G(n,x)$ and $O(n,x)$ are the transmission gain and the sensing overhead under parameters n and x, respectively, given by

$$G(n,x) = \sum_{j=1}^{\left\lceil \frac{N}{n} \right\rceil}\left[\sum_{m=x}^{M} p_j^m(n,x)r_m \overline{T_a}\right], \qquad (2.2)$$

and

$$O(n,x) = \sum_{j=1}^{\left\lceil \frac{N}{n} \right\rceil}\left[\sum_{m=x}^{M} p_j^m(n,x)r_{sum} jt_s\right]. \qquad (2.3)$$

Here, N denotes the total channel number, r_m the mth channel rate level ($m = 1, 2, ..., M$), $p_j^m(m,x)$ the probability that the channel with rate r_m is discovered in the jth round of full-parallel cooperative sensing, and $\overline{T_a}$ the average duration of channel available/idle time, whereas t_s denotes the sensing time of each cooperative SUs, and r_{sum} the sum of channel rate of all the n cooperative SUs.

To reduce the sensing overhead, the fusion center will sort all the SUs by their instant channel rate from the lowest to the highest, and then select the first n SUs as the cooperative SUs. Performance comparison among the full-parallel cooperative sensing, random sensing, and noncooperative sensing schemes is presented in Figure 2.12.

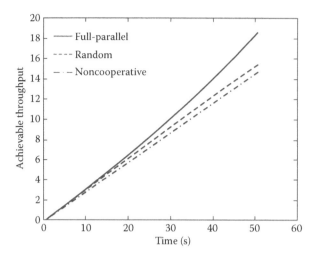

FIGURE 2.12 Comparison among the full-parallel cooperative sensing, the random sensing, and the noncooperative sensing schemes. (From R. Yu et al. Secondary users cooperation in cognitive radio networks: balancing sensing accuracy and efficiency. *IEEE Wireless Communications* 19, no. 2 (April 2012): 30–37.)

Example 2.4

To trade off the sensing accuracy and efficiency, semiparallel cooperative sensing is proposed [42]. The cooperative SUs are scheduled by the fusion center in a centralized manner. Upon receiving the sensing request from the source SU, the fusion center sends a beacon to a number of SUs to invite participation in the sensing operation. The fusion center then divides the cooperative SUs into multiple groups. The cooperative SUs in a same group are notified to sense an identical channel while each group is assigned to a different channel. All the cooperative SUs sense channels synchronously. After a sensing period, all the sensing results are sent back to the fusion center, which will determine the availability of all the sensed channels. Once an available channel is discovered, the fusion center should stop the sensing procedure. The source SU will be informed with an index of available channels.

In the semiparallel cooperative sensing, the grouping strategy will determine the levels of sensing accuracy and efficiency. For a given total number of the cooperative SUs, more SUs in a group will lead to higher accuracy. In this case, there are less groups and, hence, lower sensing efficiency. If all the cooperative SUs are allocated in the same group, the scheme becomes the sequential cooperative sensing. If each cooperative SU is individually set as a group, the scheme reduces to the full-parallel cooperative sensing. By varying the number of groups and the number of cooperative SUs in each group, the trade-off between sensing accuracy and efficiency can be adjusted to a different extent.

Let u and v denote the number of groups and the number of cooperative SUs in each group, respectively. These parameters are determined by maximizing the achievable throughput

$$T_{SPCS} = \max_{u,v}\{G(u,v) - O(u,v)\} \qquad (2.4)$$

where $G(u,v)$ and $O(u,v)$ are the transmission gain and the sensing overhead under parameters u and v, respectively, given by

$$G(u,v) = \sum_{j=1}^{\left\lceil\frac{N}{n}\right\rceil} p_j(u,v) r \overline{T_a}, \qquad (2.5)$$

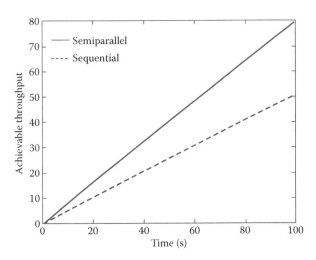

FIGURE 2.13 Comparison among the semiparallel and sequential cooperative sensing schemes. (Partially from R. Yu et al. Secondary users cooperation in cognitive radio networks: balancing sensing accuracy and efficiency. *IEEE Wireless Communications* 19, no. 2 (April 2012): 30–37.)

and

$$O(u,v) = \sum_{j=1}^{\left\lceil \frac{N}{n} \right\rceil} p_j(u,v) v \bar{r} j t_s. \qquad (2.6)$$

Here, $p_j(u,v)$ denotes the probability that the available channel is discovered in the jth round of semiparallel cooperative sensing, and $\bar{r}$ the average channel rate for SUs.

Performance comparison among the semiparallel and the sequential cooperative sensing schemes is presented in Figure 2.13.

2.4 COOPERATIVE CRN

In cooperative CRN (CCRN), SUs are able to negotiate with PUs for dedicated transmission opportunities through providing tangible service [43]. In recent years, cooperative communications and spectrum leasing between PUs and SUs have attracted attention and have been investigated separately [44–46]. In what follows, cooperative communications as well as spectrum leasing between PUs and SUs will be presented, taking into account that:

- Most existing wireless networks and devices follow legacy fixed spectrum access policies. This means that spectral bands are licensed to dedicated users and services, such as TV, cellular networks, and vehicular ad hoc networks. In this setting, only PUs have the right to use the assigned spectrum, and others are not allowed to use it, even when the licensed spectral bands are idle.
- The spectrum utilization and efficiency can be enhanced [47], if SUs can transmit in the licensed bands when PUs are absent, or if such secondary transmissions are allowed to coexist with primary transmissions in such a way that SUs cause no interference to PUs (e.g., DSA) [6].

An SU can dynamically sense the availability of unused licensed bands and transmit while spectral bands are idle. Also, it can negotiate with PUs for transmission opportunities.

2.4.1 COOPERATIVE COMMUNICATIONS BETWEEN PUs AND SUs

To gain transmission opportunities in CCRN, one or more SUs can act as relaying nodes for a PU. As a relaying node, an SU can serve to provide a multihop relay service or an additional transmission path to the destination of the PU, and the PU yields the licensed spectrum to its relaying SU for a fraction of the time in return. The primary link can be established with the help of multihop transmissions, or the receiving PU can obtain diversity gain and enhance reception by appropriately combining the signals from the direct and relayed paths. In that way, spectral efficiency and utilization are significantly improved. There are different models in which PUs and SUs can perform cooperative communications, for example, three-phase TDMA-based cooperation, two-phase FDMA-based cooperation, and two phase SDMA-based cooperation [44].

In three-phase TDMA-based cooperation, the PU transmits the primary traffic to its intended destination and the selected relaying one or more SUs in the first phase. In the second phase, the SUs relay the PU's data, whereas and the SUs transmit their own signals in the third phase. The most critical parameters of this scheme are the optimal time duration in each phase for both the PUs and SUs, and the optimal allocation of transmit power levels of the PUs and SUs for energy-efficient transmissions. Furthermore, the multiuser cooperation in the time domain may result in high overhead and collisions that degrade the cooperation performance of CCRN.

In two-phase FDMA-based cooperation, the PU uses a fraction of its licensed band for relay transmissions with an SU, and allocates the remaining resources for the SU to address secondary transmissions. As the SU can continuously transmit its own signal on a dedicated licensed band, the achievable throughput of the SU can be guaranteed. The role of the CCRN is to guarantee the PU's transmission while achieving an elastic throughput for the SU according to the properties of wireless environment.

In two-phase SDMA-based cooperation, the SU exploits multiple antennas to enable MIMO capabilities, such as spatial beamforming (see Chapter 1), to avoid interference with the PU and with other SUs in CCRN. The MIMO-CCRN framework is proposed to allow the SU to use the degrees of freedom provided by the MIMO system to concurrently relay the primary traffic, and transmit its own data at the cost of complex antenna operation and hardware requirements.

2.4.2 Spectrum Leasing

PUs typically obtain licenses to operate wireless services, such as cellular networks, by paying spectrum regulators. In this context, one approach to attaining CCRN is spectrum leasing, which adopts pricing-based incentives to motivate PUs to lease their temporarily unused spectrum to SUs in return for financial reward. In a spectrum leasing (spectrum trading) model, one challenging issue is the pricing problem. For example, spectrum providers or PUs compete with each other to lease their licensed spectrum to SUs, and SUs compete with each other to lease spectrum from PUs.

To achieve an efficient dynamic spectrum leasing protocol between PUs and SUs, some models based on economic theory have been introduced to maximize PUs' revenue and SUs' satisfaction [48]. However, there is a trade-off between these two goals. One particular form of spectrum leasing is via auctioning, which is widely used in providing efficient distribution and allocation of scarce resources [49].

The *spectrum broker* is a centralized platform that facilitates TVWS spectrum trading and its allocation to the interested operators and service providers. It can be a government-controlled body or an independent third party. The players (spectrum buyers) are supposed to be able to make use of the spectrum in flexibly assigned TVWS frequency bands, which means that their core network transceivers and mobile equipment can operate in multiple bands. The spectrum broker controls the manner in which the available resources are assigned to each user to keep the desired QoS and limited interference through appropriate mechanisms. The resources for sale in a given trading area are the available (often fragmented) frequency bands, the allowable maximum transmit power in these bands, and the period for the licensing that grants temporary exclusive rights to use the spectrum [49]. The operational goal of the broker is to achieve robust technical protection of the incumbent, QoS provisioning to the players, and spectrum trading revenue maximization.

Figure 2.14 shows the spectrum broker functional algorithm. Through its main phase 2 (Operation), the spectrum broker supports the merchant and auction modes for allocating spectrum. In the merchant mode, the base price is decided by the allocation procedure, which considers various factors that influence the value of TVWS in a given area. In the auction mode, the auctioned band has a benchmark price, then each demand has an associated price (bid), and the winning bids decide the final price. In the merchant mode, the TVWS are allocated on a first-come, first-served basis, whereas in the auction mode, the TVWS are allocated to the winning bidders. When the spectrum demand is higher than what can be offered, the auction mode should be used for maximum economic efficiency. Otherwise, the merchant mode can be used to allocate the TVWS.

By introducing an auction mechanism into CCRN, PUs can maximize their revenue through dynamic and competitive pricing based on SUs' spectrum usage demand. Due to the unique characteristics of the radio spectrum, one spectral band can be reused in different areas because of signal attenuation during propagation, which means there may be multiple winners after an auction. Another requirement in the spectrum auction model is that the auction mechanism should be quickly conducted to enable on-demand and instantaneous services of SUs, which means that the SUs' bidding should be processed immediately by PUs (or special brokers). Due to the complicated relationship between bidders and auctioneers, and

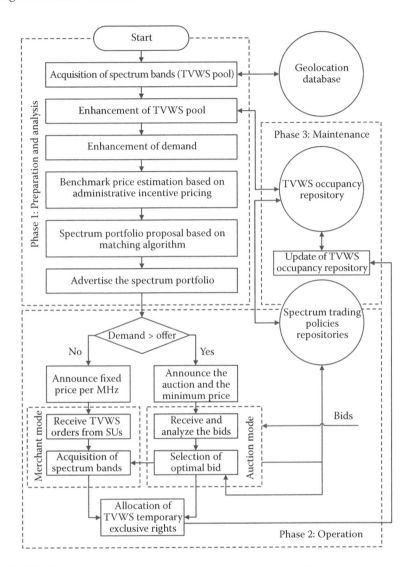

FIGURE 2.14 Spectrum broker functional diagram. (From H. Bogucka et al. Secondary spectrum trading in TV white spaces. *IEEE Communications Magazine* 50, no. 11 (November 2012): 121–129.)

the unique interference-limited characteristics, the overhead of auctions should be fully considered in the auction mechanism design. Otherwise, SUs are not willing to participate in auctions due to high overhead.

2.4.3 System Architecture for CCRN

In CCRN, the scenario consisting of one primary network with a primary BS and multiple PUs, and one secondary network with a secondary BS and multiple SUs is considered [43]. The primary BS allocates network resources (spectrum, time slots, etc.) to PUs so that PUs can access the spectral bands without interfering with each other. In addition, a dedicated control channel between the primary and secondary BSs is considered. Each SU equipped with a single CR has the knowledge of channel state information in terms of receiving signal-to-noise ratios (SNRs) of their interesting users. Moreover, SUs are assumed to have advanced signal processing functions, such as adaptive modulation, coding, and frequency agility.

To deploy user cooperation with high spectral efficiency in CCRN, the most important issue is how to exploit the available degrees of freedom in the wireless network (time, space, coding, modulation, etc.); efficient management and use of these degrees of freedom is critical to the CCRN's design and implementation. Another issue that should be taken into consideration is how to stimulate motivation for PUs and SUs to cooperate in CCRN.

As shown in Figure 2.15, two types of user cooperation models are represented in CCRN. The first one is cooperative communications between SUs and active PUs, in which an SU relays a PU's packets and obtains a transmission opportunity as a reward. The other is the cooperative spectrum leasing model among SUs, in which several SUs together lease an unoccupied licensed band from an inactive PU and establish cooperative communications with each other.

In the SU–PU cooperative communications model, an SU competes with other SUs for cooperation with an active PU. To improve the opportunity of being selected by the PU, an SU must optimize some performance

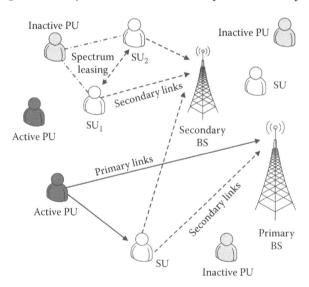

FIGURE 2.15 System architecture for CCRN.

metrics to enhance the PU's transmission as much as possible, because the PU would like to select the SU that can offer the highest gain by cooperation. Thus, the SU uses additional transmit power to forward the PU's traffic if it is selected as the relaying node by the PU. In the SU–SU cooperative spectrum leasing model, an SU would like to work with other SUs cooperatively to improve transmission performance and reduce the cost for spectrum leasing as much as possible. To fulfill this objective, an SU and its partners jointly optimize some metrics to attain transmissions in an economically efficient manner as well as satisfactory performance.

Example 2.5

As multiple SUs compete with each other to cooperate with the PU, each SU is motivated to provide as large a cooperation gain as possible to the PU subject to some constraints, such as meeting a transmission requirement and a power constraint. Because the SU uses additional transmit power to obtain the transmission opportunity, the parameters the SU needs to optimize are SUs' total transmit power P_S and power allocation factor α.

A weighted sum throughput maximization for the SU to evaluate the performance of cooperative communications was proposed by Cao et al. [43]. The optimization problem can be formulated as

$$\max_{\alpha, P_S} C_W = (1-w)C_P + wC_S, \qquad (2.7)$$

whereas

$$C_P > KC_{Pd}, \quad C_S \geq C_{ST}, \quad P_S \leq P_{SM}, \quad 0 < \alpha < 1, \quad K > 1,$$

where C_W is a weighted sum of the PU's cooperating throughput C_P, and the SU's throughput C_S, with a weighting parameter w. The objective is to maximize C_W subject to some constraints. It can be seen that C_W is a generalized metric in terms of normalized throughput, as $w = 0$ maximizes the PU's throughput, $w = 1$ maximizes the SU's throughput, and $w = 0.5$ maximizes the total throughput ($C_S + C_P$; i.e., w strikes a balance between the PU's and SU's throughputs). If the link from the PU to the primary BS is blocked by buildings, the SU provides a multihop service to the PU, and the SU may ask for a larger w for cooperation. C_{Pd} denotes the direct transmission throughput of the PU without cooperating with the SU, and K is the throughput gain that the PU tries to obtain by cooperation. The first constraint requires that by cooperating with the SU, the PU can achieve a throughput gain of at least K. The second constraint requires that the achievable through-out of the SU should meet its minimum throughput requirement C_{ST}. Otherwise, the SU is not willing to cooperate with the PU because the SU is not able to establish a connection with the secondary BS. These two constraints represent the benefits that motivate the PU and SU to cooperate with each other. The third constraint requires that the SU's transmit power should be bounded by P_{SM}. In the special case when $\alpha = 0$, the SU transmits only its own traffic without relaying that of the PU. Or the SU only helps forward the PU's traffic without transmitting its own when $\alpha = 1$, which is the case of conventional relaying communications.

This optimization problem can be solved by using primal–dual subgradient algorithms such as the method of Lagrange multipliers or its generalization the Karush–Kuhn–Tucker conditions [50]. By solving the problem, the SU determines whether itself and the PU can both obtain benefits. If so, the SU sends a response to inform the PU about the throughput that the PU can expect to achieve through cooperating with the SU. After getting responses from candidate SUs, the PU selects the SU that can provide the largest C_p as the relaying node.

During the simulation, $(1 - \alpha)$ represents the fraction of power that the SU uses to relay the PU's packets. The SU adopts the amplify-and-forward relaying mode to forward the PU's data, whereas $w = 0.3$. The channels are assumed to be Rayleigh block fading (i.e., the SNRs are exponentially distributed and invariant within each cooperation phase). The SNRs of links from the PU to PBS, from the PU to SU, from the SU to PBS, and from the SU to SBS are 7.4, 89.8, 105.7, and 108.2, respectively. As shown in Figure 2.16, the optimal power allocation factor α^* obtained by solving Equation 2.7, is achieved when the weighted sum throughput is maximized. It is shown that the total throughput $(C_S + C_P)$ is not maximal with respect to α^*. This means C_W strikes a balance between the PU's and SU's throughputs. If the total throughput is maximized, the PU's throughput is less than that of maximizing C_W, which means that the SU will not be selected as the relaying node by the PU. By introducing cooperative communications with an SU, the PU can achieve 1.9 bit/s/Hz after cooperation whereas the primary direct link without cooperation is 1.2 bit/s/Hz, and the SU can achieve a throughput of 1.6 bit/s/Hz (i.e., both the PU and SU achieve benefit by cooperation).

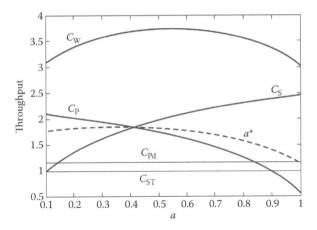

FIGURE 2.16 Effectiveness of CCRN. (From B. Cao et al. Toward efficient radio spectrum utilization: User cooperation in cognitive radio networking. *IEEE Network* 26, no. 4 (July–August 2012): 46–52.)

2.5 MULTIHOP CRNs

In multihop CRNs, the local view of available resources is not necessarily identical in all nodes due to the physical separation of nodes. The correlation between resource availability observations diminishes as peer nodes move further away from each other [51]. This observation leads to two important effects. First, resource availability information must be disseminated at least between neighboring nodes, and possibly beyond that, to ensure that P2P and E2E connectivity can be established. Second, establishment and maintenance of E2E paths, whether performed centrally or in a distributed manner, must have tight couplings with resource availability views and allocation decisions. Thus, resulting communication solutions are most likely to be based on cross-layer interactions.

The CRN environment consists of a number of primary radio networks (PRNs) that are licensed to operate over orthogonal spectrum bands and one (secondary) CRN. All networks coexist within the same geographical space. Figure 2.17 shows a conceptual composition view of a multihop CRN environment [34]. PUs that belong to a given PRN share the same licensed spectrum, whereas SUs form an opportunistic network to access the entire spectrum available to all PRNs. A characteristic of a CRN is that users must operate using relatively low transmission power to avoid degrading the performance of PUs.

MAC protocols proposed for traditional multichannel wireless networks are not well suited to the unique characteristics of CRNs. In particular, the absence of PUs in multichannel wireless networks makes their protocols

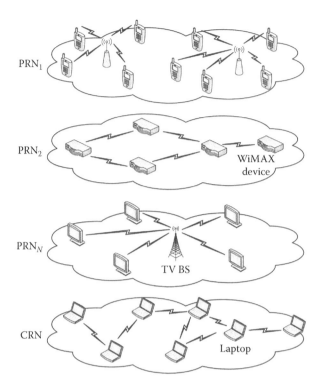

FIGURE 2.17   Multihop CRN environment.

significantly different from CRN MAC protocols. To design an efficient MAC protocol for multihop CRNs, the following attributes are required [34]:

a. The protocol should be transparent to PUs
b. The protocol should provide guarantees on PRNs' performance
c. The protocol should allow cooperation among neighboring CR users at the MAC layer to improve spectrum efficiency and fairness among them
d. The protocol should make efficient sensing and spectrum assignment decisions to explore both unused and partially used spectrum holes
e. The protocol should provide an effective distributed coordination scheme for exchanging control information without assuming a predefined dedicated control channel

Many researchers are currently engaged in developing MAC protocols related to the CRNs that attempt to address the above design requirements.

2.5.1 ROUTING IN MULTIHOP CRNs

The topology and connectivity map of the multihop CRNs are determined by the available primary RF bands and their instantaneous variations. Routing is a fundamental issue to consider when dealing with multihop CRNs. With regard to the timescale imposed by the specific primary nodes' behavior, an appropriate routing approach should be considered. The activity and holding time of the exploited primary bands by the CR determine the routing solution to use [52].

There are three separate categories defined by the primary technology used on the channel over which the CRs exploit the spectrum holes: static, dynamic, and opportunistic (highly dynamic). In a static scenario, the holding time of the used primary band offers a relatively static wireless environment. Once a frequency band is available, it can be exploited for an unlimited period. On the other hand, in a dynamic scenario, the primary band can be exploited by a cognitive user. In a highly dynamic case, a possible solution for CRs is to opportunistically transmit over any available spectrum band during the short period of the spectrum's existence. In what follows, these three categories will be discussed from the point of view for multihop CRNs. Possible routing solutions will have insight for each of them.

The three possible routing approaches are summarized in Figure 2.18. It can be seen that for every primary environment, an adequate routing solution has to be determined. How to define the boundaries that limit each approach's applicability is a challenging task. For example, choosing between a dynamic routing solution and an opportunistic approach in unstable environments is a hard decision to make. One can see that the undecided region that delimits the opportunistic approach and dynamic routing regions can be large.

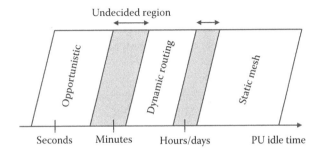

FIGURE 2.18 Three possible routing approaches in multihop CRNs.

2.5.2　STATIC MULTIHOP CRNs

When a primary frequency band is available for a duration that exceeds the communication time, static wireless networking methods defined for ad hoc and mesh networks can be adapted for CRNs [53]. The basic differences between mesh networking and CRNs are basically the dynamic and heterogeneous spectrum access and the physical capability to transmit simultaneously over multiple frequency bands. In a static environment, the dynamic dimension of the spectrum band is reduced to statically available channels [6]. The physical capability of transmitting over multiple channels can only be exploited on similar almost static bands. Selecting an E2E path over both a static channel and a dynamic one leads to path instability because failure of the dynamic spectrum band may cause the route to become inefficient. Special consideration for the detection of new arriving primary nodes over the exploited bands and the reaction it should trigger has to be included in the routing design.

Typical examples of a static CRN can be observed over satellite or analog TV bands where the bandwidth occupied by the PUs in a geographic location allows for continuous CR activity over this channel. Even a GSM or CDMA BS in a rural area, where the activity of a PU in the vicinity is very scarce, can create a static CRN.

2.5.3　DYNAMIC MULTIHOP CRNs

In dynamic multihop CRNs, the first priority is to find an available and stable path. To select a stable path that achieves acceptable performance, an option can be to accumulate the achieved throughput over many bands on every hop of the path [52]. Even if a first ineffective channel is selected, it can later be reinforced by other channels. However, selected channels must be really available and stable, and this can be ensured by including spectrum information in the path selection algorithm. One of the solutions is proposing routing metrics that capture spectrum fluctuations and less dynamic spectrum bands over unstable ones. Moreover, the computation must be fast and allow dynamic changes. Thus, routing algorithms in multihop CRNs must be less complex compared with algorithms in mesh networks.

2.5.4　HIGHLY DYNAMIC MULTIHOP CRNs

If the available time for CR activity over a primary band becomes shorter than the time needed to undergo a communication by the CRs over these bands, establishing a route for a whole flow is clearly an unthinkable solution. Furthermore, computing an E2E path cannot be considered because, in this scenario, for every sent packet the network properties may change, thus requiring a new path computation for every single transmission [52].

In such a highly dynamic and unstable environment, each sent packet may be forced to follow a different path based on primary band availability. The exploited primary bands dictate which cognitive neighbors can be observed on every channel. Therefore, opting for a complete opportunistic solution, in which every packet can be sent and forwarded over opportunistically available channels, constitutes a potential solution. Such an approach is even more interesting because the CRNs, through their intermittent channel availability, give immediate opportunistic networking possibilities. Using this feature can reduce the complexity of establishing E2E routes and increase the efficiency of the proposed solutions.

2.5.5 CHALLENGES FOR ROUTING INFORMATION THROUGH MULTIHOP CRNs

Consider the reference network model of information routing in multihop CRNs [54] shown in Figure 2.19. Secondary devices can share different spectrum bands or spectrum opportunities with PUs. Several spectrum bands $(1, ..., M)$ may exist with different capacities $C_1, C_2, ..., C_M$, and the SUs may have different views of the available spectrum bands due to inherent locality of the sensing process. The PUs are assumed motionless, whereas the SUs can change their position before and during transmission. In this scenario, the problem of routing in multihop CRNs targets the creation and the maintenance of multihop paths among SUs by deciding both the relay nodes and the spectrum to be used on each link. The additional challenge has to deal with the simultaneous transmissions of the PUs, which dynamically change the spectrum opportunities availability.

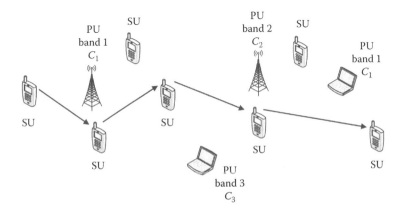

FIGURE 2.19 Information routing in multihop CRNs.

The main challenges for routing information throughout multihop CRNs include [54]: spectrum awareness, setup of "quality" routes, and route maintenance.

The *spectrum awareness* designing efficient routing solutions for multihop CRNs requires a tight coupling between the routing modules and the spectrum management functionalities such that the routing module can be continuously aware of the surrounding physical environment to take more accurate decisions. Three scenarios are possible:

a. Information on the spectrum occupancy is provided to the routing engine by external entities (e.g., SUs can have access to the TVWS database [41])
b. Information on spectrum occupancy is to be gathered locally by each SU through local and distributed sensing mechanisms [55]
c. A combined scenario of the previous two

Any routing solution designed for multihop CRNs must be highly coupled to the entire cognitive cycle of spectrum management [5].

As for the *setup of quality routes* in a dynamic variable environment, it should be noted that the actual topology of multihop CRNs is highly influenced by the PU's behavior, whereas traditional ways of measuring/ assessing the quality of E2E routes (throughput, delay, energy efficiency, etc.) should be coupled with novel measures on path stability and spectrum availability/PU presence.

In the *route maintenance/reparation* scenario the sudden appearance of a PU in a given location can render a given channel unusable in a corresponding area, thus resulting in unpredictable route failures, which may require frequent path rerouting either in terms of nodes or used channels. In this scenario, effective signaling procedures are required to restore "broken" paths with minimal effect on the perceived quality.

2.5.6 MULTICARRIER MODULATION IN MULTIHOP CRNs

Almost all proposed CRSs are based on multicarrier modulation (MCM) because multiple users can access the systems by allocating subcarriers [56]. In multihop CRNs, each node has a list of available frequency bands and must work adaptively among these frequency bands because of DSA. It is well known that two nodes cannot communicate if they work on different frequency bands. Hence, routing in multihop CRNs becomes a critical and challenging issue. Solutions for this problem mainly focus on the methods in the network layer, whose processing delay is in the order of milliseconds. The high-speed wireless channel in multihop CRNs varies on the order of microseconds due to multipath fading, the Doppler effect, and dynamic occupancy of the subchannel by PUs. Therefore, the solutions proposed in the network layer may cause heavy interference to PUs. Adopting MCM to intersection nodes of multihop, CRNs may be a promising solution in the physical layer. Because of the use of MCM, the intersection node can allocate some unused subcarriers to different information flows. Thus, all flows can be transmitted simultaneously. Considering that the access of multiple users can be implemented by the allocation of subcarriers in an MCM system, almost all the proposed spectrum overlay CRSs are based on MCM technology.

2.6 CONTROL AND COORDINATION IN DISTRIBUTED CRNs

Compared with its centralized counterpart, a distributed CRN can be a more practical choice because of its easier and faster deployment, lower system complexity, and lower cost of implementation [57]. It also offers more challenging research issues due to the lack of a central control unit. Because the control and coordination of communication over wireless channels happens mainly at the MAC layer, designing a smart and efficient MAC protocol remains a key requirement for successful deployment of any CRN. The MAC protocol should be able to adapt to the unique features of CRNs and maintain robust performance in the presence of a highly dynamic environment. It is obvious that traditional MAC protocols cannot meet the requirements for CRNs.

Distributed CRNs usually form P2P architecture among SUs. These types of networks are very similar to conventional distributed networks. However, nodes of distributed CRNs have additional features such as capability of sensing channels, and negotiating a common available channel with the intended receiver when necessary. The main characteristics of distributed CRNs are

- The channel availability for SUs at any time and location depends on the PUs' activity at that time and location
- The available channels, that is, the spectrum not being used by PUs, is generally discontinuous and may lie anywhere in the entire spectrum
- The number of available channels is a time-dependent variable, and the sets of available channels are different for various nodes
- There is no central unit to coordinate channel sensing, channel access, and synchronization among SUs

2.6.1 Distributed Medium Access Control Benefits

The distributed MAC design approach can take benefits from traditional distributed multichannel MAC protocols to address common problems in centralized as well as distributed CRN architecture. CR-specific issues are related to [57]: resource availability, interference to PU and presence of sensing error, channel sensing period and negotiation, time synchronization, QoS provisioning, multichannel hidden terminal problem, network coordination, and reconfiguration.

A CR MAC must be able to maintain its robustness and efficiency in the presence of uncertainty about the amount of radio resource, which depends on the PUs' activity. Any amount of sensing error will add uncertainty to the already dynamic nature of resource availability.

SUs opportunistically access only the spare bandwidth from the PU network. They are likely to cause some level of interference to the data transmission among PUs. As for sensing error, it cannot be completely avoided due to hardware limitations in a practical SU node. Thus, the goal of MAC design should be to minimize the level of interference as much as possible.

Sensing PU channels is necessary, but it comes with the price of increased overhead. The length and frequency of the MAC sensing phase needs to be designed carefully so that the resource utilization can be improved with a trade-off between any particular sensing strategy and its incurred overhead.

A MAC design must incorporate an efficient negotiation mechanism for proper coordination among SUs. The negotiation mechanism may need time synchronization among SUs or a common control channel (CCC), or both [58] for the initial control message exchange. The CCC is a channel allocated solely for control message exchange and is shared by many or all SUs. If not designed properly, the negotiation mechanism can consume a significant amount of resource itself due to messaging and time overhead.

Some type of time synchronization among SUs, particularly between transmitter and receiver before data transmission begins, is critical for a distributed CR MAC to function. For example, channel negotiation is a necessary functionality in a distributed CRN MAC, which is difficult to implement without time synchronization. Time synchronization is also needed for network establishment and for coordination among SUs.

QoS provisioning [59] for SUs is not possible without necessary support from the MAC protocol, which is responsible for coordinating access over the available radio resource. Such support usually comes in the form of necessary signaling, resource scheduling, and admission control mechanisms in the MAC layer.

The multichannel hidden terminal problem [60] can drastically degrade the system performance in distributed CRNs. If each node has only one transceiver, it can work either on the control channel or on a data channel. This problem needs to be tackled at the MAC layer by a proper synchronization and signaling mechanism. On the other hand, if equipped with multiple transceivers, an SU node can listen to both data and control messages simultaneously. Because an SU node can be aware of the exchange of control messages among neighboring SUs, multichannel hidden terminal problem becomes easier to handle with multitransceiver MAC. However, compared with multitransceiver MAC, single-transceiver MAC can work with SU nodes that are much cheaper and less complex to implement.

Ensuring good network coordination and reconfiguration, a negotiation mechanism may be costly in terms of overheads like control signaling, time taken for the process, and disruption to the entire network. The existence of a CCC and network-wide time synchronization may help the distributed CRN MAC solve the problem of network coordination and reconfiguration by providing a message exchange mechanism among SUs.

2.6.2 Classification of Distributed Cognitive Medium Access Control Protocols

Generally, cognitive MAC can be divided into two classes, centralized and distributed, depending on the CRN architecture [20]. In a centralized architecture, BS typically controls the spectrum access of SUs. On the other hand, distributed CRNs do not have a central unit or BS to assist SUs in spectrum sensing or spectrum access. In a distributed approach, the design of a MAC protocol largely depends on the number of transceivers needed at each SU node to support the MAC operations. Accordingly, distributed MAC solutions can be categorized into two major groups: single transceiver and multitransceiver (Figure 2.20). As previously stated, both of these design approaches have their benefits as well as drawbacks.

One of the important design assumptions for both single-transceiver and multitransceiver MACs is the presence or absence of a CCC. The presence of a

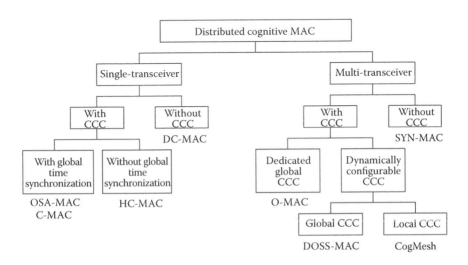

FIGURE 2.20   Classification of distributed cognitive MAC protocols.

CCC simplifies the design of the MAC protocol and offers several advantages, including better coordination among SUs and reduced collision of their data transmissions. Most of the existing distributed MACs are designed assuming the presence of a CCC. However, CCC saturation occurs when many or all SUs try to transmit control messages on the CCC at the same time. This might cause the MAC to suffer from serious performance degradation, especially in a large network or when the traffic load is high. To avoid these problems, some dynamic cognitive MACs are designed without a CCC. For example, decentralized cognitive MAC (DC-MAC) [61] and synchronized MAC (SYN-MAC) [62] are proposed to work without a CCC for single-transceiver and multitransceiver systems, respectively.

Time synchronization becomes an important issue for single-transceiver MAC designed with a CCC, that is, hardware-constrained MAC (HC-MAC) [63], opportunistic spectrum access MAC (OSA-MAC) [64], and cognitive MAC (C-MAC) [65]. In these MACs, SUs cannot remain tuned to the CCC all the time because the same transceiver is used for data transmission. Local or local time synchronization among SUs is thus needed to make sure that the SUs know when they have to tune to the CCC for control message exchanges and when they can resume transmission. Local time synchronization is usually preferred in distributed CRNs where coordination is limited to neighboring nodes.

In multitransceiver MAC, a CCC can be either dedicated or configured dynamically. In opportunistic MAC (O-MAC) [59], the CCC is assumed to be a dedicated spectrum that either belongs to the cognitive system or to an ISM band. The dedicated CCC is always available to SUs, which simplifies MAC operation. However, in practice, a dedicated CCC may not be available. Therefore, a PU data channel is dynamically selected as the CCC in dynamic open spectrum-sharing MAC (DOSS-MAC) [66] and CogMesh [67]. The dynamically selected CCC can be available either locally to a neighborhood or globally to all SUs. Cluster-based distributed CRNs usually select a local CCC in each cluster. For example, in CogMesh, a node initiates cluster formation with a locally available PU channel as the CCC. The node then becomes the leader of its cluster. The neighbors to whom the selected CCC is also available can join the cluster. On the other hand, DOSS selects a PU channel available to all SUs as a global CCC.

Example 2.6

Average length of contention phase depends on various parameters like number of SUs in the system, average number of available channels, and probability of data packet availability at SUs [57]. For given backoff window size and average number of neighbors of an SU node N_{su}, average time for one channel reservation

T_{si} can be calculated as described in [68] and is assumed to be known. If contention period T_{con} is chosen as an exact integer multiple of T_{si} (i.e., $T_{con} = nT_{si}$) where $n \geq 0$ is an integer, it can be shown that average saturation throughput S_{avg} is a concave function of n. Assuming other parameters such as total number of data channels N_{ch} and average probability of availability associated with each data channel P are known, an optimal value of T_{con} can be calculated that maximizes the throughput by formulating an optimization problem as follows:

$$\max_{n} S_{avg} = \frac{1}{T_{BI}} [N_{avg}(T_{BI} - T_{sen} - nT_{si})], \qquad (2.8)$$

where T_{BI} is the length of the backoff interval and T_{sen} is the length of the sensing phase, whereas the average number of channels reserved during the contention period T_{con} can be calculated as

$$N_{avg} = \sum_{i=1}^{N_{ch}} \binom{N_{ch}}{i} P_i(1-P)^{N_{ch}-i} \min(i,n). \qquad (2.9)$$

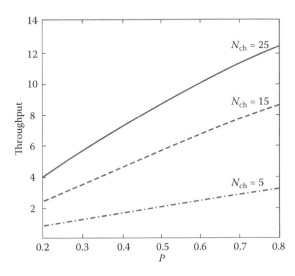

FIGURE 2.21 Saturation throughput variation with channel availability and number of channels. (From S. C. Jha et al. Medium access control in distributed cognitive radio networks. *IEEE Wireless Communications* 18, no. 4 (August 2011): 41–51.)

If n^* is optimal n, from expression of N_{avg} it can be seen that $n^* \leq N_{ch}$. Therefore, even an extensive search algorithm for optimization is feasible in this case. This is one variable optimization, and the number of needed iterations is $0 \leq n^* \leq N_{ch}$. Due to the concave nature of S_{avg}, it will increase with the increase in n. As soon as S_{avg} starts decreasing with the increase in n, the search can be stopped. The optimal T_{con} will be equal to $n^* T_{si}$.

An illustrative example of cognitive MAC performance study, in which the optimal contention phase length that maximizes the throughput was used [57], and is presented in Figure 2.21.

As expected, the maximum achievable throughput increases with the increase in number of channels and probability of channel availability.

2.7   CONCLUDING REMARKS

The cognitive radio system can be characterized as a radio system having the capability to obtain knowledge by adjusting its operational parameters and protocols. Many CRS usage scenarios are possible. CR is a paradigm for opportunistic access of licensed parts of the spectrum by unlicensed users that can provide solutions for the interference scenarios and also enhance scalability. A CR user can combine spectrum sensing and geo-location database access to determine occupancy and dynamic reconfiguration of its transceivers' parameters to avoid interference with PUs. As for cooperative spectrum sensing, it is a promising technique in CRNs by exploiting multiuser diversity to mitigate channel fading. Cooperative sensing is traditionally employed to improve sensing accuracy whereas the sensing efficiency has been largely ignored. However, sensing accuracy and efficiency have very significant effects on overall system performance. Currently, international standardization of CRS is being performed at all levels, including the ITU, IEEE, ETSI, and ECMA. Each of these organizations is considering multiple CRS deployment scenarios and business directions. Many technical and regulatory issues have been reached in opening the spectrum for more flexible and efficient use. CR technology plays a significant role in making the optimal use of scarce spectrum to support the fast-growing demand for wireless applications ranging from multimedia streaming inside homes, over healthcare applications, to public safety.

Sequential cooperative sensing mainly aims to improve the sensing accuracy in CRNs. In semiparallel cooperative sensing, the grouping strategy will determine the levels of sensing accuracy and efficiency. For a given total number of the cooperative SUs, more SUs in a group will lead to higher accuracy. If all the cooperative SUs are allocated in the same group, the scheme becomes a sequential cooperative sensing, whereas the scheme reduces to a full-parallel cooperative sensing if each cooperative SU is individually set as a group. Asynchronous cooperative sensing exploits both spatial and temporal diversities. The unique feature of asynchronous cooperative sensing is its ability to further exploit the temporal diversity of channel availability. The multiuser spatial diversity is indicated by the essential cooperation among multiple SUs. Synchronous cooperative sensing can only operate in a centralized manner whereas asynchronous cooperative sensing can operate in a centralized or decentralized manner.

In multihop CRNs, each node has a list of available frequency bands and, because of DSA, must work adaptively among these frequency bands. Hence, routing in multihop CRNs becomes an important issue. For high data rate wireless communication systems, two of the major issues are the underutilization of limited available radio spectrum and the effect of channel fading. Using DSA, CR can improve spectrum utilization. Almost all proposed CRSs are based on MCM because users can access them by allocating subcarriers.

Many technical issues still need to be addressed for successful deployment of CRNs, especially in the MAC layer. Cognitive MAC protocol design is an open area of research and will be of interest to both the industry and academia as this technology matures in the next few years. CRNs have distributed architecture because they offer ease of deployment, self-organizing capability, and flexibility in design and are believed to be more practical for future deployments compared with their centralized counterparts. In CR, identifying the available spectrum resource through spectral sensing, deciding on the optimal sensing and transmission times, and coordinating with other users for spectrum access, are the important functions of the MAC protocols.

Despite advances toward distributed CRN MAC design, a number of important design issues demand considerable further research. Among others, future research directions may include QoS support for SUs in the presence of dynamic resource availability and achieving time synchronization and network coordination without a dedicated global CCC, due to the lack of a predefined medium to exchange initial control messages to establish coordination among SUs.

3 Mobility Management in Heterogeneous Wireless Systems

The popularity of the Internet and rapid development and acceptance of wireless communications have led to the inception and development of the mobile Internet infrastructure with a goal to provide an end-to-end IP platform supporting various multimedia services. Mobile Internet applications have become popular and are being used widely. The concept of next generation networks includes the ability to make use of multiple broadband technologies and to support generalized mobility. The heterogeneity in the intertechnology roaming paradigm magnifies the mobility effect on system performance and user-perceived quality of service, necessitating novel mobility modeling and analysis approaches for performance evaluation. Because each individual network has its own characteristics (e.g., spectrum, multiaccess, and signaling) and application requirements (e.g., bandwidth, delay, and jitter) facilitating interoperability within the next generation framework raises several key design issues concerning resource management and mobility management. Adequately addressing these issues is of great importance to the successful development of future wireless environments. Seamless services require network and device independence that allow the users to move across different access networks and change computing devices. Mobility management protocols are responsible for supporting seamless services across heterogeneous networks that require connection migration from one network to another. In addition to providing location transparency, the mobility management protocols also need to provide network transparency.

3.1 INTRODUCTION

To meet the upcoming exponential growth of mobile data traffic [1], operators are deploying more network infrastructures to make mobile systems closer to users, and thus increase spectrum efficiency and spatial reuse. The availability of wireless networks is the result of low-cost deployment of local points of attachment (PoA) and the operators' short-term strategies of covering smaller geographic areas at low cost (such as deploying relay stations). The advantage of femtocells, for example, will certainly improve indoor coverage and provide reliable connectivity without the need for the cost-inefficient deployment of additional base stations. On the other hand, some dense urban areas will be served by a mix of overlapping access networks (e.g., Wi-Fi, WiMax) reaching different coverage. It is clear that mobile terminals (MTs) have been evolving from single-network interface phones to multitask devices with a number of connectivity capabilities. With the recent advances in software radio technology (see Chapter 2), most modern MTs are capable of communicating via different technologies.

In this context, heterogeneous networks (HetNets), which are composed of coexisting macrocells and low power nodes such as picocells, femtocells, and relay nodes, have been heralded as the most promising solution to provide a major performance leap [2]. However, to realize the potential coverage and capacity benefits of HetNets, operators are facing new technical challenges in mobility management, intercell interference coordination, and backhaul provisioning. Among these challenges, mobility management is of special importance [3].

In a heterogeneous environment, mobility management represents the basis for providing continuous network connectivity to mobile users roaming between access networks. The next generation in mobility management will enable different wireless networks to interoperate with one another to ensure seamless mobility and global portability of multimedia services. Mobility management affects the whole protocol stack, from the physical, data link, and network layers up to the transport and application layers. Examples include radio resource reuse at the physical layer, encryption and compression at the link layer, congestion control at the transport layer, and service discovery at the application layer. Because mobility is essentially an address translation problem, it is therefore naturally best resolved at the network layer by changing the routing of datagrams destined to the mobile node (MN) to arrive at the new PoA.

The purpose of this chapter is to survey recent research on mobility management in future generation wireless systems. The rest of this chapter is organized as follows. First, primary and auxiliary mobility management services are briefly introduced. After that, current and perspective mobility management protocols are presented. Special attention is provided to mobile IP solutions, because they are widely accepted in this research field. The concept of interdomain mobility management is introduced together with a solution for session establishment and maintenance during seamless network changes. Finally, distributed mobility management (DMM) is introduced as a new architectural paradigm for evolving wireless multimedia systems.

3.2 MOBILITY MANAGEMENT SERVICES

Mobility support is one of the major attributes of the HetNets. In this context, an open challenge is the design of mobility management solutions that take full advantage of IP-based technologies to achieve global and seamless mobility between the various access technologies and, at the same time, provide the necessary QoS guarantees. Mobility can be classified into terminal, personal, session, and service mobility [4,5]. Terminal mobility is the ability of a MT to move between IP subnets, while continuing to be reachable for incoming requests and maintaining sessions across subnet changes. Personal mobility concentrates on the movement of users instead of users' terminals, and involves the provision of personal communications and personalized operating environments. Session mobility allows user to maintain a session while changing terminals. Finally, service mobility can be defined as the ability of users to maintain access to their services even when moving and changing terminals or service providers.

Mobility management (a.k.a. management of terminal mobility) has to provide primary services like location management and handover management, as well as some auxiliary services, such as multihoming and security. The main mobility management services and corresponding procedures are presented in Figure 3.1.

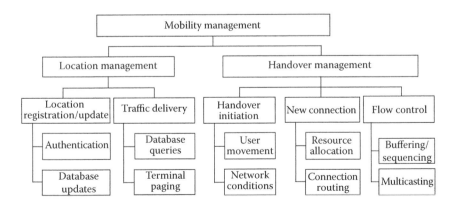

FIGURE 3.1   Main mobility management services and corresponding procedures.

3.2.1 LOCATION MANAGEMENT

Location management enables the serving network to track and locate a MT for possible connection. Location management involves handling all the information about the roaming terminals, such as original and currently located cells, authentication information, and QoS capabilities. Accordingly, it includes two major procedures:

- Location registration/update related to the periodical updating of the new MN's location on the access link to allow the network to keep track of the MN
- Traffic delivery as the ability of the network to find the MN's current location in the access network in order to deliver its data packets

In location management, the main concern is with the database architecture design and the signaling transmissions issues. Therefore, to deal with these issues, the challenge is in the location registration procedure, such as the security issue due to the MN's authentication process and delay constraints associated with static and dynamic updates in location registration. The other issue concerns the data packets' delivery procedure, such as querying delay due to the type of database architecture used (centralized or decentralized), as well as the delay constraint and paging delay cost. Furthermore, many of the issues are not reliant on specific protocols and can be applied to different networks depending on their requirements. Efficient mobility management design implies minimized signaling overhead and delay for location updating and paging procedures [6].

3.2.2 HANDOVER MANAGEMENT

Handover management is the process by which an MT keeps its connection active when it moves from one PoA to another. Moreover, when users are in the coverage area of multiple radio access networks, that is, in a heterogeneous environment, handover management can provide them connections to the best available network. This aspect of mobility management is considered in Chapter 4. At this point, it is important to note that an efficient handover management strategy implies minimum delay and packet loss during handover.

3.2.3 AUXILIARY MOBILITY MANAGEMENT SERVICES

Considering auxiliary mobility management services, multihoming has to be supported by the protocol stack, which provides the MN with network access using multiple communication technologies. Mobility and multihoming try to solve the same problem related to the session survivability in two different environments and by using different mechanisms. Moreover, nowadays, MNs are often multihomed and vice versa. Therefore, it is important to have a single protocol that manages both of them independently of the environment of deployment [7]. Also, security is a major concern for mobile networks protocol designers because MNs, which change their PoAs while roaming through different networks, present additional security risks in comparison with the fixed nodes.

3.3 MOBILITY MANAGEMENT PROTOCOLS

Mobility management protocols operate from different layers of the protocol stack with the goal of minimizing performance degradation during the handover procedure. Only the transport and application layer protocols maintain the end-to-end (E2E) semantics of a connection between communicating hosts. Most of the mobility protocols are limited to single-layer solutions only, and thus, they are transparent to the other layers [8]. In the following sections, the protocol design issues in each layer are discussed in more detail.

3.3.1 LINK LAYER MOBILITY MANAGEMENT

The seamless mobility management approaches in the link layer involve the underlying radio systems concept. Hence, mobility support by the link layer is also known as access mobility [9]. In the link layer solutions, an MT can usually change its position within the coverage area of an access router (AR). Strictly speaking, link layer solutions for mobility management are tightly coupled with specific wireless technologies and cannot be used as a general solution in heterogeneous wireless environments. Also, link layer mobility support for intersystem roaming requires additional interworking entities to help with the exchange of information between different systems.

The performance of the link layer mobility protocols can be summarized as follows [10]:

- The intersystem handover latency is high because several functions such as format transformation and address translation, user profile retrieval, mobility information related to intersystem movement recording, and authentication
- The large value of handover latency results in higher packet loss during intersystem handover
- After the intersystem handover, an MN communicates with the new system without the need for any redirection agent. Thus, the E2E delay requirement of the applications is respected

- Because an MN communicates with a new address in the new system, a transport layer connection has to be reestablished after intersystem handover. Therefore, link layer mobility management protocols are not transparent to TCP and UDP applications
- Because authentication is carried out during an intersystem handover, these procedures are mainly secure

Having these characteristics in mind, link layer mobility solutions are not solely applicable in future wireless systems, and they can be considered as merely theoretical sustenance to upper layer mobility protocols.

3.3.2 NETWORK LAYER MOBILITY MANAGEMENT

Various network layer protocols have been proposed as global or local mobility management solutions that are intended to handle the MN's mobility within the same domain or across network domains, respectively. IP mobility can be classified into two main categories: host based and network based. In the host-based category, the MN must participate in the mobility-related signaling. On the other hand, in the network-based category, the network entities are the only entities that are involved in the mobility-related signaling.

Mobile IP was proposed by the Internet Engineering Task Force (IETF), which generally focuses on MN mobility supports during its roaming across domains and redirects the MN's packets to its current domain location using typical Mobile IPv4 (MIPv4) [11]. While roaming through different foreign subnets, the MN acquires short-term addresses (care-of address; CoA) from a foreign agent (FA) in the visited network or through external assignment mechanism. The MN then registers its CoA with its home agent (HA) by sending a registration request via the FA. The HA then sends a registration reply either granting or denying the request. If the registration process is successful, any packets destined for the MN are intercepted by the HA, which encapsulates the packets and tunnels them to the FA where decapsulation takes place and the packets are then forwarded to the appropriate MN. Moreover, the corresponding node (CN) does not need to know about the MN's mobility, and it can send all packets through the HA. This procedure is presented in Figure 3.2.

It is obvious that MIPv4 is not the optimal solution to support the increasing number of users and real-time services because it suffers from extra packet E2E delay due to the routing of each packet through the HA (triangular routing), lack of addresses, and high signaling load. In addition, all on-the-fly packets, which were already tunneled to the old CoA, are lost whenever the MN moves from one FA to another, because the new FA cannot inform the old FA about this movement. Furthermore, the mobility signaling delay is very long and may vary significantly when the distance between the home network and the visited network is large.

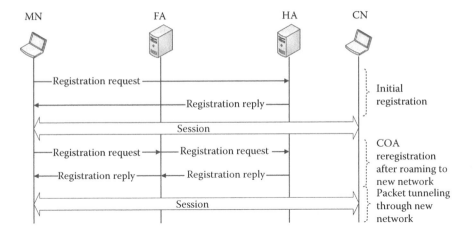

FIGURE 3.2 IP-based mobility management.

Mobile IPv6 (MIPv6) [12] is a well-known standard for global mobility support that overcomes many constraints experienced in MIPv4. MIPv6 enables a MN to move within the Internet domain without losing current data connection directly with its CN, while in MIPv4, the CN sends a data packet to the MN through the HA and FA by a longer route. MIPv6 supports MN mobility to be reachable at anytime and anywhere by its CN. This is done by providing the MN with a fixed home address provided by HA. Furthermore, if the MN is in the home network, all packets destined to it will not have to be altered and can reach through the normal routing process. Moreover, when the MN moves to a new visited network, it is assigned a temporary CoA provided by the visited network and the MN will not be reachable through its home address. Therefore, the HA is now responsible for receiving data packets that are destined for the MN. Whenever the HA receives such packets, it will tunnel it to the MN's current CoA. Therefore, MN has to update its HA on its current CoA; consequently, HA will forward all packets through a tunnel destined to the MN's home address to its current CoA at the visited network. Therefore, the data transfer between HA and MN uses the tunnel ends at the MN directly (not to the FA as in the MIPv4). Furthermore, MIPv6 introduces a route optimization operation to solve the triangular routing problem and improve network performance. The basic idea is to allow for better routing between the MN and its CN, through exchange query–response messages between the MN and CN to establish a direct and secure route. Hence, all packets can travel between the CN and MN without being intercepted by the HA. This optimization improves network reliability and security, and reduces network load.

However, despite the good standing of this protocol, it has been slow to deploy in real implementations due to some drawbacks, such as high handover latency, high packet loss, and signaling overheads [13]. In addition, the local mobility of MN is handled in the same way as global mobility, that is, when the MN moves to a new subnet, it will update its new PoA each time to HA and CN, without any locality consideration, thereby causing perceptible deterioration of real-time traffic performance. All these weaknesses led to various investigations [14] and development of other mobility enhancements that focused on MIPv6 performance improvements.

Fast MIPv6 (FMIPv6) [15] was proposed to reduce latency and minimize service disruption during handover related to the MIPv6. It uses link layer events (triggers) to improve the handover performance in terms of packet loss by anticipating the handover and tunneling the packets to the new AR until the binding update (BU) is received by the HA and CN. At the same time, the MN will advertise its presence and availability to the new AR and will start receiving data on the new CoA. This solution provides a substantial improvement of handover latency and packet loss. On the other hand, the main drawback of this solution is the precise coordination required between the MN, previous AR and new AR and high unpredictability of packets arriving at the APs. The FMIPv6 operation's steps, shown in Figure 3.3, are described as follows:

1. The MN anticipates the handover by sending the router solicitation for proxy advertisement (RtSolPr) message to the previous AR requesting the CoA.
2. The previous AR replies by sending the proxy router advertisement (PrRtAdv) message, which contains a new CoA to be used

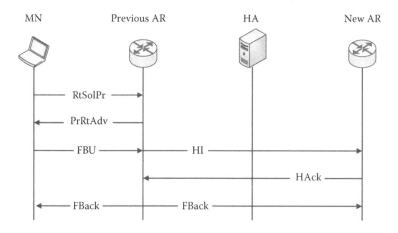

FIGURE 3.3 FMIPv6 operations.

on the new AR's link. This message can also be sent in a periodic manner.

3. The MN initiates the handover by sending a fast binding update (FBU) message with the new CoA to inform the previous AR that the packets should now be forwarded to the new AR.

4. To forward packets, a tunnel is established between the previous AR and the new AR. However, the MN does not know yet if the new CoA is unique on the new link. Therefore, the previous AR sends a handover initiate (HI) message to the new AR for address duplication check on the new link and it sets up the temporary tunnel to redirect packets between the previous AR and the new AR.

5. The new AR responds with a handover acknowledge (HAck) message if the tunnel is established successfully and if there is no address duplication.

6. After the previous AR received a HAck message, it sends a fast binding acknowledgment (FBack) to the MN through both access links. After the new AR receives both FBU and HAck messages, it starts forwarding the MN packets using the tunnel to the MN's old CoA.

This tunnel starts at the previous AR and ends at the new AR, not to the MN. This allows the MN to use its old CoA while verifying the new one. Moreover, the data packets sent by the MN from its old CoA will also be tunneled back from the new AR to the previous AR. This goes as long as the MN has verified its new CoA and updated the HA and the CN. After that, the MN will inform the new AR about its movements to its link. Then, the new AR can forward all data packets, which can be buffered during the MN's handover.

In general, FMIPv6 optimization is based on a reliable handover prediction that enables predictive configuration of the MN involved in the mobility signaling. However, this prediction relies on the link layer trigger availability and the appropriate triggering time, which affects the beginning of the handover and will determine whether fast handover optimizations will take place. Accordingly, the absence of an accurate prediction such as erroneous handover detection, and hence, early triggers, may negatively affect the seamlessness of this protocol.

Hierarchical MIPv6 (HMIPv6) [16] is a local mobility management protocol designed to reduce handover latency and signaling overheads that occur when MN frequently change PoA. It adds an indirection for locating the MN independent of where the CN and HA are located in the network topology. It tunnels packets to a mobility anchor point (MAP), which is addressed by a regional CoA. The MAP, in turn, tunnels these packets to the MN addressed by an on-link (local) CoA. Therefore, the

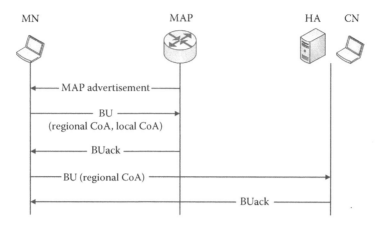

FIGURE 3.4 Hierarchical MIPv6 operations.

MN's local handover mobility information only needs to be signaled to the MAP, hence, avoiding high handover latency and BU overheads. The main operations of the HMIPv6 are shown in Figure 3.4.

To update the CN or HA about its new CoA, MN will send them a BU message through a MAP. The BU acknowledgement (BUack) from the HA/CN will be transferred back in the same way. If the link between the CN/HA and MAP is a significant distance, it means that it would take some time for the BU to travel from the MAP to the CN/HA and back. Therefore, it would make sense to have a kind of temporary HA on the MAP. In that way, the MN only needs to update the MAP as long as the same MAP is located between the MN and CN/HA. Thus, the extra time for sending a BU between CN/HA and MAP is spared. Moreover, the MN can find the MAP address from the router advertisement and form a regional CoA address from the MAP before updating the CN/HA with this CoA. After that, the CN/HA sends their packets to a regional CoA. Then, the MAP tunnels them to the MN's local CoA. In addition, the MAP can buffer the data packets destined to the MN and send them when the MN has sent the BU message through the new AR. In addition, HMIPv6 gains an advantage over MIPv6 by handling the MIP registration locally using a hierarchy of MAP, instead of the global communication handling in the MIPv6 domain.

In general, all the host-based mobility management protocols require a protocol stack modification of the MN and change its IP addresses to support its mobility within or across network domains. Consequently, it may increase the MN complexity and waste radio resources. Furthermore, some drawbacks still remain in the host-based mobility protocols (e.g., long handover latency, high packet loss, signaling overhead), which, put together, indicate the inappropriateness of these protocols to satisfy the QoS requirements for multimedia services.

Proxy MIPv6 (PMIPv6) [17] has been standardized by the IETF Network-based Localized Mobility Management (NETLMM) working group as a fundamental protocol of the homonymous category. It is based on MIPv6 and reuses some of its signaling concepts and functions. In particular, user terminals are provided with mobility support without their involvement in mobility management and signaling because the required functionality is relocated from the MN to the network. Movement detection and signaling operations are performed by a new functional entity called the mobile access gateway (MAG), which usually resides on the AR. Through standard terminal operation, including router and neighbor discovery or using link layer support, the MAG learns about MN movement and coordinate routing state updates without any mobility-specific support from the terminal. IP addresses used by nodes within localized mobility domains (LMD) are anchored at an entity called the local mobility anchor (LMA), which plays the role of local HA for the corresponding domain. Bidirectional tunnels between the LMA and MAG are set up so that the MN can keep the originally assigned address despite its location within the LMD. Through the intervention of the LMA, packets addressed to the MN are tunneled to the appropriate gateway within the domain. Upon arrival, packets are locally forwarded to the MN, which is therefore oblivious to its own mobility.

In the case when MN connects to the PMIPv6 domain through multiple interfaces and over multiple access networks, the network will allocate a unique set of home network prefixes for each of the connected interfaces. The MN will be able to configure addresses on those interfaces from the respective home network prefixes. However, if a handover is performed by moving its address configuration from one interface to another, and if the LMA receives a handover hint from the serving MAG about the same, the LMA will assign the same home network prefixes that it previously assigned before the handover. The MN will also be able to perform a handover by changing its PoA from a previous MAG to a new one using the same interface and will be able to retain the address configuration on the attached interface.

As illustrated in Figure 3.5, the overall PMIPv6 signaling flow includes an initial attachment phase and a handover procedure phase. Once a MN attaches to a MAG module and sends router solicitation (RtrSol) message for the first time, the MAG and the LMA exchange proxy binding update (PBU) and proxy binding acknowledgement (PBA) messages. The LMA sends an address assigned to a MN via a PBA message, and also sets up a bidirectional tunnel with the MAG for the MN to be able to communicate with a CN.

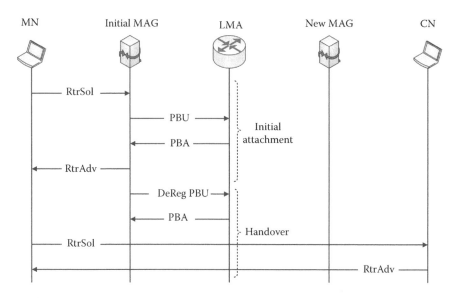

FIGURE 3.5   PMIPv6 operations.

The second phase describes the MN movement from the initial MAG to a new MAG until it can resume sending/receiving data packets to/from its CN. When the initial MAG detects the MN's movement away from its access link to the new MAG, it sends a deregistration PBU (DeReg PBU) message to the LMA with a zero value for the PBU lifetime. Upon receiving this request, the LMA will identify the corresponding mobility session for which the request was received, and accepts the request after which it waits for a certain amount of time to allow the MAG on the new link to update the binding. However, if it does not receive any PBU message within the given amount of time, it will delete the binding cache entry. Upon detecting the MN on its access link, the new MAG will signal the LMA to update the binding state. Finally, the serving MAG will send the router advertisements (RtrAdv) containing the MN's home network prefixes.

In PMIPv6, all the data traffic sent from the MN gets routed to the LMA through a tunnel between the MAG and the LMA. The LMA forwards the received data packets from the CN to the MAG through a tunnel. After receiving the packets, the MAG at the other end of the tunnel will remove the outer header and forward the data traffic to the MN. Furthermore, PMIPv6 is a localized mobility management protocol that shortens the signaling update time and reduces the disruption period. Therefore, the PMIPv6 handover can be relatively faster than the MIPv6 by using the link layer attachment information. However, PMIPv6 still suffers from communication interruptions due to the link layer handover, which basically, depending on the underlying technology used, needs some time to complete [13]. Consequently, all data packets sent during this handover period are lost. Moreover, enhancing the seamlessness of the PMIPv6 handover is still needed to support the QoS of real-time sensitive services and multimedia applications as well as the interaction with MIPv6 to support global mobility. Recently, the 3GPP and WiMAX forum have envisaged employing PMIPv6 for interworking with heterogeneous wireless access networks [18].

Considering the reduction of handover latency and data loss in PMIPv6, the fast handovers for PMIPv6 (FPMIPv6) procedure is proposed [19]. This standard specifies some necessary extensions for FMIPv6 to support the scenario when the MN does not have IP mobility functionality and hence is not involved with either MIPv6 or FMIPv6 operations. Moreover, FPMIPv6 does not require any additional IP-level functionality on the LMA and the MN running in the PMIPv6 domain.

FPMIPv6 can operate in predictive and reactive modes. In the predictive mode (if the MN detects a need for handover), the MN initiates handover procedures by transmitting an indication message to the previous MAG. In reactive mode, when the MN requires handover, the MN executes network re-entry to the new MAG. Then, the new MAG initiates handover procedures before the MN informs the necessity of handover to the previous MAG. Predictive (initiated over previous MAG) and reactive (initiated by new MAG) FPMIPv6 operations are presented in Figure 3.6.

FPMIPv6 can reduce the PMIPv6 handover delay by allowing the MN to begin forwarding packets as soon as it detects a new link. FPMIPv6 operates on the assumption that each of the MAGs has a database that has information (e.g., PoA identification and proxy CoA) of all the other MAGs

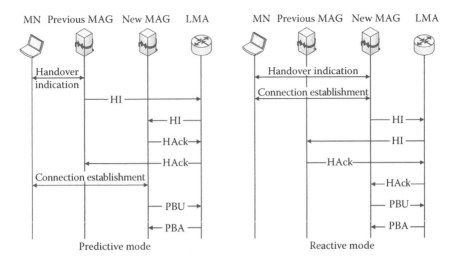

FIGURE 3.6   Fast handovers operations for PMIPv6.

that exist in the same network. However, if the new MAG is located in a HetNet, then there may be no way to obtain the proxy CoA of the new MAG in advance, which is a problem that occurred when conducting handover between heterogeneous access networks. Especially when a real-time multimedia packet stream needs handover management support, information on the new MAG that is located in the HetNet needs to be known in advance as the QoS parameters on the target network side may need to be negotiated. Due to these reasons, even with FPMIPv6 implemented, it may be difficult to conduct seamless handover to the new MAG. Therefore, an enhanced FPMIPv6 scheme is proposed to reduce the packet-forwarding delay and E2E data transmission delay by eliminating traffic aggregation problems through improved coordinated data-path switching and also by using shorter data-paths instead of longer IP-tunneled data-paths [20].

Example 3.1

Commonly, signaling packets are small and approximately the same size. Therefore, compared with packet transmission time, the propagation delay becomes an important criteria in mobility protocols performance analysis. To present performance analysis of FPMIPv6 and enhanced FPMIPv6, it can be assumed that the previous and new PMAG are at equal distance with the LMA. Therefore, the message propagation delay of PBU (T_{PBU}), PBA (T_{PBA}), fast PBU (T_{FPBU}), fast PBA (T_{FPBA}), handover packet-forwarding address request (T_{HPAR}), and corresponding response (T_{HPAP}) can all be set equal to the propagation delay between the LMA and MAG ($T_{LMA-MAG}$) because these messages are exchanged between the previous MAG and LMA or the new MAG and LMA. Because signaling packets exchanged between MAGs need to be routed through the LMA, the propagation delay for HI (T_{HI}), HAck (T_{HAck}), and $T_{MAG-MAG}$ are all equal to $2T_{LMA-MAG}$. The connection establishment time (T_{conn}) can be set to $2T_{MAG-MN}$, where T_{MAG-MN} is the propagation delay between the MAG and MN, and the transmission delay of forwarded data packets (T_{df}) is set to 10 ms. T_{MAG-MN} is set to 0.5 ms and $T_{LMA-MAG}$ is set as a variable where various values were tested during the simulation. The handover indication message propagation delay ($T_{h.ind}$) is equal to T_{MAG-MN}.

For the predictive mode, the handover signaling delay ($T_{HO,pre}$) and data packet-forwarding delay ($T_{DF,pre}$) of FPMIPv6 are expressed, respectively, as

$$T_{HO,pre} = T_{h,ind} + T_{HI} + T_{HAck} + T_{conn} + T_{PBU} + T_{PBA}, \quad (3.1)$$

$$T_{DF,pre} = 2T_{df} + T_{MAG-MAG} + T_{MAG-MN}. \tag{3.2}$$

On the other hand, handover signaling delay ($T_{E-HO,pre}$) and data forwarding delay ($T_{E-DF,pre}$) for enhanced FPMIPv6 are expressed, respectively, as

$$T_{E-HO,pre} = T_{h,ind} + 2T_{FPBU} + 2T_{FPBA} + 2T_{conn} + (T_{HPAR} + T_{HPAP}), \tag{3.3}$$

$$T_{E-DF,pre} = 2T_{df} + T_{LMA-MAG} + T_{MAG-MN}. \tag{3.4}$$

In the reactive mode, the handover signaling delay ($T_{HO,rea}$) and data packet-forwarding delay ($T_{DF,rea}$) of FPMIPv6 are expressed, respectively, as

$$T_{HO,rea} = T_{conn} + T_{HI} + T_{HAck} + T_{PBU} + T_{PBA}, \tag{3.5}$$

$$T_{DF,rea} = 2T_{df} + T_{MAG-MAG} + T_{MAG-MN}. \tag{3.6}$$

Signaling delay ($T_{E-HO,rea}$) and data packet-forwarding delay ($T_{E-DF,rea}$) for reactive mode of enhanced FPMIPv6 are expressed, respectively, as

$$T_{E-HO,rea} = T_{conn} + 2T_{FPBU} + 2T_{FPBA}, \tag{3.7}$$

$$T_{E-DF,rea} = 2T_{df} + T_{LMA-MAG} + T_{MAG-MN}. \tag{3.8}$$

The total time-delay of the FPMIPv6 procedure is defined as $T_{FPMIPv6,x} = T_{HO,x+TDF,x}$ and the total time-delay of the enhanced FPMIPv6 procedure is defined as $T_{E-FPMIPv6,x} = T_{E-HO,x+TE-DF,x}$, where x = pre or x = rea when representing predictive or reactive mode, respectively. Based on the above assumptions and relations, it can be concluded that $T_{FPMIPv6,x} = (T_{EFPMIPv6,x} + 3T_{LMA-MAG})$; therefore, $T_{FPMIPv6,x} > T_{E-FPMIPv6,x}$. Figure 3.7 shows the results obtained through simulation of total handover delay for predictive and reactive modes with varying $T_{LMA-MAG}$ [20].

Obvious delay reductions, particularly in reactive mode, can be regarded as a consequence of reduction in the data forwarding and the handover delay of the enhanced FPMIPv6 scheme. Because data packets are buffered at the LMA in the enhanced FPMIPv6, no IP tunneling is required between the previous and new MAG. These advantageous properties make the enhanced FPMIPv6 scheme more suitable for heterogeneous mobility compared with FPMIPv6 operations.

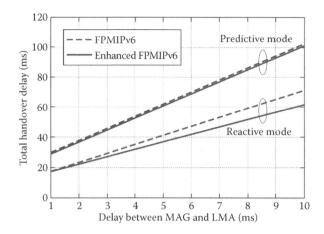

FIGURE 3.7   Total handover delay for FPMIPv6 and enhanced FPMIPv6. (Partially from J.-M. Chung et al., *IEEE Wireless Communications* 20, no. 3 (June 2013): 112–119.)

The Network Mobility (NEMO) Basic Support protocol [21] extends MIPv6 to support the movement of a whole network (mobile network), by the mobile router taking care of the mobility management (i.e., mobility signaling and tunnel setup) on behalf of the nodes of the network, called mobile network nodes (MNNs). The IP addresses of these nodes belong to the mobile network prefix (MNP) that is anchored at the HA of the mobile router. Regarding mobility, the NEMO is a client-based solution because it is also based on mobility functionality in the MN, a router in this case. The main purpose of this protocol is to address the requirement of transparent Internet access from vehicles.

NEMO inherits the limitations of MIPv6 as well as having its own drawbacks. Although NEMO seems to fit well in the context of terrestrial transport systems, it has not been designed to support the dynamics and special characteristics of vehicular communication networks (VCNs) [22]. The current version, as defined by the standard, does not incorporate a route optimization mechanism, and that affects its performance in vehicular scenarios.

Because packets sent by CNs reach the mobile network through one or more bidirectional tunnels between the HA and the mobile router, the route traversed by packets may be suboptimal when the mobile network and CN are in the same network (or topologically close) that is far away from the HA. A suboptimal route results in inefficiencies such as higher E2E delay, additional infrastructure load, susceptibility to link failures, etc. [23]. Moreover, requirement of all packets from or to the mobile network to pass through HA creates a bottleneck. Header overhead is another issue associated to the problems of suboptimal route. As a packet passes through each tunnel, it is encapsulated resulting in increased packet size. Encapsulation results in header overhead that decreases bandwidth efficiency, and increases the chance of fragmentation. Moreover, encapsulated packets are also decapsulated as often as the number of encapsulations. Encapsulation and decapsulation require additional processing at the HA and mobile router. Handover of a mobile router is similar to that of a MIPv6 node. When a mobile router moves from one network to another, it has to discover an AR to obtain a CoA, and register with the HA. This procedure results in a delay that interrupts ongoing connections. The problem of handover delay reduction is not unique to NEMO, and has been adequately addressed for MIPv6.

An interesting extension to PMIPv6, specially designed for public transport system communications, is proposed by Soto et al. [24]. NEMO-enabled PMIPv6 (N-PMIPv6) fully integrates mobile networks into LMDs. Concerning this approach, users can obtain connectivity either from fixed locations or mobile platforms (e.g., vehicles) and can move between them while keeping their ongoing sessions. N-PMIPv6 architecture exhibits two remarkable characteristics:

1. It is a totally network-based solution; therefore, no mobility support is required in the terminals
2. The handover performance is improved, both in terms of latency and signaling overhead

A MN is able to roam not only between fixed gateways (i.e., MAGs as in conventional PMIPv6) but also between moving gateways (called mMAGs, which are also able to roam within the domain), without changing the IPv6 addresses they are using. Detailed signaling operations for N-PMIPv6 are shown in Figure 3.8.

A mMAG behaves as a MN from the viewpoint of fixed gateways because moving gateways roam between different fixed gateways while keeping the same IP address. Moreover, a mMAG behaves as a regular gateway from the MN's perspective, and extends the localized domain by providing attached terminals with IPv6 prefixes of the domain, and by forwarding their packets through the LMA. An additional bidirectional tunnel between the moving gateway and the LMA is used to hide the network topology and avoid changing the particular prefix assigned to the terminal while roaming within the same domain.

3.3.3 HYBRID IP-BASED MOBILITY MANAGEMENT SOLUTIONS

Although each of the IP-based mobility support protocols is designed to be used independently, there are circumstances in which two or more of them

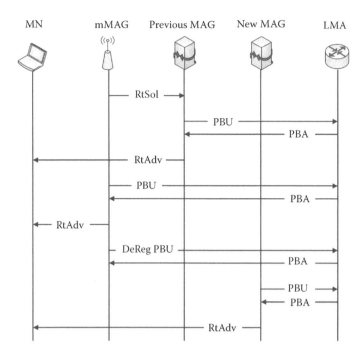

FIGURE 3.8   NEMO-enabled PMIPv6 operations.

can be combined. In most cases, the combination is the result of individual actions of the different actors involved in the scenario (e.g., users and operators), with each of them deploying a solution to fulfill its own requirements [25]. For example, a client-based solution can be set up by a user requiring global mobility, but then the user's MN could visit a network in which the operator has deployed a network-based solution to provide mobility support to its visiting nodes. On the other hand, the combination can also be planned to get together different functionalities, for example, network mobility and host mobility.

The basic combinations do not require modifying the individual protocols. Although they are used together, they are not aware of each other and they do not have explicit mechanisms to cooperate, so there is no increased complexity because new functionality implemented in the involved nodes does not exist. In what follows, some representative hybrid solutions are presented.

3.3.3.1   MIPv6 + PMIPv6

A MN uses MIPv6 to obtain global mobility support (i.e., it can roam to any visited network while keeping global reachability and session continuity). On the other hand, an operator deploys PMIPv6 to offer local mobility support (within the domain) without requiring any support from the MNs. In this scenario, a MIPv6 node may visit the PMIPv6 domain. The operation of MIPv6 in the MN, when visiting a PMIPv6 access network, is the same as when visiting any other foreign network. Initially, after attaching to the domain, the MN obtains an IP address (to be used as its CoA), and registers it in its global mobility agent (i.e., the HA). Because the address used in the PMIPv6 domain remains the same while roaming within this domain, movements are transparent to the mobility management software in the user terminal (i.e., MIPv6). Furthermore, the terminal can also move to an access network outside the domain while keeping ongoing sessions. This is done by the terminal getting another temporal address from the new access network and using MIPv6 to keep its global mobility agent (i.e., the HA) updated with its new location.

In this case, 24 bytes of additional overhead are added in the entire path between the MN and the CN, due to the use of MIPv6, plus an IPv6 tunnel (40 bytes) between the LMA and the MAG where the MN is attached (due to the use of PMIPv6). It is important to note that, out of the overall overhead, only the 24 bytes added by MIPv6 are present in the wireless access [25].

3.3.3.2   NEMO + PMIPv6 (+MIPv6)

A mobile router uses a NEMO protocol to acquire global mobility support for itself and the corresponding network. A node inside the mobile network can be a normal IP node without mobility support if it is not going to move away from the mobile network. It can also be a node with MIPv6-based global mobility support, able to roam to other networks. In addition, an operator deploys PMIPv6 to offer local mobility support without requiring any support from visiting nodes (hosts/routers).

Concerning this scenario, two different tunnels are involved to enable the communications of the mobile network: one between the mobile router and its HA (due to the use of NEMO), and another between the LMA and the MAG serving the mobile router (due to the use of PMIPv6). Consequently, there are up to 80 additional bytes of overhead in some wired segments of the path (when both tunnels are present), and up to 40 bytes in the wireless access (due to the use of NEMO), although not in the last wireless hop between the user terminal (i.e., the MNN) and its AR. A third overhead component (24 bytes in a route-optimized mode) is required if the terminal attached to the mobile network is itself a MIPv6 node outside its home network.

A particularly relevant example of this scenario is the provision of Internet connectivity in public transportation systems where users benefit from seamless access using mobility-unaware devices while the network mobility support (i.e., the mobile router) takes care of managing the mobility on behalf of the terminals. Some of the access networks can also provide PMIPv6 support. In this situation, where NEMO and PMIPv6 protocols are combined, when the mobile router enters the localized domain, it gets a temporal address (to be used as its CoA) from the domain and registers this address in its global mobility agent (i.e., HA), binding the MNPs managed by the mobile router to its current location (i.e., CoA). Because this newly acquired IPv6 address is provided by the PMIPv6 domain, it does not change while the mobile network roams within the localized domain, and therefore its movements are transparent to the NEMO protocol. Moreover, the mobile network is able to roam not only within the localized domain but also outside the domain, thanks to the NEMO operation, which provides global mobility support. A user terminal will not be able to leave the mobile network without breaking its ongoing sessions unless this terminal has MIPv6 support.

3.3.3.3   MIPv6 + N-PMIPv6

This scenario is very similar to the MIPv6 + PMIPv6 combination. A MN uses MIPv6 to obtain global mobility support (i.e., it can roam to any visited network while keeping global reachability and session continuity). In addition, an operator deploys N-PMIPv6 to offer local mobility support enabling local roaming without requiring any support from user terminals. With N-PMIPv6, this local mobility domain is composed of fixed and moving MAGs. In this scenario, a MIPv6 terminal may visit the N-PMIPv6 domain. A user terminal can both move within a localized domain without changing its IP address and can also leave the domain without breaking any ongoing communications by acquiring a new temporal address from the new access network and using MIPv6 to register this temporal address in its global mobility agent. The difference with the MIPv6 + PMIPv6 combination is that here the localized domain integrates both fixed and moving MAGs, so that a user terminal is able to roam between fixed and mobile access infrastructure within the domain without involving/requiring any IP mobility support in the terminal (thanks to the N-PMIPv6 protocol). Whenever the terminal changes its location within the domain, the new access gateway (fixed or mobile) will update the terminal's location in the LMA. In this hybrid solution, three overhead components are required: one (24 bytes) between MN and CN (due to the use of MIPv6 in a route-optimized mode), an IPv6 tunnel (40 bytes) between the LMA and the fixed MAG, and a second tunnel between the LMA and the mMAG.

An example of this scenario can also be a public transportation system, in which mobility-unaware devices would get Internet access not only while moving or while waiting at the station platforms but also while roaming between fixed and mobile access infrastructure. Additionally, the use of MIPv6 would also enable a MN to roam outside the localized domain, for example, when leaving the public transportation environment.

3.3.3.4 NEMO + N-PMIPv6 (+MIPv6)

In this combination, as in the previous one, an operator deploys N-PMIPv6 to support local mobility, enabling local roaming (within the domain) without requiring any support from the user terminals. However, the operator also deploys NEMO mobile router capabilities in the mMAGs, which enable the corresponding mobile networks to be able to move outside the localized domain while keeping ongoing sessions. This can be a common configuration if the mobile network needs to move out of a domain (e.g., a vehicle leaves the N-PMIPv6 localized domain deployed in a city and connects to another network operator). Using the N-PMIPv6 protocol, the localized domain integrates both fixed and moving MAGs, so that a user terminal is able to roam between fixed and mobile access infrastructures within the domain without changing IP address. The terminal can also be connected to a mMAG that moves outside the localized domain, and thanks to the use of NEMO functionality, this movement will be transparent to terminals in the mobile network, that is, they will not need to change their IP addresses. The terminal can also use MIPv6 to obtain global mobility, that is, to be able to roam outside the access infrastructure over N-PMIPv6 and the mobile networks created by using NEMO.

The most effective way of deploying this scenario is by colocating the global mobility agent of the mobile router functionality (i.e., the HA) and the LMA in the same node so they can share the same range of addresses. With this configuration, the localized domain also becomes the home network (domain) of the global mobility support. Therefore, when the mobile network is at the home domain, packets addressed to a MN attached to this network are forwarded through the LMA, as in the N-PMIPv6 simple case. This means that when the mobile network is away from its home domain, a bidirectional tunnel is created between the mobile router and the HA, which is used to forward all the traffic from or to terminals connected to the mobile network. In case the mobile network moves out of its home domain, the mobile router can no longer act as a mMAG, either because the visited domain is not an N-PMIPv6 localized domain or because the mMAG lacks the appropriate security associations with the LMA of the visited domain. When the mobile network is not at its home domain, a user terminal moving away from the mobile network would need to change its IP address, thus breaking ongoing sessions unless the MN has its own MIPv6 support. In this combination, a node in the network has to combine LMA (N-PMIPv6) and HA (NEMO) functionality.

When a mobile network is attached to a mMAG (located in home N-PMIPv6 domain), and assuming a deployment scenario in which the

LMA and the HA are colocated, two IPv6 tunnels are required. The first one is between the LMA and the fixed MAG, and the second one between the LMA and the mMAG. If the user terminal is a MN running MIPv6 (which is outside its home network), an additional overhead component (24 bytes) is required due to the use of MIPv6 in route-optimized mode.

Considering the presented hybrid solutions, it can be concluded that they can have an important effect on the overhead and handover delay, leading to performance penalties that can be significant in certain cases. Comprehensive experimental evaluation of hybrid mobility management solutions regarding handover delay is conducted by de la Oliva et al. [25].

Example 3.2

To select the most suitable mobility management solution, the protocol provided by the network and the MN's preference of the mobility management should be considered [26]. For example, during the authentication process, MAG finds the MN's preference from the user's profile. When the MN's preferred protocol matches that provided by the access network, the agreed protocol is selected. Otherwise, the MN's preference has higher priority. If the MN has no preference, the network takes the responsibility of evaluating the performance of basic MIPv6 and the hybrid MIPv6 + PMIPv6 scheme and selects the appropriate protocol.

To evaluate the performance of basic MIPv6 and the hybrid scheme, the path latency related to them is probed by MAG. During the path probing, the MAG sends out two probing messages to home LMA/HA for multiple times. The first one is sent through the visited LMA and then redirected to the home LMA/HA and the related round-trip time (RTT) is denoted as RTT_{hybrid}. The other probing message is directly sent to the home LMA/HA and the related RTT is denoted as RTT_{mip}.

The average RTT of the MIPv6 path $(\bar{z}_n)$ after path probing for n times can be calculated as

$$\bar{z}_n = \alpha RTT_{mip}(n) + (1-\alpha)\bar{z}_{n-1} \qquad (3.9)$$

Parameter α reflects the significance of past events in the calculation of the weighted average. For example, if $\alpha = 0.8$, then the most recent value $\bar{z}_{n-1}$ contributes to the calculated $\bar{z}_n$ value with a 20% weighting. In a similar manner, the average RTT for the hybrid MIPv6 + PMIPv6 scheme can be calculated and denoted as $\bar{t}_n$.

When the path latency of the basic MIPv6 is much smaller than that of the hybrid scheme and MN hands over with low frequency, the performance of the basic MIPv6 is better. On the other hand, when the latency of basic MIPv6 is not much smaller than that of the hybrid scheme and MN hands over with high frequency, the performance of the hybrid scheme is better. The protocol selection criterion presented as

$$\begin{cases} \dfrac{\overline{t}_n - \overline{z}_n}{N_h} < H_t, & \text{select hybrid scheme} \\[2ex] \dfrac{\overline{t}_n - \overline{z}_n}{N_h} \geq H_t, & \text{select basic MIPv6} \end{cases} \tag{3.10}$$

can be used to adaptively select the better protocol according to network conditions and mobility parameters. Here, N_h is the handover frequency, whereas H_t is the quality threshold, which can be used to determine which protocol should be selected.

3.3.4 TRANSPORT LAYER MOBILITY MANAGEMENT

Transport layer protocols are based on the E2E mechanism as an alternative to MIP-based solutions. These protocols achieve mobility management using end hosts that have been allocated to take care of mobility without altering the network infrastructure. Well-established transport layer protocols, such as TCP and UDP, provide application E2E communication services as well as reliable data delivery, and congestion and flow control. However, these transport protocols do not target wireless network or mobility management support. Therefore, several approaches have been designed and directed toward transport layer performance improvement in wireless networks and support mobility management.

The advantages of transport layer mobility include inherent route optimization (triangle routes never occur), no dependence on the concept of a home network or additional infrastructure, the possibility to support seamless handovers as well as location management if the MN has multiple interfaces, and the ability to pause transmissions in expectation of a mobility-induced temporary disconnection. The main drawback of transport layer approaches is the requirement for cooperation between layers compared with the network layer approach, such as location management handling [27].

The stream control transmission protocol (SCTP) [28] is accepted as a general-purpose transport protocol that operates over a potentially unreliable connectionless packet service, such as IP. Although it inherits many TCP functions, it also incorporates many attractive features such as multihoming, multistream, and partial reliability. The SCTP-based approach uses multihoming for implementing mobility management. The multihoming feature allows a SCTP to maintain multiple IP addresses. Among those addresses, one address is used as the primary address for current transmission and reception, whereas other (secondary) addresses can be used for retransmissions. These IP addresses are exchanged and verified during the association initiation, and are considered as different paths toward the corresponding peer. The multihoming feature of SCTP provides a basis for mobility support because it allows a MN to add a new IP address, while holding an old IP address already assigned to it. Other applications of multihoming, such as load balancing over multiple paths, are not supported by the standard SCTP. Indeed, simultaneous data transfers over multiple paths may cause packet reordering leading to congestion–control problems because SCTP adheres strictly to the TCP congestion–control algorithm, which is not designed to support multihoming [4].

Despite this limitation, SCTP multihoming seems to be an interesting protocol feature that may easily be leveraged to provide transport-layer handover to end-user applications. However, when considering standard SCTP multihoming support for transport-layer handover, it is very important to note that only the primary path is used for data transmission, whereas all other available paths can handle retransmissions only. Then, the decision of changing the primary path relies mainly on the failover mechanism. Another important consideration about SCTP multihoming support is the absence of mechanisms to dynamically change the set of IP addresses specified for an active association. Thus, in a mobile network scenario, if an association has already been established for a given IP address, and a new PoA with a different IP address becomes available, there is no possibility to include it in the association and switch the primary path over to the new network connection.

The SCTP multihoming feature and dynamic address reconfiguration (DAR) extension [29] can be used to solve the mobility management problem in HetNets by dynamically switching between alternate network interfaces. Using this solution, also referred to as mobile SCTP (mSCTP), to enable seamless handover has many advantages including simpler network architecture, improved throughput and delay performance, and ease of adapting flow/congestion control parameters to the visiting networks. mSCTP mobility management procedure is performed through the following basic steps [13]:

1. The MN and CN nodes exchange IP address lists, which are valid for the communication to establish an mSCTP association. Moreover, only one source or destination pair is selected as a primary path to send the data, whereas the remaining pairs will be kept untouched, which serve in the backup situation.

2. When the MN receives other domain network advertisements and before the MN moves to the new domain network, it obtains a new IP address (not included in the exchange list). Then, the MN must communicate its CN using specific control messages (i.e., address configuration chunks) regarding this new IP address.
3. Once the new IP address has been added to the association, the MN may decide, depending on an appropriate moment, such as receiving strong signal strength measurement obtained from the link layer, that it will move to the new network domain.
4. The MN switches the primary path associated with the old IP address to the new IP address.
5. The MN communication session continues without disruption to the new IP address.
6. As further mobility session transmissions continue, unnecessary IP addresses can be removed from the exchange IP address list.

With mSCTP, the primary path may be announced to the receiver's end point during association initialization and changed whenever it is needed during the association lifetime. When adding/removing an IP address to/from an association, the new address is not considered fully valid until the address configuration acknowledgment message is received. Changing the primary address may be combined with the addition or deletion of an IP address. However, only addresses already belonging to the association can be set as the primary, otherwise the set primary address request is discarded. mSCTP preserves the same congestion–control rules as standard SCTP, and logically, a lot of research performed recently on SCTP can be useful for mSCTP development [4].

3.3.5 APPLICATION LAYER MOBILITY MANAGEMENT

The application layer approach achieves mobility management support by taking care of node mobility without altering its IP stack and does not rely on the underlying technology. The application layer mobility management schemes, unlike other approaches in transport and network layers, do not change the MT's kernel. Another advantage is that the application layer, architecturally the highest layer of operation, can function across HetNets.

Session initiation protocol (SIP) [30] is a major application layer signaling protocol for establishing, modifying, and terminating multimedia sessions. These sessions include IP telephone calls, multimedia distribution, multimedia conferences, and other similar services. SIP invitations used to create sessions carry descriptions that allow participants to agree on a set of compatible media types. SIP uses elements called proxy servers to help route requests to the user's current location, authenticate and authorize users for services, implement provider call-routing policies, and provide features to users. SIP also provides a registration function that allows users to upload their current locations for use by proxy servers.

With minor modifications, SIP is capable of handling terminal, session, personal, and service mobility. Moreover, SIP is free of many of the drawbacks of network or transport layer mobility, such as suboptimal routing and protocol stack modification. The fact that SIP has been accepted as the signaling standard by 3GPP, and is also capable of providing mobility support, made it the representative application layer mobility solution. Also, the SIP-based approach is widely accepted as a seamless handover management solution in HetNets [31–33].

SIP exploits knowledge about the traffic at a higher layer to benefit real-time flows. This scheme is quite similar to MIPv6 and is especially advantageous for multimedia traffic, as it reduces E2E delay by allowing a CN to directly communicate with the MT's CoA, but does not require direct traffic tunneling through the HA. The main entities in SIP are user agents, proxy servers, and registrar servers. A user is generally identified using a SIP uniform resource identifier (URI), for example, sip:user@domain.

Signaling messages in typical SIP infrastructure are shown in Figure 3.9. Mobile users register their SIP ID and IP address into a registrar server (that may be colocated with a proxy server) using REGISTER message. When an MN decides to establish a session, it sends an INVITE message to its local proxy server. The INVITE message is routed to the proxy server of the destination domain according to the URI of the destination. Destination proxy server queries the location service of registrar server to find the IP address that is mapped to the destination URI and forwards the INVITE message to the destination user agent.

Although SIP is primarily the protocol for personal mobility, it can also be used for terminal mobility. When an MN moves to a new subnet during a session, it obtains a new IP address, which is to be informed to the CN via a re-INVITE message containing the callID of the session. The CN continues the session with the MN using its new IP address. The MN also sends a REGISTER message to its home registrar server to update its new location.

Several variations of enhanced SIP-based schemes to reduce the handover delay have been proposed in open literature [34,35]. Based on recent comprehensive research, it became evident that SIP is not a totally independent solution to support terminal mobility because it does not take into account the effect of lower layer triggers and information on handover detection and decision. The most perspective approach to overcome this issue is related to the integration of media independent handover (MIH)

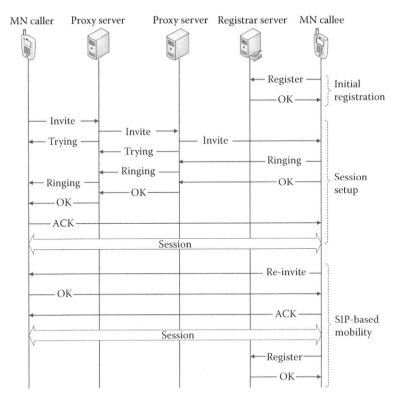

FIGURE 3.9   Signaling messages in SIP-based infrastructure.

functionalities and associated services (see Chapter 4) into SIP mobility management framework [36,37].

3.4   INTERDOMAIN MOBILITY MANAGEMENT

Interdomain mobility management is related to the multihoming capability of a terminal, where users not only have access to services anywhere at any time from any network but also consider using different access networks simultaneously through several wireless interfaces [38]. It allows mobile users more flexibility in terms of ubiquitous access, resiliency, reliability, and bandwidth aggregation. The multihoming terminal should be able to maintain multimedia applications, especially delay-sensitive ones, with minimal disruption during a handover.

The concept of interdomain mobility management also means that one domain for each access network is considered. Even if it is easier for users to have all the subscriptions at the same operator, it is more advantageous to use different operators compared with roaming. Although all operators do not offer this possibility nowadays, the use of separate operators provides a wider coverage area.

3.4.1 SHIM6 PROTOCOL

The Shim6 protocol [39] specifies a network layer shim approach for providing locator agility below the transport protocols so that multihoming can be provided for IPv6 with failover and load spreading properties, without assuming that a multihomed site will have a provider independent address prefix, which is announced in the global routing table. Moreover, Shim6 brings a clear split between the identifier and locator functions of an IP address, which is very helpful in mobile environments, especially for multihomed terminals that have to manage as many IP addresses as access networks. The first IP address a terminal uses to communicate is its identifier, called the upper layer identifier (ULID). The identifier or ULID of a Shim6 terminal remains unchanged for the upper layers, even if the active IP address changes. The locators correspond to the remaining set of IPv6 addresses that are associated with the MT. ULID is also used as a locator at the beginning of the communication. The mapping between ULID and locators is performed in the Shim6 sublayer [40].

Shim6 can easily be adopted in a multihoming MT context. Observing two communicating MNs (MN and CN), Shim6 first initiates a context establishment exchange between these terminals to exchange their available sets of IP addresses. At this step, it also establishes a security association to identify these hosts safely.

The locator change during an active communication is achieved thanks to a reachability protocol (REAP) [41], which implements failure detection and locator pair exploration functions. The failure detection function reveals disconnections in the current path by sending periodic keep alive messages. If the current path fails, a locator pair exploration procedure is launched to select an alternative one. It represents the shortest available path between hosts from the list of locators exchanged at the context establishment. After switching to the new locator pair, the ULIDs remain unchanged. At the arrival of each packet at the Shim6 layer, a comparison is made to determine if the ULID corresponds to the current locator. If they are different, a mapping is performed to set up the correct locator and then keep the changes invisible to the application.

3.4.2 INTERDOMAIN SESSION MANAGEMENT

Integration of Shim6 protocol in the IP multimedia subsystem (IMS) architecture seems to be a perspective approach for interdomain session management. The main goal of this concept is to allow a user to change the access network seamlessly while maintaining application QoS [38]. Shim6 can manage interface switching in a seamless and secure way, whereas IMS supports real-time session negotiation and management, guaranteeing an expected QoS level to the ongoing sessions. With such a combination and the fact that Shim6 makes the interface change transparent, the implementation of a SIP-based proxy is needed to handle the session renegotiation procedure. This proxy is capable of managing the registration and session establishment.

Once the new location has been determined, Shim6 sends a notification message to inform the proxy about the address change so that this entity can start session initiation in the new location. After session establishment, the proxy sends back a notification response to Shim6 so that it can update the locators and switch to the new domain. Considering the interface switching as a handover process, it can be observed as being reactive or proactive. Shim6 originally works in a reactive mode. In fact, the MT first detects the failure and switches to an alternative path afterward. It does not imply any change in the basic concepts (Shim6 and REAP) concerning mobility management. This method involves a handover delay and can cause significant packet loss. On the other hand, in the proactive mode, the handover delay is significantly reduced because the session establishment and resource allocation are anticipated.

Example 3.3

If it is assumed that the access networks belong to independent IMS domains, MN needs to be subscribed to both operators and registered with each IMS. As a case study, an environment in which MN has two interfaces to access IMS_1 and IMS_2 is considered. The MN obtains two IP addresses from the corresponding IMSs. First, the MN connects to IMS_1 and after a while, a handover occurs to the new access network.

Figure 3.10 illustrates the signaling flow for session establishment and its maintenance during terminal movement in the reactive mode. The session established has a callID with $callID_1$ value. After that, the MN initiates a ShimM6 context establishment to benefit from multihoming. At this step, the ULIDs and locators are identical. Sometime later, the MN moves and changes its access network and connects to IMS_2. The interface change is managed by Shim6 as previously described, and the SIP proxy receives a notification message. At this step, the ULIDs remain the same, but the locators are different and correspond to the IP addresses of the new access network (in IMS_2). This change is transparent to the application but should be considered to reestablish a session

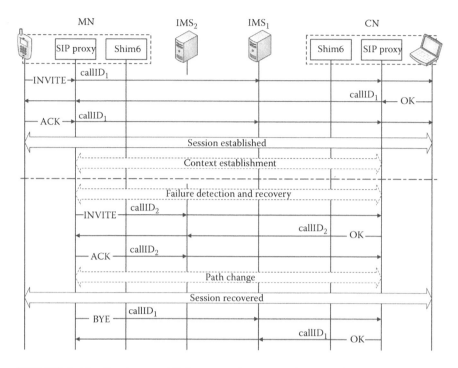

FIGURE 3.10 Session establishment and recovery in a reactive mode with two independent IMSs.

according to the new location of the MN and maintain the appropriate QoS parameters. The SIP proxy initiates the session establishment according to the new location in IMS$_2$ through a second interface. The new session has callID$_2$ as callID value. To hide this change from the application and preserve the continuity of the session, the SIP proxy has to handle the callID change. From the application view, the session is established with callID$_1$, whereas from the network view, the new session has a callID$_2$ value. The application is not aware of the new session establishment.

Once the new session is established, a notification response is sent back to Shim6. At this point, Shim6 enables the use of the new locators, so the traffic is redirected to the new location IMS$_2$. The previous session in IMS$_1$ is explicitly ended to avoid any negative effects on delay and allocated resources (i.e., SIP proxy sends a BYE message with callID$_1$). In the reactive mode, the media loss begins when the Shim6 detects the failure and ends when the Shim6 path change ends. It is obviously visible from the user point of view. Afterward, when the MN ends the session, the BYE message has callID$_1$, and the SIP proxy changes this to callID$_2$ to end the session properly and in the appropriate network.

In the same way, the OK message is sent with the callID$_2$ value. When the CN's proxy receives this message, it changes the callID$_2$ to callID$_1$ and forwards the message to the application to complete the BYE procedure. From the application view, the session does not change. All modifications are hidden by the SIP proxy to maintain the session as active.

3.5 DISTRIBUTED MOBILITY MANAGEMENT

Traditional hierarchical structures of mobile systems employ heavily centralized mobility management protocols, notably PMIPv6 and dual-stack MIPv6, to handle network-based and host-based mobility schemes, respectively. The centralized approach might have been reasonable at the time the existing mobile networks were designed. However, it brings several obvious limitations in handling a large volume of mobile data traffic, due to the involvement of the centralized mobility anchors (HA, LMA, etc.) in handling mobility signaling and routing for all registered MNs. These drawbacks are identified by Chan [42] as

- *Suboptimal routing.* Because the (home) address used by an MN is anchored at the home link, traffic always traverses the central anchor, leading to paths that are, in general, longer than the direct one between the MN and its CN. This is exacerbated by the current trend in which content providers push their data to the edge of the network, as close as possible to the users, for example, in deploying content delivery networks. With centralized mobility management approaches, user traffic will always need to go first to the home network and then to the actual content source, sometimes adding unnecessary delay and wasting operator resources.
- *Scalability problems.* Existing mobile networks have to be dimensioned to support all the traffic traversing the central anchors. This poses several scalability and network design problems because central mobility anchors need to have enough processing and routing capabilities to deal with all the users' traffic simultaneously. Additionally, the entire operator's network needs to be dimensioned to be able to cope with all the users' traffic.
- *Reliability.* Centralized solutions share the problem of being more prone to reliability problems because the central entity is potentially a single point of failure and vulnerability to attacks.

To address these issues, a new architectural paradigm, called DMM, is being explored by both research and standards communities. DMM introduces the concept of a flatter system architecture in which mobility anchors are placed closer to the user, distributing the control and data infrastructures among the entities located at the edge of the access network. Critically, DMM introduces the ability of an MN to move between mobility anchors, something that is not possible with any of the present centralized approaches. Additionally, it is worth noting that by removing all the difficulties posed by the deployment of a centralized anchoring and mobility approach, adoption of DMM is expected to be easier [43]. Currently, both leading standardization bodies in the field of mobility management, IETF and 3GPP, have their own initiatives related to the DMM framework.

3.5.1 IETF FRAMEWORK FOR DMM

At this moment, the IETF is the main driver in the DMM standardization process. The DMM WG [44] was chartered to address the emerging need for mobile operators to evolve existing IP mobility solutions toward supporting a distributed anchoring model. The IETF first identified the requirements that DMM architecture should meet, and it is currently analyzing existing practices for the deployment of IP mobility solutions in a distributed environment. The main goal of this analysis is to identify what can be achieved with existing mobility solutions and which functions are missing to meet the identified DMM requirements. In terms of solution space, three main classes of solutions can be identified: client (MIPv6)-based, network (PMIPv6)-based and routing-based DMM.

In the case of client-based distributed mobility [45], multiple HAs are deployed at the edge of the access network to distribute the anchoring. The basic concept is that an MN no longer uses a single IP address anchored at a central HA, but configures and uses an additional address at each visited access network. The MN uses the locally anchored address to start new communications, while maintaining reachability for those IP addresses that are still in use by active communications. This requires the MN to bind each of the active (home) addresses with the locally anchored address currently in use, which is actually playing the role of CoA in these bindings. Session continuity is guaranteed by the use of bidirectional tunnels between the MN and each one of the HAs anchoring in-use addresses as shown in Figure 3.11. Here, MN initially attaches to the distributed anchor HA/AR$_1$ and configures the IPv6 address HoA$_1$ to communicate with a CN$_1$. If MN moves to HA/AR$_3$, a new locally anchored IPv6 address (HoA$_2$) is configured and used for new communications, for example, with CN$_2$. The continuity of the session with CN$_1$ is provided by a tunnel set up between the MN and HA/AR$_1$.

This deployment model does not require changes in the protocol behavior of the network entities. However, it requires extensions and additional intelligence on the MN side because it has to manage multiple addresses simultaneously, select the right one to use for each communication, keep

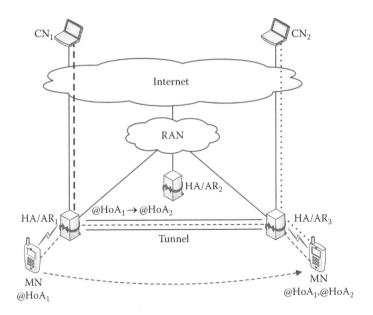

FIGURE 3.11 Client-based DMM concept.

track of those addresses that need mobility support, and perform the required maintenance operations (i.e., binding, signaling, and tunneling) [43]. Handover latency of MIPv6-based DMM is improved from MIPv6 due to the localized and distributed mobility management being introduced, and depends strictly on the underlying network topology [46]. Because of this fact and good performance in terms of packet delay, client-based DMM seems suitable for supporting real-time video applications that can tolerate some amount of delay (e.g., video-on-demand with a large buffer) and also non-real-time applications (e.g., video downloading).

Considering the network-based distributed mobility solutions, two subclasses can be identified: solutions with a fully distributed model and solutions with a partially distributed model. The difference between fully and partially distributed approaches has to do with whether or not the control plane and data plane are tightly coupled. In the fully distributed model, mobility anchors are moved to the edge of the access network, and they manage both control and data planes. For instance, in a fully distributed model and using PMIPv6 terminology, each AR implements both LMA and MAG functions, and for each user, the AR could serve as an LMA (thus anchoring and routing the local traffic for a given user) or as an MAG (thus receiving the tunneled traffic from the virtual home link of the given user). In the partially distributed model, the data and control planes are separated, and only the data plane is distributed. In this sense, the operations are similar to 3GPP networks in which the control plane is managed by the mobility management entity (MME), whereas the data plane is managed by the corresponding gateways. An example of the operation of a generic network-based DMM concept is shown in Figure 3.12.

The split of the control plane and data plane allows the mobility anchors to optimally route the data traffic while relying on a single central entity to retrieve the localization of the connected MNs. There are proposals to extend the PMIPv6 protocol to achieve this by either maintaining legacy with current PMIPv6 deployments or proposing new extensions and changing the way PMIPv6 functions are implemented. Among the ones of the first category, Korhonen et al. [47] propose implementing local routing at the MAG. In fact, the PMIPv6 binding signaling exchange is extended to allow the MAG to defend a set of IP addresses, thus routing traffic directly through the Internet. To access the operator services, the MN can still use the IP address anchored at the LMA. This, however, requires the management of several IP addresses at the MN and the selection of a specific IP address for a certain service. Alternatively, the logical entity of a centralized mobility database to maintain users' localization information and allow the setup of on-demand tunneling when a specific service requires seamless mobility support is introduced by Bernardos et al. [48]. Taking into account the qualitative analysis carried out by Shin et al. [46], the PMIPv6-based DMM seems most suitable for real-time interactive video applications (e.g., videoconferencing and gaming) due to its low handover latency.

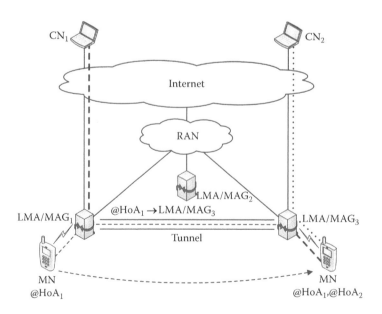

FIGURE 3.12   Network-based DMM concept.

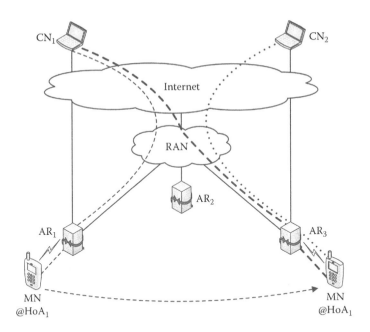

FIGURE 3.13   Routing-based DMM concept.

The routing-based distributed mobility proposal [49] follows a completely different philosophy. In this case, when the MN attaches to an AR, it obtains an IP address that is then internally advertised within the domain using an intradomain protocol (e.g., border gateway protocol; BGP). When the MN moves, the new AR first looks up an IP address using the MN's host name obtained during the authentication. If found, the new AR performs a reverse lookup to confirm that some AR has actually assigned the IP address to the MN's host name. If this is confirmed, a routing update is performed. The new AR creates a BGP update message containing the MN's IP address and sends the message to its peers to announce the new route. An example of routing-based concept is shown in Figure 3.13.

However, routing-based DMM has some limitations in terms of handover latency (limited by the intradomain routing convergence) and scalability (i.e., storms of routing updates) [43]. This approach would be the most efficient way to support non-real-time video applications due to its superior performance in packet routing. On the other hand, the scope of its usage seems limited to less mobile users due to its high signaling overhead [46].

3.5.2   3GPP EFFORTS TOWARD FLEXIBLE AND DYNAMIC MOBILITY MANAGEMENT

Although no DMM-specific efforts are ongoing in 3GPP, several developments indicate a trend toward more flexible and dynamic mobility management. Namely, 3GPP continues to make ongoing efforts to alleviate the traffic load on the mobile core network. The prevalent schemes are local IP address (LIPA) and selected IP traffic off-load (SIPTO) [50].

LIPA enables an IP-capable MN connected via a femtocell to access other IP-capable entities in the same residential or enterprise network, without the user plane traversing the mobile operator's core network. LIPA is achieved by collocating a local gateway with the femtocell and enabling a direct user plane between the local gateway and the femtocell. On the other hand, SIPTO enables a mobile operator to off-load certain types of traffic at a network node close to the MN's PoA to the access network. This is achieved by selecting a set of gateways (i.e., serving gateway and packet gateway) geographically or topologically close to the PoA. For 3GPP Rel. 12, there is currently a work item called LIPA mobility and SIPTO at the local network (LIMONET) [51], which aims to provide mobility support for LIPA between femtocells within the same local gateway and also enable SIPTO at the local network. Although SIPTO and LIPA allow off-loading traffic from the network core in ways that seem similar to the DMM approaches, the LIMONET concept provides only localized mobility support (within a small geographical region).

Another important aspect of the 3GPP approach to mobility management is the role of the MME. Based on the key supported functions (e.g., MT reachability, selection of the appropriate gateway, and management of bearers associated with the MT), it should be clear that the role of MME in such evolved networks is an important consideration. For instance, the MME can play a role in the selection of the distributed anchor.

3.6   CONCLUDING REMARKS

Many mobility support protocols have been and are still being developed by several standardization bodies, in particular by the IETF. Each protocol provides a different functionality (e.g., terminal mobility or network mobility) and requires operations in different network entities. The evaluation of mobility management schemes clearly shows that the network-based mobility protocol, PMIPv6, is the most promising solution to improve the mobile communications performance on localized networks and it is expected to fulfill most of the service requirements in the wireless networks.

The current trend in the evolution of mobile communication networks is toward terminals with several network interfaces that get ubiquitous Internet access by dynamically changing access network to the most appropriate one. Handovers between different access networks will become more common. In this situation, the different solutions are going to coexist to provide seamless mobility. There is no general solution because each of them addresses different requirements.

As a flexible architectural paradigm, DMM offers perspective solutions for future multimedia wireless systems. It can be introduced into real environments in a gradual and additive fashion, either complementing or replacing existing functions, depending on the operator's needs. Moreover, it evolves those features of the 3GPP systems that are already being deployed as spot solutions to the bandwidth crunch. According to the comparative analysis in terms of overhead, handover latency, and packet delay, it can be concluded that client-based and network-based DMM is more suitable for efficient multimedia content delivery compared with routing-based approaches, and can better support delay-sensitive and delay-tolerant multimedia services, respectively.

4 Network Selection in Heterogeneous Wireless Environment

An important characteristic of next generation wireless networks is the compositeness of the communication model. The combination of different wireless technologies and architectures can be used for providing a large variety of multimedia services for users to access from "any place and any time." In such a heterogeneous environment, handover management is the essential issue that supports seamless mobility for users. Handover management, as one of the components of mobility management, controls the change of the mobile terminal's point of attachment (PoA) during active communications. Handover management includes mobility scenarios, metrics, decision algorithms, and procedures. Major challenges in heterogeneous handover management are seamlessness and automation aspects in network switching. Because users can always be connected through the optimal radio access network (RAN), it is necessary to develop an adequate mechanism for its selection. Because some other parameters must be taken into consideration besides the traditional received signal strength (RSS) and signal to interference and noise ratio (SINR), it is possible that the problem can be pointed out from the aspect of multicriteria analysis.

4.1 INTRODUCTION

Growing consumer demands for access to multimedia services from anywhere at any time is accelerating technological development toward the integration of heterogeneous wireless access networks. This next generation of wireless multimedia systems represents a heterogeneous environment with different RAN technologies that differ in terms of their general characteristics such as coverage, bandwidth, security, cost, and QoS. However, differences can exist among networks with the same architecture. For example, two wireless local area networks (WLANs) based on the same standard can be different in terms of security and QoS as well as the cost of service.

In this kind of environment, handover management is the essential issue that supports the mobility of users from one system to another. Handover management, as one of the components of mobility management, controls the change of the mobile terminal's PoA during active communications. Handover management includes mobility scenarios, metrics, decision algorithms, and procedures [1].

In homogeneous networks, handover is typically required when the serving PoA becomes unavailable due to user's movement, whereas the need for vertical (heterogeneous) handover can be initiated for convenience rather than connectivity reasons. Major challenges in vertical handover management are seamlessness and automation aspects in network switching. These specific requirements can refer to the always best connected (ABC) concept [2]. ABC represents a vision of fixed and mobile wireless access as an integral and challenging dimension in developmental paradigm of the next generation of wireless networks. It is a strategic goal to define important advancements that happen and are predicted in technologies, networks, user terminals, services, and future business models that include all these issues while realizing and exploiting new wireless networks. On the other hand, because users can always be connected through the optimal RAN, it is necessary to develop an adequate mechanism for its selection. Because some other parameters must be taken into consideration besides the traditional RSS and SINR, it is possible that the problem can be pointed out from the aspect of multiple criteria analysis [3].

Network selection is one of the most significant challenges for next generation wireless heterogeneous networks, and thus it draws researchers' and standardization bodies' attention. International Telecommunication Union's (ITU) concept of Optimally Connected, Anywhere, Anytime proposed in M.1645 [4] states that future wireless networks can be realized through the coalition of different RANs. According to such a scenario, the heterogeneity of access networks, services, and terminals should be fully exploited to enable higher utilization of radio resources. The main objective is to improve overall networks performances and QoS perceived by users. 3GPP is defining an Access Network Discovery and Selection Function (ANDSF) [5] to assist mobile terminals in vertical handover between 3GPP and non-3GPP networks, covering both automated and manual selection as well as operator and user management.

IEEE 802.21 TG is developing standards to enable handover and interoperability between heterogeneous link layers [6]. This standard defines the tools required to exchange information, events, and commands to facilitate handover initiation and handover preparation. The IEEE 802.21 standard does not attempt to standardize the actual handover execution mechanism. Therefore, it is equally applicable to systems that employ mobile IP at the network layer as to systems that use SIP at the application layer (see Chapter 3).

In the network selection scenario, users are always trying to seamlessly access high-quality multimedia service at any speed, any location, and at any time by selecting the optimal network. Therefore, ensuring a specific QoS is the objective in the process of network selection. A great number of techniques related to the handover initiation and optimal access network selection are proposed in the open literature. The suggested techniques use different metrics and heuristics for solving the abovementioned problems. Unfortunately, the currently proposed vertical handover techniques do not completely satisfy the demands from the covering technologies' point of view, as well as for analyzing the parameters' adequacy, implementation of complexity, together with invoking all the entities to the access network selection.

This chapter starts with a synopsis of the handovers in a heterogeneous environment. Additionally, Media-Independent Handover (MIH; IEEE 802.21), which is designed for vertical handover, is briefly described. After that, the definition and systematization of the decision criteria for network selection is provided. Next, the influences of users' preferences together with the service requirements are envisaged. Finally, existing studies on network selection using cost function, multiple attribute decision making, fuzzy logic, artificial neural networks, etc., are systematically discussed.

4.2 HANDOVER FRAMEWORK IN HETEROGENEOUS ENVIRONMENT

The area serviced by each PoA can be identified as its cell. The dimensions and profile of every cell depend on the network type, transmission and reception power, as well as size of each PoA. Usually, cells of the same network type are adjacent to each other and overlap in such a way that, for the majority of time, any mobile terminal is within the coverage area of more than one PoA. Cells of HetNets, on the other hand, are overlaid within each other. Therefore, the key issue for a mobile host is to reach a decision from time to time as to which PoA of which network will handle the signal transmissions to and from a specific host and hand off the signal transmission if necessary.

Handover (handoff) is the temporal process of maintaining a user's active session (call) when a mobile terminal changes its PoA to the access network. It is possible to classify handovers based on several factors as shown in Figure 4.1. In a heterogeneous environment, besides network type, many other factors constitute handover categorization including the administrative domains involved, number of connections, and frequencies engaged. Also, necessity and user control can be observed as characteristic factors for NGWS.

The type of network environment is the most common classification factor. Depending on the access network that each PoA belongs to, the handover can be either horizontal or vertical [7,8]. A horizontal (intrasystem)

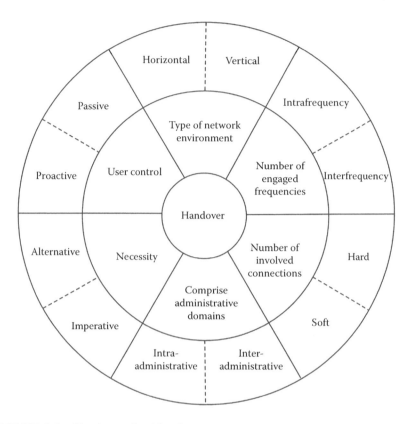

FIGURE 4.1 Handover classification.

handover takes place between PoAs supporting the same network technology, for example, the changeover of signal transmission between two base stations of a cellular network. On the other hand, a vertical (intersystem) handover occurs between PoAs supporting different network technologies, for example, the changeover of signal transmission from an IEEE 802.11 AP to an overlaid IEEE 802.16 network. Vertical handoffs can be further distinguished into upward vertical handover and downward vertical handover. An upward vertical handover roams an overlay with a larger cell coverage and lower bandwidth, and a downward vertical handover roams an overlay with a smaller cell coverage and larger bandwidth. Downward vertical handovers are less time-critical because a mobile device can always remain connected to the upper overlay.

Considering the number of engaged frequencies, handover can be observed as

- Intrafrequency handover process between PoAs operating on the same frequency. This type of handover is characteristic of CDMA networks with FDD.

- Interfrequency handover process between PoAs operating on different frequencies. This type of handover is present in CDMA networks with TDD.

Handovers can also be classified as hard and soft, depending on the number of connections involved. In the case of a hard handover, the radio link to the old PoA is released at the same time a radio link to the new base station is established. Here, the data does not have to be duplicated and, therefore, the data overhead is minimized [9]. However, excessive service interruptions can result in an increased demands dropped rate. Hard handovers are used by systems such as GSM and GPRS in which time division multiple access (TDMA) and frequency division multiple access (FDMA) are applied. Hard handover is also the compulsory method to maintain active sessions in the IEEE 802.16-based networks. Contrary to hard handovers, in a soft handover, a mobile node (MN) maintains a connection with no less than two PoAs in an overlapping region and does not release any of the signals until it drops below a specified threshold value. Soft handovers are possible in situations in which the MN is moving between cells operating on the same frequency. They can be used to extend the time needed to take a handover decision without any QoS degradation. However, because the data are transmitted to all links, frequent soft handovers can result in an increased data overhead. The CDMA systems use soft techniques because, in these networks, a MN may communicate with more than one coded channel, which enables it to communicate with more than one PoA [10]. A softer handover is a special case of a soft handover, in which the mobile terminal switches connections over radio links that belong to the same PoA.

An administrative domain is a group of systems and networks operated by a single organization of administrative authority [8]. Administrative domains play a significant role in heterogeneous wireless networks, and consequently, the classification of handovers in terms of intra-administrative and interadministrative handovers is a crucial issue. Intra-administrative handover represents a process in which the mobile terminal transfers between different networks (supporting the same or different types of network interfaces) managed by the same administrative domain. On the contrary, interadministrative handover is a process in which the mobile terminal transfers between different networks managed by different administrative domains.

The factor of necessity becomes relevant in handover classification with the introduction of heterogeneous wireless networks. In some situations, it is necessary for the mobile terminal to transfer the connection to another PoA to avoid disconnection. In that case, imperative (forced) handover is initiated. This type of handover is triggered by physical events regarding network interfaces' availability. In other situations, the transfer of connection is optional and may or may not improve the QoS. In that case, handover time is not critical, and alternative (optional) handover can be initiated. These handovers are triggered by the users' policies and preferences.

Handover can furthermore be classified as proactive or passive. In a proactive approach, the mobile terminal user is allowed to decide when to initiate the handover. The handover decision can be based on a set of preferences specified by the user. Proactive handoff is expected to be one of the radical features of 4G wireless systems. On the other hand, passive handover occurs in cases when the user has no control over the handover process. This type of handover is the most common in 2G and 3G wireless networks.

4.2.1 SEAMLESS HANDOVER

In one of the revolutionary drivers for next generation wireless multimedia systems, technologies will complement each other to provide ubiquitous high-speed connectivity to mobile terminals [11]. To satisfy these demands, it will be necessary to support seamless handovers of mobile terminals without causing disruption in their ongoing sessions. Whereas wired networks regularly grant high bandwidth and consistent access to the Internet, wireless networks make it possible for users to access a variety of services even when they are moving. Consequently, seamless handover, with low delay and minimal packet loss, has become a crucial factor for mobile users who wish to receive continuous and reliable services. One of the key issues that aid in providing seamless handover is the ability to correctly decide whether or not to initiate handover at any given time. A handover scheme is required to preserve connectivity as devices move about, and at the same time curtail disturbance to ongoing transfers. Therefore, handovers must exhibit low latency, sustain minimal amounts of data loss, as well as scale to large networks. An efficient handover algorithm can achieve many desirable features by trading off different operating characteristics. Some of the major desirable handover features are summarized in Table 4.1.

4.2.2 HANDOVER MANAGEMENT FRAMEWORK

The handover management process can be described mainly through three phases [12]:

- Handover initiation (system discovery, information gathering)—
  This phase involves collecting all required information from
  available (candidate) RANs. This information is then used for

TABLE 4.1
Desirable Handover Features

Feature	Description
High reliability	Handover algorithm is reliable if session (call) achieves good quality after handover execution. Many factors can help in determining the potential QoS of a candidate PoA (RSS, SIR, SNR, etc.).
Interference prevention	Co-channel interference is caused by devices transmitting on the same channel. This is usually caused by a neighboring detrimental source that is operating on the same channel. Interchannel interference, on the other hand, is caused by devices transmitting on adjacent channels.
Seamless mobility	A handover algorithm should be fast so that the mobile user does not experience service degradation or interruption. Service degradation may be due to a continuous reduction in signal strength or increasing co-channel interference.
Load balancing	This feature is extremely important for all cells. This helps to eliminate the need for borrowing channels from neighboring cells that have free channels, which simplifies cell planning and operation, and reduces the probability of new demand blocking.
Reducing number of handovers	A high number of handover attempts may result in more delay in the processing of requests, which will cause signal strength to decrease over a longer period to a level of unacceptable quality. In addition, the demand may be dropped if a sufficient link quality is not achieved. Superfluous handovers should be prevented, especially when the current PoA is able to provide the desired QoS without interfering with other network elements.

identifying the need for handover. Such information is collected periodically or when an event occurs. Depending on the instance that triggered the initiation, the handover can be either network-initiated (forced handover dependent from network condition, load balancing, operator policy, etc.) or terminal-initiated (requested handover dependent on users' preferences, services' requirements, etc.).

• Handover decision (network selection)—This phase determines whether and how to perform the handover by selecting the most suitable RAN and transfer instructions to the execution phase. During the vertical handover procedure, network selection is a crucial step that affects communication. An incorrect or slow handover decision may degrade QoS and even break off current communication. To make an accurate decision, this phase takes advantage of algorithms that, considering the information available, perform an evaluation process to obtain the best choice for handover execution.

• Handover execution (transfer connection)—This phase follows access network selection and leads to channels changing state, conforming to the details resolved during the decision process. Because the converged core network is IP-based, the mobile IP and multiple care-of address (CoA) registrations solutions will be adopted. If handover fails, the network selection attempts to select another RAN, and the list of available networks will be updated.

Handover process control (decision mechanism) can be located in a network entity or in the mobile terminal, and usually involves some sort of measurements and information about when and where to perform the handover. In network-controlled handover (NCHO), the network entity has the primary control over the handover. In mobile-controlled handover (MCHO), the mobile terminal must take its own measurements and make the handover decision on its own. When information and measurements from the MN are used by the network to decide, it is referred to as a mobile-assisted handover (MAHO), similar to 2G mobile systems [1,11]. When the network collects information that can be used by the terminal in a handover decision, it is a network-assisted handover (NAHO), which is characteristic for IEEE 802.16 systems [13].

Especially for heterogeneous wireless networks, user-assisted handover (UAHO) has to be developed as an alternative, proactive approach. On the user–terminal relation, handover initiation, optimal link determination, and active connection maintenance are provided [3,14].

4.3   MEDIA-INDEPENDENT HANDOVER

The main purpose of the IEEE 802.21 [15] standard is to enable handovers between heterogeneous access networks, without service interruption, hence, improving the user's experience of mobile terminals. Many functionalities required to provide session continuity depend on complex interactions that are specific to each particular technology. Supporting seamless intertechnology mobility, handover is a key element to help operators manage and thrive from the heterogeneity. Operators who have the ability to switch a user's session from one access technology to another can better manage their networks and better accommodate the service requirements of their users. For example, when the quality of an application running on one network is poor, the application can be transferred to another network where there may be less congestion, fewer delays, and higher throughput. Operators can manage multiple interfaces to balance traffic loads more appropriately across available RANs, improving system performance and capacity.

The IEEE 802.21 standard defines a MIH framework [15] that can significantly improve handover between HetNets technologies. Here, media refers to the method/mode of accessing a communication system. This standard defines the tools required to exchange information, events and commands to facilitate handover initiation and handover preparation. The IEEE 802.21 standard does not attempt to standardize the actual handover execution mechanism. Therefore, the MIH framework is equally applicable to systems that employ mobile IP at the network layer such as systems that employ SIP at the application layer.

The main purpose of the IEEE 802.21 standard is to enable handover between heterogeneous technologies. Its contributions are centered on:

- A framework that enables seamless handover between heterogeneous technologies
- The definition of a new link layer service access points (SAPs), and
- The definition of a set of handover-enabling functions that provide the upper layers with the required functionality to perform enhanced handovers

The framework that enables seamless handover between heterogeneous technologies is based on a protocol stack implemented in all devices involved in the handover process. The defined protocol stack aims to provide the necessary interactions among devices for optimizing handover decisions.

SAPs define both media-independent and media-specific interfaces. SAP offers a common interface for link layer functions and is independent of any specific technology. For each of the technologies in IEEE 802.21, SAP is mapped to the corresponding specific technology. Some of these mappings are included in the draft standard. Functions concerning the definition of a set of handovers provide the upper layers, for example, mobility management protocols such as mobile IP [16], with the functionality to perform enhanced handovers. The secondary goals are service continuity, handover-aware applications, QoS-aware handovers, network discovery, network selection assistance, and power management [15].

This standard presents a framework that supports a complex exchange of information aiming to enable seamless handover among heterogeneous technologies. Service continuity is defined as the continuation of the service during and after the handover procedure. It is very important to avoid the need to restart a session after handover. From the point of view of handover-aware applications, it is important to take into account that IEEE 802.21 provides applications with functions for participating in handover decisions. For example, a voice application may decide to execute a handover during a silent period to minimize service disruption. The services defined by IEEE 802.21 provide information on networks that are potential handover targets, report events, and deliver commands related to handover. These services speed up handovers while helping to retain E2E connectivity.

Furthermore, this framework provides the necessary functions to make handover decisions based on specific criteria. For example, a user can decide to hand over to a new network that guarantees the desired QoS level. Network selection assistance is the process of making a handover decision based on several factors such as QoS level, traffic condition, policies, security, etc. The IEEE 802.21 framework only provides the necessary functions to assist network selection but does not make handover decisions, which are left to the higher layers.

Finally, power management can also benefit from the information provided by this standard. For example, power consumption can be minimized if the user is informed of network convergence maps, optimal link parameters, or idle modes.

A media-independent framework is a more scalable and efficient method of addressing intertechnology handovers. To address handovers, each access technology requires only a simple extension to ensure interoperability with all other access technologies. The complexity of this approach grows on the order of N access technologies and scales more efficiently than a media-specific approach. This is the approach adopted by the IEEE 802.21 standard, which defines a common set of MIH services that interact with the higher layers of the protocol stack. Then, each access technology requires only one media-specific extension to ensure interoperability with the common IEEE 802.21 framework. IEEE 802.21 provides internetworking with IEEE 802 systems and between IEEE 802 and non-IEEE 802 systems (i.e., 3GPP systems).

Recently, important modifications emerged on both standardization bodies and manufacturers that clearly indicate that IEEE 802.21 will soon take off and be implemented in real networks. For example, InterDigital, British Telecom, and Intel have recently deployed experimental test beds with MIH functionalities to optimize seamless handover. Also, some open source implementations of MIH protocol are emerging to be used by developers.

4.3.1 IEEE 802.21 GENERAL ARCHITECTURE

The architecture of the IEEE 802.21 standard can be described through layer structures and their interactions over nodes and network levels [15]. Figure 4.2 shows a logical diagram of the general architecture of the different nodes supporting IEEE 802.21.

As can be observed, all IEEE 802.21-compliant nodes have a common structure surrounding a central entity called the media-independent handover function (MIHF). The MIHF acts as an intermediate layer between the upper and lower layers whose main function is to coordinate the exchange of information and commands between the different devices involved in making handover decisions and executing handovers. From the MIHF perspective, each node has a set of MIHF users, typically mobility management protocols, that use the MIHF functionality to control and gain handover-related information. The communications between the MIHF and other functional entities such as the MIHF users and lower layers are based on a number of defined service primitives that are grouped in following SAPs [15,17]:

- MIH_SAP: this interface allows communication between the MIHF layer and higher-layer MIHF users
- MIH_link_SAP: this is the interface between the MIHF layer and the lower layers of the protocol stack
- MIH_net_SAP: this interface supports the exchange of information between remote MIHF entities

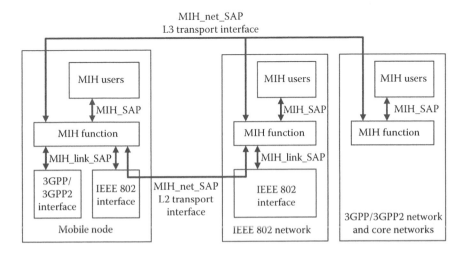

FIGURE 4.2 IEEE 802.21 general architecture.

It should be remembered that all communications between the MIHF and lower layers are done through the MIH_link_SAP. This SAP has been defined as a media-independent interface common to all technologies, so the MIHF layer can be designed independent of the technology specifics. However, these primitives are then mapped to technology-specific primitives offered by the various technologies considered in IEEE 802.21. Standard specification does not mandate a specific programming language for representing the primitive and requires implementers of the MIHF to define specific application programming interfaces (APIs) in terms of their chosen programming language [18].

4.3.2 MIH SERVICES

IEEE 802.21 defines three primary types of MIH services to facilitate intertechnology handovers [15,18–21]: media-independent event service (MIES), media-independent command services (MICS), and media-independent information services (MIIS). These services allow MIHF clients to access handover-related information as well as deliver commands to the link layers or network. As for MIH services, they can be delivered asynchronously or synchronously. Events generated in link layers and transmitted to the MIHF or MIHF users are delivered asynchronously, whereas commands and information generated by a query/response mechanism are delivered synchronously.

These primary services are managed and configured by a fourth service called the management service [19]. This service consists of MIH capability discovery, MIH registration, and MIH event subscription. Through the service management primitives, MIHF is capable of discovering other MIHF entities. Registration can also be performed to obtain proper service from a remote entity. By providing a standard SAP and service primitives to the higher layers, the MIHF enables applications to have a common view across different media-specific layers. Media-specific SAPs and their extensions enable the MIHF to obtain media-specific information that can be propagated to the MIH users using a single media-independent interface.

The MIES provide link layer triggers, measurement reports, and timely indications of changes in link conditions. Generally speaking, MIES define events that represent changes in dynamic link characteristics such as link status and link quality. In general, MIES can support two types of events:

- Link events, which are media-dependent information exchanged between the lower layers and the MIHF, and
- MIH events containing media-independent information exchanged between the MIHF and the MIH users

MIH users subscribe to receive notifications when events occur. This registration may be for a local or a remote MIHF. Local events are subscribed by local MIHF and are contained within a single node. Remote events are subscribed by a remote node and are delivered over a network by MIH protocol messages. Furthermore, MIES can be observed as

- MAC and PHY state change events (e.g., link up and link down events)
- Link parameters events generated by a change in the link layer parameters (synchronously—a parameters report on a regular basis or asynchronously—reporting when a specific parameter reaches a threshold)
- Link handover events (PoA changing indication)
- Link transmission events (information of the transmission status of higher-layer protocol data units by the link layer)

The MICS allows an MIH user to control the behavior of the lower layers. This includes turning the interface on or off, performing scanning, or changing the PoA. This service also allows an MIH user to request instant status updates and to configure thresholds for event generation. The mobility management protocols combine dynamic information regarding link status and parameters, provided by the MICS with static information regarding network states, network operators, or higher layer service information provided by the MIH service, to help in the decision making. Through remote commands, the network may force a terminal to hand over, allowing the use of network-initiated handovers and network-assisted handovers. A set of commands are defined in the specification to allow the user to control lower layer configurations and behavior. Similar to MIES, there are two types of commands:

- Link commands issued by the MIHF to the lower layers, and
- MIH commands issued by MIH users to the MIHF

It can be noted that link commands are always local and destined to one interface whereas MIH commands may be for a local or remote MIHF. Furthermore, MIH commands may contain actions regarding multiple links.

The IEEE 802.21 standard also defines MIIS to provide information about surrounding networks (network topology, properties, and available services) without connecting to them. In other words, the MIIS defines a set of information elements, their information structure, and representation as well as a query–response-based mechanism for information transfer. The MIIS provides a framework to discover information useful for making handover decisions. The information provided by this service can be divided into the following groups:

- General information (overview about the networks covering a specific area such as network type, operator identifier, etc.)
- RAN-specific information (e.g., security characteristics, QoS information, revisions of the current technology standard in use, cost, roaming partners, and supported mobility management protocol)
- PoA-specific information (comprises aspects like MAC address, geographical location, data rate, channel range, etc.)
- Additional information (e.g., vendor-specific information)

The MN should be able to discover whether the network supports IEEE 802.21 using a discovery mechanism or information obtained by MIIS through another interface [15]. It is also important that the MN is able to obtain MIIS information even before the authentication in the PoA is performed to be able to check the security protocols, QoS support, or other parameters before performing a handover.

MIH services framework and the communication flows between local and remote MIHF entities are shown in Figure 4.3. It is assumed that the events, commands, and information service queries are initiated by the local entity only and are propagated as remote events, commands, and information service query–response to the remote entity.

Dotted lines represent the remote events, commands, and information service query and response. The MIHF protocol facilitates communication between peer MIHF entities through the delivery of MIH protocol messages. The MIH protocol defines message formats including a message header and message parameters. These messages correlate with the MIH primitives that trigger remote communication.

Example 4.1

The IEEE 802.21 specification describes high-level MIHF reference models for IEEE 802 and the 3G networks family, and shows how existing primitives or protocols can be mapped to the media-specific link SAPs. Two example deployment scenarios, in which IEEE 802.21 can improve the user's experience by facilitating seamless handover across heterogeneous access technologies, are presented by Taninchi et al. [18].

A handover scenario between wireless metropolitan area network (WMAN) and WLAN managed by the same operator is shown

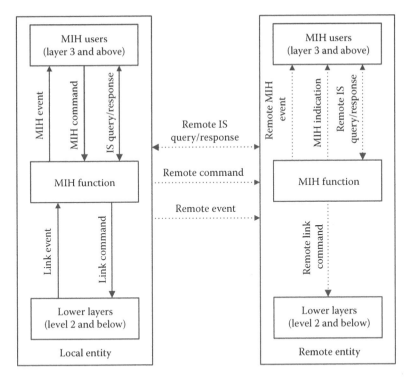

FIGURE 4.3 Communication flows between local and remote MIHF entities.

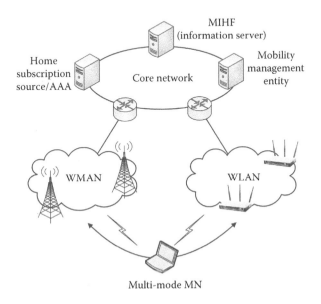

FIGURE 4.4 Handover scenario between WMAN and WLAN.

in Figure 4.4. The IEEE 802.21 provides a media-independent framework to support handover preparation and initiation. MIH services can be integrated in the common core network elements and mobile devices to facilitate such handovers.

Although a MN is within WMAN, it can query the information server to obtain available WLAN information without activating and directly scanning through the IEEE 802.11 interface. This can significantly increase battery power saving. Using the information provided by the information server, the MN activates its IEEE 802.11 interface, confident that an appropriate access network is available. Then, the MN can associate and authenticate with the WLAN while the session is active through the IEEE 802.16 interface. The MIES allows new links to be discovered and qualified before handover, whereas MIH commands can be used to begin handover processes. Use of the MIH service allows much of the time-consuming work associated with handover initiation and preparation to be completed before the handover takes place. This significantly reduces handover latency and losses.

A handover scenario between WMAN and wireless wide area network (WWAN) is represented in Figure 4.5. The WMAN operator has a roaming relationship with the WWAN operator. Placement of IEEE 802.21 entities can occur in either or both of

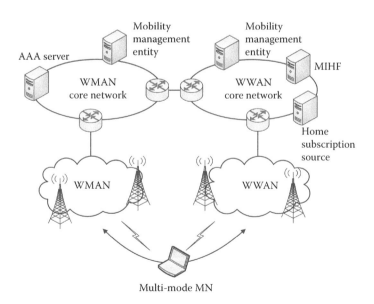

FIGURE 4.5 Handover scenario between WMAN and WWAN.

the provider core networks. In this scenario, it is assumed that MIH functionality resides within the 3GPP core network.

The MIH can be used to discover the presence and character-istics of available WMANs, whereas the MICS can instruct the mobile device to activate its IEEE 802.16 interfaces only when WMAN coverage is available. Carrying this information over the 3GPP RAN allows the mobile device to discover and access avail-able WMANs without activating its IEEE 802.16 interface, eliminat-ing the need to actively scan, and preserve the battery life. Efforts are required in 3GPP and IEEE standardization bodies to incorpo-rate such MIH services to support this type of interworking.

Example 4.2

Figure 4.6 provides an example call flow or messaging diagram to illustrate how IEEE 802.21 MIH services can facilitate transition-ing from a high-mobility low–data rate network to a low-mobility high–data rate one. For this example, the mobile controlled mobil-ity model is assumed first. That is, the mobility decision is made by an IEEE 802.21-enabled MN. Furthermore, it is assumed that at point 1, the MN is initially traveling at high velocity and is con-nected to the IEEE 802.16m access network. At point 2, the mobil-ity management decision logic on the MN decides that it needs to prepare for a possible handover to an IEEE 802.11ac network. It may realize that its velocity is decreasing. Or the user or one of the applications running wants to explore the possibility of getting higher data throughput from an 802.11ac network.

The MN then uses IEEE 802.21 information service primitives (3, 4) to query the network-based IEEE 802.21 handover informa-tion database. It does this to both identify and get information on available WLANs in its vicinity. Example information that may be available on these networks includes identity/owner, service agreement, base station locations, QoS grades, and data rate. It is assumed in the following that there are two IEEE 802.11 networks available in the MN's nearness. After obtaining information on the two candidate networks, the MN (5) uses its own criteria to evalu-ate the suitability of each as the handover target network. For example, a candidate can be eliminated for lack of an adequate SLA.

In addition, the MN may also decide to check the resource status of each candidate before making a choice. The MN (6, 7) uses IEEE 802.21 MIH command service messages to query each candidate. In the query, the MN indicates to each candidate the

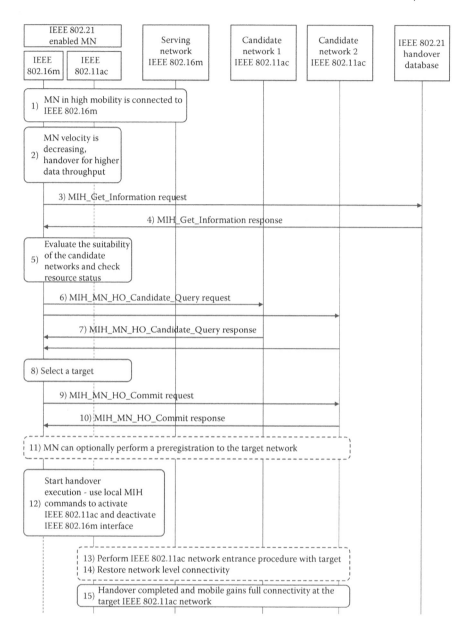

FIGURE 4.6   Example of IEEE 802.21 handover procedure.

minimal set of resources necessary for sustaining its ongoing sessions after a handover. Taking into account resource status at each
candidate, the MN (8) decides which candidate will be the handover target network. For this example, it is assumed that the MN
selects candidate network #2. The MN (9, 10) then notifies the
selected target network with IEEE 802.21 MIH_MN_HO_Commit

primitives, requesting the resource reservation at the target network.

At this point, the MN (11) may also perform some preregistration procedures (e.g., preauthentication, and IP address preallocation) at the target network, to reduce the overall handover latency. At point 12, the MN decides to start execution of the handover to the target, and invokes IEEE 802.21 commands to activate its IEEE 802.11 interface and deactivate its IEEE 802.16m interface. After this, the handover enters the execution phase. Most remaining procedures (13–15) are defined in either IEEE 802.11 or Internet Engineering Task Force (IETF) specifications.

4.4   VERTICAL HANDOVER DECISION CRITERIA

Handover criteria represent the qualities that are measured to give an indication of whether or not a handover is needed and select optimal available network. In general, the conventional single criteria-based algorithms cannot react easily to the changing environmental conditions and the possible accumulated human knowledge [22]. Usually, they cannot cope with the different viewpoints and goals of the operators, users, and QoS requirements, which make them inefficient for a multidimensional problem, such as network selection. The network selection is considered to be a complex problem because of the multiple mix of static and dynamic, and sometimes conflicting criteria (parameters) involved in the process [23].

The decision criteria that may be considered in the network selection process can be subjective or objective with minimizing or maximizing nature. From the origin point of view, they can be classified into four categories [1,23]:

- *Network-related criteria* express the technical characteristics and performance of the RANs in terms of certain performance indicators (technology, coverage, link quality and capacity, throughput, load, pricing scheme, etc.). These criteria are referred to the network and not to the service provisioning
- *Terminal-related criteria* include information about the end user's terminal characteristics and current status (supported interfaces, velocity, hardware performances, location-based information, energy consumption, etc.)
- *Service-related criteria* concern the QoS offered to the end users through a series of metrics. This is the most commonly used category in network selection problems because it involves basic QoS metrics of packet switched networks (delay, jitter, error ratio, and loss ratio)

- *User-related criteria* are mostly subjective and express certain aspects of the end user's satisfaction (user's profile and preferences, cost of service, quality of experience, etc.). Because of the imprecision they bear, these criteria are often expressed in linguistic terms.

RSS is the most widely used criterion because of its measure simplicity and close correlation to the link quality. There is a close relationship between RSS readings and the distance from the mobile terminal to its PoA. Traditional horizontal handover techniques are basically analyzing metrics through the variants of comparing RSS of the current PoA and candidate network PoA. In combination with threshold and hysteresis, RSS metrics represent a satisfying solution for a homogeneous network environment. In a heterogeneous environment, the RSS metric is not a sufficient criterion for initiating a handover, but in combination with other metrics it can be applied as a compulsory condition.

The following network parameters are particularly important for vertical handover decision (network selection) and because of this, they are most commonly found in the open literature. In terms of *network conditions*, available bandwidth is the most commonly used indicator of traffic performances and transparent parameter for the current and future users of the multimedia services. This is the measure of per user bandwidth allotted by the network operator, which is dynamically changeable according the utilization of the network. The maximum theoretical bandwidth is closely related to the link capacity (total number of channels) [24]. Transition to a network with better conditions and higher performance would usually provide improved QoS and it is especially important for delay-sensitive applications.

QoS parameters such as delay, jitter, error ratio, and loss ratio can be measured to decide which network can provide a higher assurance of seamless service connectivity. The *QoS level* can objectively be declared by the service provider based on ITU-T recommendation Y.1541 [25] and specified parameters. By declaring the QoS level in this way, a complex examination of QoS parameters by users and the additional load of user's terminals and other network elements can be avoided while this criterion becomes more transparent to the user. An example of QoS level mapping was proposed by Bakmaz et al. [26] and is presented in Table 4.2.

When the information exchanged is confidential, a network with high encryption is preferred. The *security level* concept is similar to service level in QoS management [27]. Security level is a key piece of information within a security profile and is used to determine whether data should be allowed for transfer by a particular network or not.

TABLE 4.2

QoS Levels Mapping Based on ITU-T Recommendation Y.1541

Application	Transfer Delay	Delay Variation	Loss Ratio	Error Ratio	QoS Class	QoS Level
Real-time, jitter sensitive, high interactive	<100 ms	<50 ms	$<10^{-3}$	$<10^{-4}$	0	Q
Real-time, jitter sensitive, interactive	<400 ms	<50 ms	$<10^{-3}$	$<10^{-4}$	1	Q-1
Transaction data, highly interactive	<100 ms	*	$<10^{-3}$	$<10^{-4}$	2	Q-2
...	...	...	...	...	...	...
Traditional	*	*	*	*	N	1

*Unspecified.

The *cost of service* can significantly vary from provider to provider, but in different network environments. In some cases, cost can be the deciding factor for optimal network selection and it includes the traffic costs and the costs of roaming between HetNets. In some contexts, cost of service is in tight relation with the available bandwidth and QoS level, but in a heterogeneous environment, this criteria is a fast time differentiable function depending on many other parameters. Pricing schemes adopted by different service providers is crucial and will affect the decisions of users in network selection [28].

The specified criteria do not represent an exhaustive list and are possible choices that can be used as input information for the decision mechanism. Different approaches may use only a subset of the parameters, or may include additional parameters.

After the definition of the convenient parameters, the question that often arises is how to transfer the metrics information from the network entities to the user's multimode terminals. Through the End to End Reconfigurability (E²R) project, concepts and solutions for a cognitive pilot channel (CPC) were developed [29]. It was concluded that CPC will be able to bring enough information (e.g., proposed parameters) for network selection to the terminals when users are proceeding to either initial connection or handover.

A uniform set of parameters for each candidate RAN has to be provided as input to the decision algorithm to form the basis for the network selection process. Depending on the type of architecture, protocol in use, and whether it is a centralized or decentralized decision, different information will be available in different forms and accuracy levels [23]. For example, for a decentralized approach, the mobile device can collect the network condition information as statistics, usually represented by mean values of previous sessions, or can estimate network bandwidth, for example, through the use of IEEE 802.21 Hello packets. A mobile station can collect authentication, routing, and network condition (e.g., available throughput, average delay, and average packet loss) information through advertisement Hello packets sent by a gateway node. This information can be collected from the link layer by using the IEEE 802.21 reference model [6]. Another option would be to predict the future network state based on past history. For example, many QoS parameters (e.g., availability and utilization) of different PoAs can be predicted depending on usage pattern statistics based on location (e.g., home, office, or airport) or time-related statistics (e.g., peak/off-peak hours, working days/weekends, or holidays). The accuracy in collecting network state information is very important because the selection of the optimal value network depends on it. However, a trade-off between accuracy and overhead has to be taken into account because keeping accurate estimates for the more dynamic parameters, which depend on their frequency of change and can be data intensive, adds to signaling, processor, and memory burden.

Another form of data prediction is the estimation of probabilities. This procedure is related to Markov chains and is usually associated with a limited set of decision parameters, such as SINR, blocking probability, and termination probability. One representative approach, related to a traffic model with handover call termination probability estimation, was proposed by Shen and Zeng [30]. Finally, in the presence of imprecise information, network selection can be performed using sequential Bayesian estimation, which relies on dynamic QoS parameters estimated through bootstrap approximation [31]. The bootstrap method is a computer-based, nonparametric approach in which no assumptions are made on the underlying population from which the samples are collected and allows for an estimation of the probability distributions of critical QoS parameters from acquired data. This concept can be employed to provide reliable inference from incomplete network condition information.

Network selection criteria are mainly represented in the form of a decision (performance) matrix

$$\mathbf{D} = \left\| x_{ij} \right\|_{m \times n}, \tag{4.1}$$

where x_{ij} represents performance of ith RAN (for $i = 1, \ldots, m$), related to the jth criteria ($j = 1, \ldots, n$). Here, m is a set of available RANs, and n is a set of observed criteria. To compare the criteria of different values and different measurement units, normalization is treated as a necessary step for most of the network selection techniques. In the normalization process, the starting matrix (Equation 4.1) moves into a normalized matrix

$$\mathbf{R} = \left\| r_{ij} \right\|_{m \times n,} \tag{4.2}$$

where r_{ij} is defined as normalized performance rating, obtained as

$$r_{ij} = \begin{cases} x_{ij} / \sum_{i=1}^{m} x_{ij}, \\ x_{ij} / \sqrt{\sum_{i=1}^{m} x_{ij}^2}, \\ \left(x_{ij} - \min_{i}(x_{ij}) \right) / \left(\max_{i}(x_{ij}) - \min_{i}(x_{ij}) \right). \end{cases} \tag{4.3}$$

For applications of normalization methods, refer to the articles by Bakmaz et al. [14] and Bari and Leung [32].

4.5   CRITERIA WEIGHTS ESTIMATION

Once the decision criteria have been determined, the next step is to define their importance, that is, the weight of each one in the final outcome. Weights are differentiated based on context because each user or application type may bear different requirements. The users' preferences and service requirements play an important role in the decision mechanism and they may be used to weight the involved criteria.

In certain cases encountered in the literature, the weights of the selection criteria are defined through the derivation and the analysis of questionnaires that capture the user's overall perception of a service. However, these approaches depend only on the user's feedback to determine the relative weights and thus cannot be considered precise because the user's perception and opinion is subjective [22]. Obviously, this approach is adequate when subjective criteria are considered. However, in the case of objective criteria, no accurate results can be guaranteed. For example, some of the existing weighted solutions obtain the weights through questionnaires on the users' and services' requirements. Other solutions integrate a user interface in the mobile terminals to collect their preferences. An important aspect is to find a trade-off between the cost of involving the user and the decision mechanism. One solution for minimizing the user interaction may be implementing an intelligent learning mechanism that can predict the users' preferences over time.

4.5.1 SUBJECTIVE CRITERIA WEIGHTS ESTIMATION

In several network selection mechanisms, the use of the analytical hierarchy process (AHP) for subjective criteria weight estimation has been proposed [33–36]. This method is considered to be a well-known and proven mathematical process. AHP is defined as a procedure to divide a complex problem into a number of deciding factors and integrate the relative dominances of the factors with the alternative solutions to find the optimal one [37]. AHP criteria weights estimation procedure is carried out in three steps:

1. Structuring a problem as a decision hierarchy of independent elements
2. Comparing each element to all the others within the same level through a pairwise comparison matrix. The comparison results are presented in a square matrix form

$$\mathbf{A} = \left\| a_{ij} \right\|_{n \times n},\tag{4.4}$$

where n is the number of observed criteria, and $a_{jj} = 1$ (criteria is compared with itself), $a_{ij} > 1$ (element i is considered to be more important than element j), $a_{ij} < 1$ (element j is considered to be more important than element i), and $a_{ij} = 1/a_{ji}$
3. Normalization and calculation of the relative weights using the relation

$$w_j = \sum_{j=1}^{n} a_{ij} / \sum_{i=1}^{n} \sum_{j=1}^{n} a_{ij},\tag{4.5}$$

where it is obvious that $\sum_{j=1}^{n} w_j = 1$.

The AHP method can be used alone for network selection [34], or in combination with other methods [33,38]. Despite its popularity, the conventional AHP is often criticized for its inability to adequately handle the inherent uncertainty and imprecision associated with the mapping of the decision-maker's perception to exact numbers.

Example 4.3

Considering the set of network selection criteria $\{\alpha, \beta, \gamma, \delta\}$, according to Equation 4.5, it is possible to calculate relative weights related to the comparison matrix. The assumed criteria comparison and obtained weight coefficients are presented in Table 4.3.

4.5.2 OBJECTIVE CRITERIA WEIGHTS ESTIMATION

As an exact and objective approach, the *entropy method* was proposed by Bakmaz et al. [14]. This method is based on the relation

$$e_j = [-1/\ln(m)] \cdot \sum_{i=1}^{m} [r_{ij} \ln(r_{ij})], \; j \in \{1, \dots, n\}. \tag{4.6}$$

Deviation within each criterion $d_j = 1 - e_j$, leads to the weight coefficients determination

$$w_j = d_j / \sum_{i=1}^{n} d_i, \tag{4.7}$$

in the case when the user equally prefers all the parameters (objective approach), or

$$w_j = d_j w_j^* / \sum_{i=1}^{n} d_i w_i^*, \tag{4.8}$$

if the user determines the subjective weights w_j^*.

TABLE 4.3

Criteria Pairwise Comparison and Obtained Weights

	α	β	γ	δ	Σ	w_j
α	1	3	2	9	15	0.485
β	1/3	1	1/2	3	4.83	0.156
γ	1/2	2	1	6	9.5	0.307
δ	1/9	1/3	1/6	1	1.61	0.052
Σ	1.94	6.33	3.67	19	30.94	1

TABLE 4.4

Entropy Weights Determination

	α	β	γ	δ
Starting (input) Parameters Values				
RAN$_1$	20	3	2	2
RAN$_2$	16	3	2	1
RAN$_3$	18	2	3	1
Normalized Matrix				
RAN$_1$	0.639	0.640	0.485	0.184
RAN$_2$	0.511	0.640	0.485	0.592
RAN$_3$	0.575	0.426	0.728	0.592
Entropy Weights Calculation				
e_j	0.862	0.851	0.849	0.848
d_j	0.138	0.149	0.151	0.152
w_j	0.234	0.253	0.256	0.258

After objective or subjective criteria weights estimation, in general, the normalized matrix (Equation 4.2) moves into a weighted matrix

$$\mathbf{V} = \left\| w_j r_{ij} \right\|_{m \times n} = \left\| v_{ij} \right\|_{m \times n}. \tag{4.9}$$

Example 4.4

Considering the set of network selection criteria {α, β, γ, δ}, and a set of three available RANs, according to Equations 4.6 and 4.7, it is possible to calculate weights applying the entropy method (Table 4.4).

4.6 ALTERNATIVE RANKING TECHNIQUES

The ranking process, through which the optimal choice is pointed out, is based on the input of the previous two steps, as presented in Figure 4.7. This process can be considered as the core phase of the vertical hand-over management because it is in charge of evaluating and deciding the most appropriate network choices to fulfill both the systems' and the users' requirements, thus providing the desired seamless communication. Due to

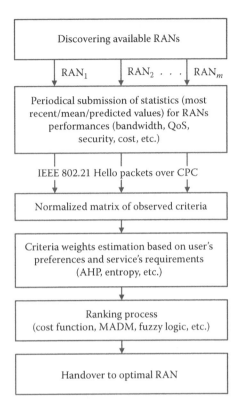

FIGURE 4.7   General model of network selection process.

the different possible strategies and the numerous parameters involved in the process, researchers have tried many different techniques to find the most suitable network selection solution [23].

These techniques are usually called vertical handover decision (VHD) algorithms. Recently, many VHD techniques were proposed in the open literature. Current techniques do not fully comply with requirements related to the technology coverage, adequacy of the analyzed parameters, complexity of implementation, and integration of all the entities in the selection of appropriate RAN. Consequently, because of its complexity, they are only representing theoretical solutions, which are currently the focus of many researchers but, for a number of deficiencies, are not yet applicable in real environment [38,39].

Performance analysis of the network selection techniques can be performed through the determination of mean and maximum handover delays, number of handovers, number of handovers failed due to incorrect decisions, handover failure probability, resource utilization, etc. [40].

Handover delay refers to the duration between the initiation and completion of the handover process. It is related to the complexity of the applied heuristic. Reduction of the handover delay is especially important for delay-sensitive voice and multimedia sessions. Reducing the number of handovers is usually preferred because frequent handovers can cause wastage of network resources. A handover is considered superfluous when a mobile terminal coming back to the previous PoA is needed within a certain time duration ("ping-pong" effect), and such handovers should be minimized.

A handover failure occurs when the handover is initiated, but the target network does not have sufficient resources to complete it, or when the mobile terminal moves out of the coverage area before the process is finalized. In the first case, the handover failure probability is related to the resource availability (e.g., channel availability) of the target network, whereas in the second case, it is related to terminal mobility.

Resource utilization is defined as the ratio between the mean amount of utilized resources and the total amount of resources in a system. In the case of efficient channel utilization, the ratio between the mean number of channels that are being served and total number of channels in a system is taken into account.

For efficient network selection strategy, the following important issues have to be fulfilled [41]:

- Only considerable parameters must be analyzed.
- Equilibrium among users' preferences, services' requirements, and networks' performance must be achieved.
- The technique has to be reliable and transparent to the user.
- The algorithm has to minimize handover latency, blocking probability, and number of superfluous handovers.
- Flexible and suitable implementation in real environments is necessary.

4.6.1 Cost Function Techniques

The simplest approach for network ranking indicates that an optimal RAT is selected based on a cost function, which is a function of the selected criteria. Cost function is a measurement of the benefit obtained by handover to a particular network. Usually, the cost of a network can be considered as the inverse of its utility, but the form of this inversion is related with the way to combine multiple attributes.

A general form of cost function for the network selection problem was proposed by McNair and Zhu [11]. It integrates a large number of attributes, corresponding weights, and network elimination factors through the relation

$$CF_i = \sum_k \left(\prod_j E_{ij}^k \right) \sum_j f_j^k \left(w_j^k \right) N \left(Q_{ij}^k \right),$$ (4.10)

where $N\left(Q_{ij}^k\right)$ is the normalized QoS parameter Q_{ij}^k, representing the cost in the jth criteria to carry out service k over network i, $f_j^k\left(w_j^k\right)$ is the weighting function of criteria j for service k, and E_{ij}^k is the network elimination factor of service. The network elimination factor is represented by values of one or infinity, to reflect whether current network conditions are suitable for the requested services. For example, if a RAN cannot guarantee the delay requirement of certain real-time service, its corresponding elimination factor will be set to infinite. Thus, the corresponding cost becomes infinite, which eliminates the corresponding RAN. The algorithm of the introduced technique is shown in Figure 4.8.

The fundamental benefit of cost function usage and handover-independent initiation for different services is reduced failure (blocking) probability. However, weight coefficients estimation techniques are not discussed by the authors. Although cost functions have been widely used, currently, the

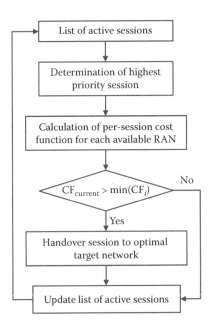

FIGURE 4.8 Example of cost function based vertical handover algorithm.

need for more intelligent and dynamic schemes has been increased because more criteria must be taken into account.

4.6.2 Multiattribute Decision-Making Techniques

Multiple criteria decision-making (MCDM) problems, although very different, have some characteristics in common:

- Conflicts among criteria are possible (low-cost conditions decrease network's performances)
- Criteria contain nonmeasurable units (cost and bandwidth)
- The goal is to design an optimal, or to select the best, alternative

Based on the third characteristic, MCDM can be divided into multiple objective decision-making (MODM) and multiple attributes decision-making (MADM) algorithms. MODM consists of conflicting set goals that cannot be achieved simultaneously, whereas MADM deals with the problem of selecting an alternative from a set of alternatives that are characterized in terms of their attributes. In fact, the attributes are parameters or performance factors with an influence on the selection whereas alternatives are characterized by more attributes with the reliable level of successfulness. Usually, the goal is not explicit, that is, it is defined and often can be represented as the satisfactions' maximization.

In the network selection problem in heterogeneous environment, the specified candidate network is characterized by attributes such as bandwidth, losses, delay, cost of service, etc. The number of available networks is finite, whereas the resolution space is discrete. In this way, the problem can be characterized in the MADM category.

MADM algorithms can be divided into compensatory and noncompensatory ones [32]. Noncompensatory algorithms (e.g., dominance, conjunctive, disjunctive, or sequential elimination) are used to find acceptable alternatives that satisfy the minimum cutoff. On the contrary, compensatory algorithms combine multiple attributes to find the best alternative. Most MADM algorithms that have been studied for the network selection problem are compensatory algorithms [42], including simple additive weighting (SAW), multiplicative exponential weighting (MEW), gray relational analysis (GRA), technique for order preference by similarity to an ideal solution (TOPSIS), and others. These methods are often called "soft" optimization techniques compared with standard mathematical optimization methods such linear and dynamic programming [43], or game theory [44].

The SAW technique (a.k.a. the weighted sum method) [45] is one of the most widely used MADM methods in the network selection related open literature. The basic concept of SAW in this context is to obtain a weighted sum of the normalized form of each parameter over all candidate networks. Normalization is required to have a comparable scale among all parameters. For m available RANs, and n observed parameters, the VHD function for ith RAN is obtained as

$$\text{SAW}_i = \sum_{j=1}^{n} w_j r_{ij}. \tag{4.11}$$

Application of SAW-based VHD function in the process of evaluating the qualitative performance of potential target networks is proposed by Nasser et al. [8]. By using normalization and weight distribution methods, VHD function determines a network quality factor as

$$Q_i = \frac{w_c(1/C_i)}{\max(1/C_1,...,1/C_m)} + \frac{w_s S_i}{\max(S_1,...,S_m)} + \frac{w_p(1/P_i)}{\max(1/P_1,...,1/P_m)} + \frac{w_d(D_i)}{\max(D_1,...,D_m)} + \frac{w_f(F_i)}{\max(F_1,...,F_m)} , \tag{4.12}$$

where C_i is the cost of service, S_i security, P_i power consumption, D_i network conditions, and F_i network performance, whereas the w_c, w_s, w_p, w_d, and w_f, respectively, are weights for each of the criteria that are proportional to the significance of a parameter to the VHD function. These weights are obtained from the user via a user interface. The RAN with the highest Q_i is the preferred network. If the newly detected network receives a higher Q_i, vertical handover takes place; otherwise, the MT remains connected to the current network. High overall throughput and the user's satisfaction can be regarded as major advantages of this technique. An identical approach is proposed by Tawil et al. [46], the difference being in terms of the observed criteria.

MEW (a.k.a. the weighted product method) [45] uses multiplication among attribute values that are raised to the power of the attribute importance, whereas the normalization procedure is not obligatory. VHD function can be represented in the form

$$\text{MEW}_i = \sum_{j=1}^{n} r_{ij}^{w_j}. \tag{4.13}$$

The results obtained by Nguyen-Vuong et al. [47] indicate the inaccuracy of the SAW method and the benefits of using MEW as a network selection technique.

TOPSIS [48] is one of MADM techniques based on a concept that the chosen alternative should have the shortest distance from the ideal possible solution and the longest from the worst possible solution, in which distances are determined with certain values for p ($1 \leq p < \infty$) from the Minkowski's metric. The Minkowski's row distance p, between two points, $V = (v_1, v_2, \ldots, v_n)$ and $U = (u_1, u_2, \ldots, u_n) \in R^n$, can be defined as

$$L_p = \left(\sum_{i=1}^{n} |v_i - u_i|^p \right)^{1/p},$$ (4.14)

where the case $p = 1$ is used, very often (Manhattan distance, the first row metric), and $p = 2$ (Euclidean distance). On the boundary case, when $p \to \infty$, Chebyshev's distance is obtained (L_∞ metrics).

Considering the weighted matrix defined by Equation 4.9, the ideal and worst solutions are representing the sets

$$A^+ = \left\{ \left(\max_{1 \leq i \leq m} v_{ij} \Big| v_{ij} \in V^{\max} \right); \left(\min_{1 \leq i \leq m} v_{ij} \Big| v_{ij} \in V^{\min} \right) \right\} = \left\{ v_1^+, \ldots, v_n^+ \right\},$$

$$A^- = \left\{ \left(\min_{1 \leq i \leq m} v_{ij} \Big| v_{ij} \in V^{\max} \right); \left(\max_{1 \leq i \leq m} v_{ij} \Big| v_{ij} \in V^{\min} \right) \right\} = \left\{ v_1^-, \ldots, v_n^- \right\},$$ (4.15)

where $V^{\max}$ is the set of the larger-the-better criteria and $V^{\min}$ is the set of the smaller-the-better criteria. Euclidean distances of all alternatives, in relation to the ideal and worst solution, can be calculated from

$$D_i^+ = \sqrt{\sum_{j=1}^{n} \left(v_{ij} - v_j^+ \right)^2}, \quad D_i^- = \sqrt{\sum_{j=1}^{n} \left(v_{ij} - v_j^- \right)^2}, \quad i = 1, \ldots, m.$$ (4.16)

Lastly, the ranking of networks can be done through the relative closeness (RC) to the ideal solution in the form

$$RC_i = \frac{D_i^-}{D_i^- + D_i^+}, \quad RC_i \in (0,1).$$ (4.17)

A novel modified synthetic evaluation method (M-TOPSIS) based on the concept of the original TOPSIS with optimized ideal reference point in the D^+D^- plane is proposed by Ren et al. [49]. It can solve the problems of the original TOPSIS, such as rank reversals and evaluation failure when the alternatives are symmetrical.

A network selection algorithm based on the TOPSIS method (Figure 4.9) was proposed by Bakmaz et al. [14]. The criteria considered in the decision matrix are available bandwidth (B), QoS level (Q), security level (S), and cost of service (C). It can be noted that the computational complexity involved in calculating Euclidean distances and entropy weights used in TOPSIS is very low.

The proposed solution is evaluated using numerical examples. Through simulation studies, MADM is envisaged as a promising tool especially when the TOPSIS method is used because of high sensitivity to users' preferences and the parameter values. This solution is realistic and not very complex to implement in mobile terminals and other network elements.

The results of research from Bakmaz et al. [41,50] extend merit function (measure of network quality) obtained as relative closeness to the ideal solution (Equation 4.17) with a predefined hysteresis, the goal of which is to minimize the influence of undesirable effects of frequent superfluous

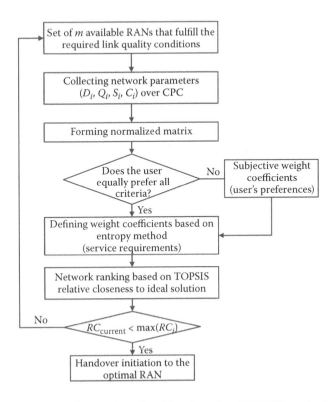

FIGURE 4.9 Network selection algorithm based on TOPSIS method.

handovers and can also be used in admission control mechanisms if the need to control or influence the network selection process exists. A handover is considered superfluous when a mobile terminal back to the previous PoA is needed within a certain time duration ("ping-pong" effect). The larger the hysteresis value, the smaller the number of handovers; however, there is a longer handover initiation delay. On the other hand, the smaller the threshold value, the shorter the handover initiation delay but the larger the number of handovers. Handover initiation delays lead to an increase in the call dropping probability, especially in the case of highly mobile users. Moreover, frequent handovers can cause an increase in signaling overhead and in the network load. Therefore, the determination of the hysteresis value is very important in terms of mobility performance.

Different from TOPSIS, GRA uses only the ideal solution to calculate the ranking coefficient alternative (network) i, given by

$$\text{GRA}_i = \frac{1}{\sum_{j=1}^{n} w_j \left| v_{ij} - V_j^+ \right| + 1}. \tag{4.18}$$

Song and Jamalipour [33], develop a network selection mechanism for an integrated WLAN and 3G cellular systems. The proposed scheme comprises two parts. The first applying an AHP to estimate the relative weights of evaluative criteria set according to users' preferences and service applications, whereas the second adopts GRA to rank the network alternatives with faster and simpler implementation than AHP. The design goal is to provide the user the best available QoS at any time. This method mathematically presents a complex solution and unnecessarily takes into account a large number of parameters (delay, jitter, response time, bit error rate, etc.) only for 3G and WLANs. Processing a large number of parameters leads to increased computational time, while the terminal and infrastructure network elements are additionally loaded. Thus, this model is interesting from theoretical point of view, but not adequate for a direct implementation.

4.6.3 FUZZY LOGIC TECHNIQUES

Users usually think in terms of linguistic descriptions, so giving these descriptions some mathematical form helps exploit their knowledge. Fuzzy logic [51] utilizes human knowledge by giving the fuzzy or linguistic descriptions a definite structure. Fuzzy logic has found successful applications in real-time control, automatic control, data classification, decision analysis, expert systems, time series prediction, robotics, and pattern recognition.

A fuzzy set is a class of objects with a continuum of grades of membership, which is characterized by a membership function assigning to each object a grade of membership ranging between zero and one. Fuzzy set is considered as an extension of the classic notion of a set. In the classic set theory, the membership of elements in a set is assessed in binary terms, which means that either it belongs or it does not belong to the set. On the contrary, the fuzzy set theory permits the gradual assessment of membership using a membership function valued within [0,1]. The classic set is usually called *crisp set* in the fuzzy logic theory.

There are different ways to use the fuzzy logic theory in network selection. Some studies use it alone as the core of the selection scheme, whereas some use fuzzy logic with recursion (learning techniques), whereas some combine fuzzy logic with MCDM algorithms.

An elementary framework for fuzzy logic–based network selection, without combining with any other technique, is proposed by Kher et al. [52], as shown in Figure 4.10, eliminating the recursion part. In the proposed scheme, three input fuzzy variables are considered (the probability of a short interruption, the failure probability of handover to radio, and the size of unsent messages).

At the beginning of the procedure, the fuzzy variables are fuzzified and converted into fuzzy sets by a singleton fuzzifier. Then, based on the fuzzy rule (IF–THEN) base, the fuzzy inference engine maps the input fuzzy sets into output fuzzy sets by the algebraic product operation. Finally, the output fuzzy sets are defuzzified into a crisp decision point.

Because some dynamic factors change frequently, the recursion can be used to combine the latest information with previous ranking results to obtain the latest rank. The recursion procedure can be a simple recursion without any further operation or certain learning procedure, such as neural networks or learning techniques. The second fuzzy logic-based scheme proposed by Kher et al. [52] gives a simple recursion that considers the requirements of both operator and user. The rank produced by the fuzzy

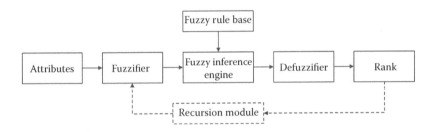

FIGURE 4.10   Fuzzy logic–based network selection procedure.

module is fed back to this module, so that it can produce a new rank when some factors change. The advantage of this method is that the designer is no longer obliged to define rules, which are generated automatically by inserting a series of training data sets to the system.

The network selection solution proposed by Kassar et al. [53] represents an interesting and promising solution while combining the heuristics of the fuzzy logic systems and MCDM (Figure 4.11). In the process of handover initiation, the proposed technique uses fuzzy logic analyzing the criteria such as RSS, bandwidth (B), network coverage (NC), and terminal velocity (V). The gathered information, depending on its availability, is fed into a fuzzifier and converted into fuzzy sets. A fuzzy set contains varying degrees of membership in a set. The membership values are obtained by mapping the values retrieved for a particular variable into a membership function. After fuzzification, fuzzy sets are fed into an inference engine, in which a set of fuzzy rules (81 predefined rules) are applied to determine whether handover is necessary. Fuzzy rules utilize a set of IF–THEN rules and the result is YES, Probably YES, Probably NO, or NO (Table 4.5).

By application of the AHP method and Saaty's scale on criteria such as cost of service, preferred interface, battery status, and QoS level, the optimal access network is determined. On the other hand, by applying

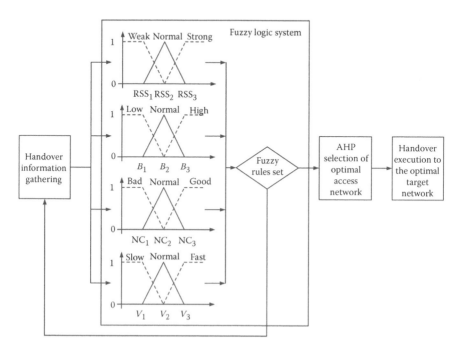

FIGURE 4.11   Network selection technique based on fuzzy logic and MCDM.

TABLE 4.5
Examples of Fuzzy Rules

Rule	RSS	B	NC	V	Handover is Needed?
1	Weak	Low	Bad	Slow	YES
...	...	...	...	...	...
58	Strong	Low	Normal	Slow	Probably YES
...	...	...	...	...	...
69	Strong	Normal	Normal	Fast	Probably NO
...	...	...	...	...	...
81	Strong	High	Large	Fast	NO

fuzzy logic in the decision-making process, the number of unnecessary handovers is reduced, as well as the signaling traffic and handover delays. The inflexibility of the effect of user preferences on the system is the basic shortcoming of the applied AHP method, which can possibly be exceeded by using some MADM techniques, for example, TOPSIS.

4.6.4 Artificial Neural Network Techniques

Artificial neural networks (ANNs) can be most adequately characterized as computational models with particular properties such as the ability to adapt or learn, to generalize, or to cluster data and the operation of which is based on parallel processing. ANN consists of a pool of simple processing units that communicate by sending signals to each other over a large number of weighted connections [54]. A set of major aspects of a parallel distributed model can be distinguished:

- A set of processing units (neurons, cells)
- A state of activation for every unit, which is equivalent to the output of the unit
- Connections between the units
- A propagation rule, which determines the effective input of the unit from its external inputs
- An activation function, which determines the new level of activation based on the effective input and the current activation
- An external input (bias, offset) for each unit
- A method for information gathering (the learning rule)
- An environment within which the system must operate, providing input signals and, if necessary, error signals

ANNs are of interest to researchers because they have the potential to treat many problems that cannot be handled by traditional analytic approaches.

Backpropagation neural networks are the most prevalent ANN architectures because they have the capability to "learn" system characteristics through nonlinear mapping.

A network selection technique based on ANNs was proposed by Nasser et al. [55]. An applied feedforward neural network topology, which consists of input, hidden, and output layers, is shown in Figure 4.12. The input layer is made of the h nodes representing different criteria for optimal network selection, whereas the hidden layer consists of the n nodes that represent the available access networks. The output layer is formed by a node that generates the identification of the optimal access network. For the training process, an error backpropagation algorithm is used. During the simulation, the authors adopted the same VHD function as that used by Nasser et al. [8].

In this study, all neurons use a sigmoid activation function. Random values, which serve as weights, are generated for all connections from input to hidden (v_{hi}) and from hidden to output layers (w_{ij}). In addition, biases are assigned random values at the hidden nodes (θ_i) and the output node (τ_i). The activation functions at the hidden layer are calculated as

$$b_i = f\left(\sum_{i=1}^{n} a_h v_{hi} + \theta_i\right),\qquad(4.19)$$

whereas activation values at the output layer are calculated as

$$c_j = f\left(\sum_{i=1}^{n} a_h w_{ij} + \tau_j\right).\qquad(4.20)$$

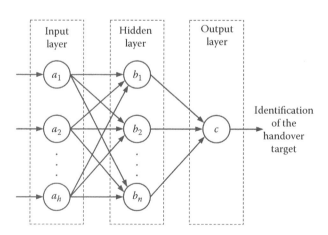

FIGURE 4.12 Example of ANN topology for network selection.

In both Equations 4.19 and 4.20, $f(x)$ represents the logistic sigmoid threshold function, $f(x) = 1/(1 + e^{-x})$.

The simulations have shown the high accuracy and reliability of the model while selecting the optimal network. The shortcomings of the algorithm are reflected in the complexity of the system and in increased handover delays due to the training process.

Example 4.5

Applying the TOPSIS-based algorithm proposed by Bakmaz et al. [14], it is possible to evaluate the influences of traffic parameters on the network selection process. In the analyzed case, besides the usual criteria (QoS level [Q_i], security level [S_i], and cost of service [C_i]), classic grade of service (GoS) parameters are pointed out by Bakmaz et al. [50]. The influence of criteria (D_i), that is, network conditions for ith RAN, can be analyzed through link capacity (s_i), the link capacity and traffic ratio (s_i/a_i), as well as the ratio of link capacity and loss obtained using Erlang formula (s_i/B_i) [56,57], link capacity and traffic losses ratio ($s_i/[a_iB_i]$), and finally, through the available bandwidth ($s_i - a_i[1 - B_i]$). Input criteria values for the three available RANs are given in Table 4.6.

The influence of the traffic intensity variations in one of the networks (a_1) to the merit function (calculated as relative closeness to the ideal solution) of all three RANs, for differently defined network conditions D_i, are shown in Figures 4.13, 4.14, and 4.15, respectively.

When network conditions are presented only by link capacity ($D_i = s_i$), merit functions do not depend on traffic intensity. The values of level of security criteria (S_i) as well as cost of service (C_i) have influence based on the fact that MF_3 has the highest value, although RAN_3 does not have the highest link capacity. By invoking traffic parameters in D_i, that is, $D_i = s_i/a_i$ and $D_i = s_i - a_i(1 - B_i)$,

---

TABLE 4.6

Input Criteria Values

RAN$_i$	s_i	a_i	Q_i	S_i	C_i
RAN$_1$	20	1–15	3	2	2
RAN$_2$	15	12	3	2	1
RAN$_3$	18	10	2	3	1

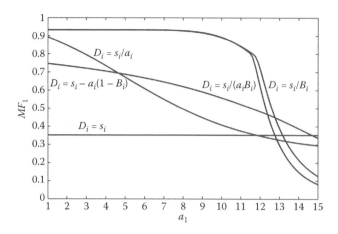

FIGURE 4.13   Influence of traffic intensity variation to the RAN_1 merit function.

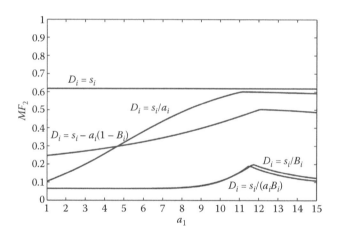

FIGURE 4.14   Influence of traffic intensity variation to the RAN_2 merit function.

with the increment of a_1, successive decrease of MF_1 appears, whereas MF_2 and MF_3 increase successively.

The most interesting point is the phenomenon concerning MF_i, when the network conditions $D_i = s_i/B_i$ and $D_i = s_i/(a_iB_i)$ are used. With a sufficient high value a_1, which produces loading per channel like loading in RAN_3 ($a_1/s_1 \approx a_3/s_3$), which is lower compared with the loading per channel RAN_2 (a_2/s_2), the effect of rapid variation for MF_1 and MF_3 exists. This property can affect the network selection process, and takes us a step closer to the always best connected and served (ABC&S) concept.

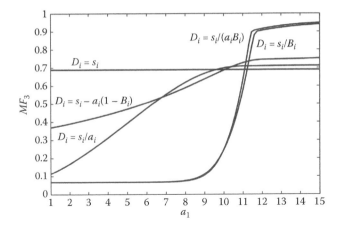

FIGURE 4.15   Influence of traffic intensity variation to the RAN_3 merit function.

4.7   CONCLUDING REMARKS

Following the principles of heterogeneous networking, a mobile user/terminal may choose among multiple available connectivity alternatives based on the criteria related to networks' performances, users' preferences and services' requirements. The scope of this chapter is to address the

TABLE 4.7

A Comparative Survey of Representative Network Selection Techniques

	Technique			
Characteristics	**Cost Function** [11]	**MADM** [8,14,32,33]	**Fuzzy Logic** [52]	**Neural Networks** [55]
Computational complexity	Low	Medium	High	High
Implementation complexity	Low	Medium	High	High
Reliability	Low	High	Medium	Very high
Number of superfluous handovers	High	Conditionally reduced	Reduced	Reduced
Handover latency	Negligible	Acceptable	Acceptable	Significant due to the training process

issue of network selection in a heterogeneous wireless environment. The main challenges involved in the network selection process are pointed out, and a significant collection of relevant approaches encountered in the open literature is presented. Theoretically acceptable and interesting decision techniques are discussed, emphasizing the use of tools such as fuzzy logic and MADM, which may lead to more efficient and objective decisions. A qualitative comparison of the analyzed network selection techniques in terms of functionality and complexity of implementation is briefly presented in Table 4.7.

Current proposed network selection techniques in heterogeneous environment require more significant challenges to be overcome before they can successfully be deployed in real systems. Because of that, research in this area is still a challenging issue.

5 Wireless Mesh Networks

Mobile communications are driven by converged networks that integrate technologies and services. The wireless mesh environment is envisaged to be one of the key components in the converged networks of the future, providing flexible high-bandwidth backhaul over large geographical areas. Although single-radio mesh nodes operating on a single channel suffer from capacity constraints, multiradio mesh routers using multiple non-overlapping channels can significantly alleviate the capacity problem and increase the aggregate bandwidth available to the network. Wireless mesh networks (WMNs) are anticipated to resolve the limitations and to significantly improve the performance of traditional wireless architectures. They will deliver wireless services for a large variety of applications in personal, campus, and metropolitan areas. With the advances in wireless technologies and the explosive growth of Internet, designing efficient WMNs has become a major task for network operators. Joint design and optimization of independent problems such as routing and link scheduling have become one of the leading research trends in WMNs. From this point of view, cross-layer approach is expected to bring significant benefits to achieve high system utilization. Existing routing schemes that are designed for single-channel multihop wireless networks may lead to inefficient paths in multichannel WMNs. On the basis of topological differences, various network architectures are possible for wireless mesh environment. Such architectures could affect wireless characteristics differently. WMNs promise to expend high-speed connectivity beyond what is possible with the current wireless infrastructures. However, their unique architectural features leave them particularly vulnerable to most of the security threats. Despite recent advances in mesh networking, many research challenges, including quality of service (QoS) provisioning, remain in all protocol layers.

5.1   INTRODUCTION

Wireless mesh network (WMN) technologies have been researched and developed as key solutions to improve the performance and services of different wireless environments such as wireless personal area network (WPAN), wireless local area network (WLAN), wireless metropolitan area networks

(WMAN), and cellular multihop networks [1]. WMNs are considered as significant components in next generation mobile systems [2,3]. Compared with ad hoc, sensor, and infrastructure-based mobile networks, WMNs represent flexible and not resource-constrained quasi-static network topology.

WMN comprises the mesh routers (MRs), mesh clients (MCs), and the mesh backbone infrastructure. Each node operates not only as a host but also as a router, forwarding packets on behalf of other nodes that may not be within direct wireless transmission range of their destinations. WMNs are dynamically self-organized and self-configured, with the nodes in the network automatically establishing and maintaining mesh connectivity among themselves. This characteristic brings many advantages to WMNs such as low up-front cost, easy network maintenance, robustness, and reliable service coverage. Also, the gateway/bridge functionalities in MRs enable the integration of WMNs with various existing wireless network environments described in Chapter 1. Thus, through an integrated WMN, the users of existing networks can be provided with otherwise unsupported services of these networks.

MRs self-organize and establish rich radio mesh connectivity in a way that has never been possible within purely wired networks. Because of these features, WMNs are being considered for a wide variety of application scenarios such as backhaul for broadband networking, building automation, intelligent transport systems, sensor systems, health and medical systems, security and public safety systems, emergency/disaster networking, and so on. These wide ranges of applications have different technical requirements and challenges in the design and deployment of mesh architectures, algorithms, and protocols. WMN is also one possible way for cash-strapped Internet service providers (ISPs), carriers, and others to roll out robust and reliable wireless broadband service access in a way that needs minimal up-front investments [1]. Using the capability of self-organization and self-configuration, WMNs can be deployed incrementally, one node at a time, as needed. As more nodes are installed, the reliability and the connectivity for the users increase accordingly. Today, WMNs continue to receive significant interest as a possible means of providing seamless data connectivity, especially in urban environments.

Performance analysis and improvement of protocol stack, particularly the MAC protocol, is an important research topic for WMNs. IEEE 802.11x is one of the most influential WLAN standards, and its basic MAC protocol is called the *distribution coordination function* (DCF). DCF is based on carrier sense multiple access with collision avoidance (CSMA/CA), which is a typical contention-oriented protocol. CSMA/CA is widely used in test beds and simulations for WMN research [4]. The IEEE 802.16 standard allows for backhauling by providing an optional mesh connectivity mode in addition to the inherent point-to-multipoint (P2MP) connectivity. The key difference between P2MP and mesh is that communication in P2MP is based on direct connection between the base station (BS) and the subscriber stations (SSs), whereas in mesh mode, multihop communication is allowed and traffic can be routed through other SSs or occur directly between SSs [5].

WMNs combine wired and wireless networks with wireless routers as backbone and mobile solutions as users. Wireless routers communicate with one another, from a backbone of the wireless network, connect wired networks, and conduct multihop communications to forward mobile station traffic to/from wired networks. Mobile users' traffic is carried over wireless routers and reaches wired networks. Each mobile user also acts as a router behaves, as a router forwarding packets for other mobile users. Wireless routers can have multiple different interfaces to access heterogeneous wireless networks [6].

QoS provisioning is a challenging issue in WMNs, which promise to support a variety of traffic types. They should satisfy the requirements of both delay-sensitive multimedia applications, such as video streaming and delay-tolerant services such as web browsing, and smoothly handle bursty traffic over the Internet. In addition, they may need to deal with different types of traffic simultaneously. As a consequence, various QoS classes are defined according to traffic types and their requirements in terms of delay, jitter, error and loss probability, and so on. Users have strong preference for wireless access, and hence, the increase in link capacity can drive a wave of innovations that increase the total demand and high-speed access [7,8].

5.2   WMN ARCHITECTURE

WMNs have become a practical solution providing community broadband Internet access services [9]. WMN architecture can take different topologies based on the structural design of its components with respect to one another and to the network environment. Variations in the WMNs architecture pose fundamental difference in the physical characteristics and network performance, leading to the various results in the performed studies. Also, the types of architecture are important for WMNs because of the influence on routing management and applications. Various architectures can be combined to build more complex hybrid structures and can be customized to fulfill specific requirements set by clients [10].

In WMN infrastructure, access mesh routers (AMRs) provide Internet access to MCs by forwarding aggregated traffic to MRs, known as relays, in a multihop fashion until a mesh gateway (MG) is reached. MGs act as bridges between the wireless mesh infrastructure and the Internet or other network environments. AMRs and MGs can be observed as edge MRs with specific functionalities. Typical (baseline) WMN architecture is illustrated in Figure 5.1.

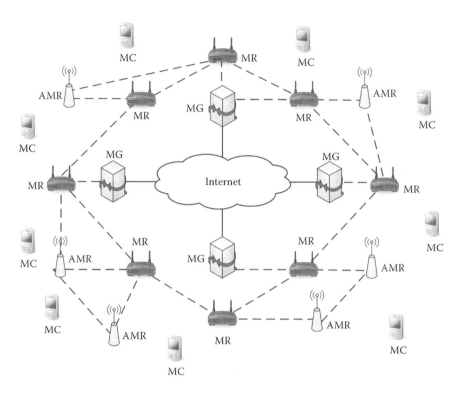

FIGURE 5.1   An example of a typical WMN architecture.

MCs can access the network through MRs as well as by directly mesh-ing with other MCs. The routing capabilities of clients provide improve-ment connectivity and coverage inside the WMN. Each infrastructure node can be equipped with multiple radios. Each radio is capable of accessing multiple orthogonal channels (aka multiradio multichannel transmissions). In multiradio multichannel networks [11], simultaneous communications are possible by using noninterfering channels, which have the potential of significantly increasing the network capacity.

Three mainstream architectures for WMNs can be identified [10]: cam-pus mesh, downtown mesh, and long-haul mesh. The difference is in the topology, the geographical location and physical orientation of the equip-ment with respect to one another and to the network environment. The backbone is a group of different types of wireless MRs organized in circu-lar, ad hoc, or longitudinal fashion, depending on whether the architecture is campus mesh, downtown mesh, or long-haul mesh, respectively. MRs are often fixed and stable and have access to unlimited power supply. They use proactive routing protocols such as open shortest path first (OSPF) [12]. The access network is a group of clusters, each containing several MCs. These clusters are highly mobile and unstable. They use temporary sources of power and on-demand and ad hoc routing protocols, such as ad hoc on-demand distance vector (AODV) [13].

5.2.1 CAMPUS MESH NETWORK ARCHITECTURE

The network in campus mesh architecture is generally easy to deploy, monitor, manage, and upgrade [10]. In this architecture, a limited number of buildings are located in a campus environment, with generally good line-of-sight (LoS) and a central management and administration unit. Wireless MRs can be installed on existing campus infrastructure. The number of MCs in such environments is usually fixed, whereas MCs have little or no mobility. The entire network is usually under a single administration and is controlled by an ISP. As for traffic, it can be easily monitored, and the intensity of exchanged traffic can be easily predicted during different periods. This gives a more static and predictable network requirement. Also, it is easy to monitor and control different aspects of network management, such as routing, congestion, and interference control. Typical campus mesh architecture is shown in Figure 5.2.

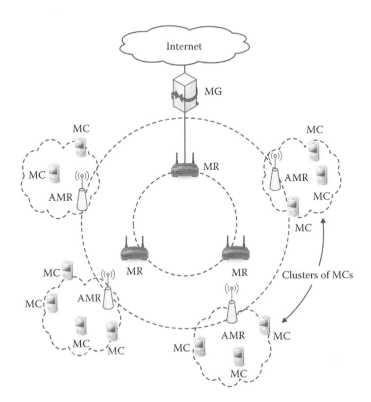

FIGURE 5.2 Campus mesh architecture.

It can be seen that there are two rings of wireless MRs in the backbone, that is, inner and outer rings. The outer ring represents AMRs connecting MCs to the backbone. The backbone MRs are in the inner ring, with no direct connection to the access network. Some of the inner-ring MRs can also act as AMRs. The inner ring has several major functions, such as the following:

- To act as a redundant array of routers or a backup path in case of congestion or disconnection
- To provide multipath routing options to the AMRs, and
- To collect traffic and forward it to MGs for Internet connectivity

5.2.2 Downtown Mesh Network Architecture

In the downtown topology-based mesh architecture [10], many low and tall buildings are scattered over several blocks in a downtown environment (Figure 5.3). In this type of architecture, LoS is not adequate, and towers

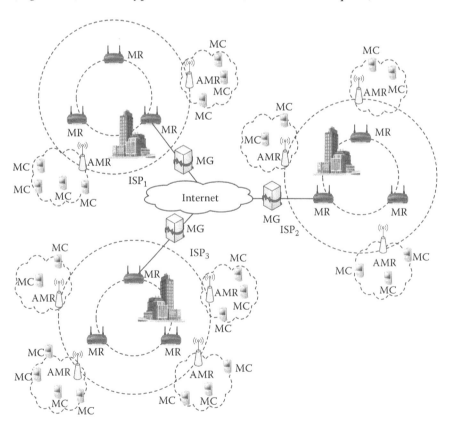

FIGURE 5.3 Downtown mesh architecture.

are not available (accessible) in many locations. The number of MCs varies with time, and they tend to change their locations frequently around the downtown area. The network can be under different administrations, managements, and ISPs. This introduces more technical and billing difficulties, such as roaming as well as network sharing among ISPs.

This type of architecture is different from the campus mesh. Namely, in a campus mesh, most of the traffic is generated by users on the campus. The number of permanent users and the type of operation, application, and usage are known to the administration over time. In a downtown environment, it is hard to predict real-time multimedia traffic demands. Some factors, such as the number of temporary users, the type of traffic they generate, the time and day they are passing by, and so on, can very well change the fluctuations of the traffic intensities. Exchanged traffic is highly bursty, depending on different client operations and different periods. Different ISPs provide different types of services to their clients. This makes it extremely difficult for them to coordinate with one another. Thus, downtown mesh architecture requires a more advanced high-capacity network and costly equipment for deployment, and tight coordination among ISPs is required.

5.2.3 LONG-HAUL MESH NETWORK ARCHITECTURE

Long-haul mesh architecture [14] is a perspective networking solution because it eliminates the need for extensive and costly infrastructure, as required by traditional wired and wireless technologies. In the long-haul mesh environment depicted in Figure 5.4, there are no buildings around, but there is a long set of MRs along a highway inside a city or in suburban areas, where there is no infrastructure in place or it is difficult and expensive to deploy one. The MRs could be as far apart as their transmission

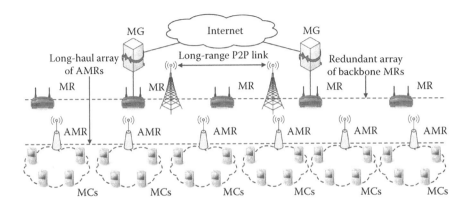

FIGURE 5.4 Long-haul mesh architecture.

range allows. An array of redundant backbone MRs runs along with the main backbone of AMRs.

For long LoS, using single powerful unidirectional antennas between each pair of adjacent routers is allowed. Backbone MRs can be deployed on the other side of the roadway for redundancy. The major difference between long-haul and the other two architectures consists of the structure itself, the type of equipment, and the lack of multipath routing in the long-haul environment.

5.3 CHARACTERISTICS OF WMNs

On the basis of their characteristics, WMNs are generally considered as a type of ad hoc networks because of the lack of wired infrastructure [1]. Although ad hoc networking techniques are required by WMNs, the additional capabilities necessitate more sophisticated algorithms and design principles for the WMN realization. More specifically, instead of being a type of ad hoc networking, WMNs aim to diversify the capabilities of ad hoc networks. Consequently, ad hoc networks can actually be considered as a subset of WMNs. The differences between WMNs and ad hoc networks can indicate some substantial characteristics.

The wireless backbone provides large coverage, connectivity, and robustness in the wireless domain. However, the connectivity in ad hoc networks depends on the individual contributions of end users, which may not be reliable.

WMNs support conventional clients that use the same radio technologies as an MR. This is accomplished through a host-routing function available in MR. WMNs also enable the integration of various existing networks through gateway/bridge functionalities in routers. The integrated wireless networks through WMNs resemble the Internet backbone because the physical location of network nodes becomes less important than the capacity and network topology.

In ad hoc networks, end-user devices also perform routing and configuration functionalities for all other nodes. However, WMNs contain MRs for these functionalities. Hence, the load on end-user devices is significantly decreased, which provides lower energy consumption and high-end application capabilities to possibly mobile and energy-constrained end users. Moreover, the end-user requirements are limited, which decrease the cost of devices that can be used in WMNs.

MR can be equipped with multiple radios to perform routing and access functionalities. This enables the separation of two main types of traffic in the wireless domain. While routing and configuration are performed between MRs, the access to the network by end users can be carried out on a different radio. This significantly improves network capacity. On the other hand, in ad hoc networks, these functionalities are performed in the same channel, and as a result, the performance decreases.

Because ad hoc networks provide routing using the end-user devices, network topology and connectivity depend on the users' mobility. This imposes additional challenges on routing protocols as well as on network configuration and deployment.

The fundamental characteristics of WMNs can be summarized through the following issues [1]:

- *Multihopping.* To extend the coverage range of current wireless networks without sacrificing the channel capacity, the mesh-style multihopping is indispensible [15,16]. It achieves higher through-put without sacrificing effective radio range via shorter link distances, less interference between the nodes, and more efficient frequency reuse. Another objective is to provide non-LoS connectivity among the users without direct LoS links.

- *Support for ad hoc networking and capability of self-forming, self-healing, and self-organization.* WMNs enhance the network performance because of flexible network architecture, easy deployment and configuration, fault tolerance, and mesh connectivity. Here, multipoint-to-multipoint communication is applied [3]. Because of these features, WMNs have low up-front investment requirements, and the network can grow gradually as needed.

- *Mobility dependence on the type of mesh nodes.* MRs usually have minimal mobility, whereas MCs can be stationary or mobile nodes. Because of MC's mobility in WMNs, mobility management is required for efficient and correct routing. A perspective wireless mesh mobility management mechanism based on location cache approach is proposed by Huang et al. [17].

- *Multiple types of network access.* In WMNs, both backhaul access to the Internet and P2P communications are supported. In addition, the integration of WMNs with other wireless networks and providing services to end users can be accomplished through WMNs.

- *Dependence of power consumption constraints on the type of mesh nodes.* MRs usually do not have strict constraints on power consumption. However, MCs may require power-efficient protocols. The MAC or routing protocols optimized for MRs may not be appropriate for MCs such as sensors because power efficiency is the primary concern for wireless sensor networks.

- *Compatibility and interoperability with existing wireless networks.* WMNs built based on Wi-Fi technologies must be compatible with IEEE 802.11 standards in the sense of supporting both mesh capable and conventional Wi-Fi clients. Also, WMNs need to be interoperable with other wireless networks such as ZigBee, WiMAX, LTE, and so on (see Chapter 1).

5.4   WMN PERFORMANCE IMPROVEMENT

Commercial and academic deployments of WMNs in real environments are beginning to demonstrate some advantages such as large covering areas, low cost of backhaul connections, small energy consumption, and non-LoS connectivity [18]. Earlier WMNs deployments have been linked to a number of problems mainly related to connectivity (lack of coverage, dead spots, obstructions, etc.) as well as performance problems (low throughput and/or high latency). Network performance is highly affected by wireless interference and congestion, causing considerable frame losses and higher delays.

Performance problems occur for many reasons, such as multipath interference, traffic slowdown due to congestion, large cochannel interference due to poor network planning, or poorly configured MCs or AMRs. Topology-aware MAC and routing protocols can significantly improve the performance of WMNs. Many solutions in the context of WMN's performance improvement have been proposed. These contributions can be classified into two general classes [9]: fixed topologies and unfixed topologies (Figure 5.5).

Approaches based on fixed topologies aim at better exploiting and using network resources. They can improve the channel reuse and routing protocols/metrics together with possible admission control mechanisms. They assume a given topology, that is, the position and the type of all mesh nodes are decided beforehand. On the other hand, approaches based on unfixed topologies are subdivided into two groups. The first group (partial design) encompasses all approaches that attempt to optimize network performance by optimally selecting the position and type of each mesh node given a different set of predeployed nodes. The second group is more generic and uses more complex techniques to design a network from scratch. It requires the consideration of many factors before network deployment, such as clients' coverage, optimal placement of MGs for better throughput and less delay/ congestion, and adequate resource amount per node.

Partial design can be observed through fixed gateway and unfixed schemes [19,20]. In the fixed gateway subcategory, the WMN design problem is viewed as the problem of looking for strategic locations to optimally place the APs or the MRs given a set of positioned MGs and a set of connectivity, coverage, and cost constraints to satisfy. The unfixed gateway scheme is related to MG's location problem and their number minimization, as well as the AP-MG path length while satisfying the AP Internet

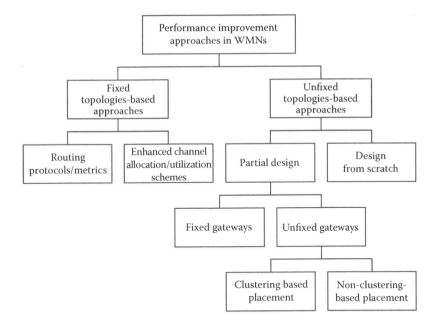

FIGURE 5.5 Classification of approaches for WMN performance improvement.

demand. MGs constitute IP traffic sinks/sources to WMNs, and consequently, a WMN can be unexpectedly congested at one or more of them if MGs are not adequately located. Basically, the placement of these mesh nodes determines the hop length of the communication paths, the amount of congestion, and the availability bandwidth to and from the Internet.

MG placement schemes are categorized into clustering-based placement and non-clustering-based placement classes. Placing MGs based on a clustering approach has several benefits, including the tight relationship between the resulting MG placement and the network throughput. When the network is partitioned into clusters, then independently of the network size, each node can send to nearby MGs within a fixed radius. Consequently, all nodes in a cluster have a bounded distance (in terms of the number of hops) to reach a gateway. Depending on whether the placed MGs are following a tree structure or not, the research studies related to the clustering-based placement can be subdivided into tree-based and non-tree-based approaches.

In tree-based clustering, the generating clusters are represented by trees rooted at the MGs [19]. These techniques have several benefits, such as low routing overhead and efficient flow aggregation. Nevertheless, they suffer from reliability degradation because a tree topology uses a smaller number of links than a mesh topology, where there are at least two nodes with two or more paths between them to provide redundant paths. Also, topologies restricted to tree structures may require, under the link capacity constraint, a larger number of MGs and thus may increase the network deployment cost.

As for non-tree-based clustering, the sole contribution proposes that it models the MG placement problem as a combinatorial optimization problem [20]. Two algorithms, self-constituted gateway algorithm (SCGA) and predefined gateway set algorithm (PGSA), are proposed. Both algorithms make use of a genetic search heuristic [21] to search for feasible configurations coupled with a modified version of Dijkstra's algorithm [22] to look for paths with bounded delays. In the PGSA, the number of MGs (initially set to one) is iteratively incremented by one until a feasible configuration is obtained. On the other hand, the number of MGs in the SCGA is set up dynamically when needed. The design problem solved by both search algorithms does not consider bounded delay in terms of communication hops. Here, delay is seen as the ratio of packet size over link capacity.

An approach to non-clustering-based placement is to study the MG position for throughput optimization in WMNs using a grid-based deployment scheme [23]. Specifically, for a given mesh infrastructure and several MGs, it is important to investigate how to place the gateways in the corresponding infrastructure to achieve optimal throughput. First, the throughput optimization problem for a fixed mesh network has to be mathematically formulated. After that, an interference-free scheduling method has to be proposed to maximize the throughput. The basic idea is to sort the links based on some specific order. Then, it is possible to process the requirement for each link in a greedy manner. The solution can be used as an evaluation tool to decide on the optimal MG placement scheme. The proposed approach achieves better throughput in the grid scheme than in random and fixed schemes.

A good planning task of a WMN essentially involves a careful choice of the installation's locations, an optimal selection of the network nodes type, and a good decision on a judicious channel/node interface assignment, while guaranteeing the users' coverage, wireless connectivity, and traffic flows at a minimum cost. In optimization terms, this is translated into determining the following [9]:

- The optimal number of wireless routers required to cover the area under consideration
- The optimal number of MGs for the efficient integration of WMNs with Internet
- The optimal initial channel assignment, and
- An optimal number of wireless interfaces per router

while taking into account all physical and financial constraints of the network provider.

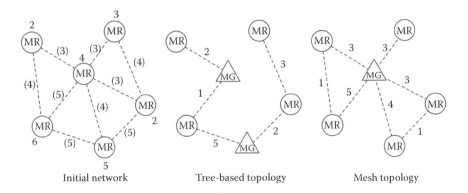

FIGURE 5.6   Effect of network topology on MG deployment.

Example 5.1

The effect of network topology on MG deployment is presented in Figure 5.6. Every potential link, presented by a dashed line, is associated with a link capacity (the value between brackets), and a traffic demand is associated to every MR. It can be seen that a tree-based topology tends to deploy more MGs than a mesh topology (two MGs instead one MG) because of the link capacity constraint.

5.5   ROUTING IN WMNs

Routing is one of the key components for data delivery in a WMN [24]. As previously stated, backbone MRs are usually stationary, while MCs roam among them. Consequently, they can be permanently power supplied. As mobility and energy saving are no longer issues, WMN routing considers link-quality metrics such as capacity or error probability. To improve network performance, the design of a routing protocol must take into account the lossy characteristic of a wireless channel. In multihop scenarios, performance depends on the routing protocols to properly choose routes, given the current network conditions. Assuming that the common-case application in WMNs is Internet access, traffic is usually concentrated on links close to the MGs.

5.5.1   FUNDAMENTAL ROUTING METRICS

The routing metrics present criteria to judge the "goodness" of a path in routing algorithms. The most typical routing metrics for multihop networks is the hop count [6]. However, this metric cannot capture the quality of a path in a wireless environment.

The first metric proposed for WMNs is the expected transmission count (ETX) [25]. It is the expected number of transmissions a node requires to successfully transmit a packet to a neighbor. To compute the ETX, each node periodically broadcasts probes containing the number of received probes from each neighbor. The number of received probes is calculated at the last T time interval in a sliding window fashion. Node A computes the ETX of the link to node B by using the delivery ratio of probes sent on the forward d_f and reverse d_r directions. The ETX considers both directions because of data ACK frame transmission. Delivery ratios are, respectively, the fraction of successfully received probes from A announced by B and the fraction of successfully received probes from B at the same T interval. The ETX of link AB can be defined as

$$\text{ETX} = \frac{1}{d_f \times d_r}. \tag{5.1}$$

The total ETX of a path is the summation of ETX of all links on the path. The chosen path is one with the lowest sum of ETX along the route to the destination.

The minimum loss (ML) metric [26] is based on probing to compute the delivery ratio. ML finds the route with the lowest E2E loss probability. ML multiplies the delivery ratios of the links in the reverse and forward directions to find the optimal path. The use of multiplication reduces the number of route changes, improving network performance.

There are two shortcomings in the implementation of ETX, that is,

a. Broadcasts are usually performed at the network basic rate, and
b. Probes are smaller typical data packets

Thus, unless the network is operating at low rates, the performance of ETX becomes low because it neither distinguishes links with different bandwidths nor considers data packet size. To cope with these issues, the expected transmission time (ETT) is invoked. This is the time a data packet requires to be transmitted successfully to each neighbor. ETT adjusts ETX to different physical rates and packet sizes. There are two main approaches to computing ETT. In the first one, ETT is presented as the product between the ETX and the average time a single packet requires to be delivered, that is,

$$\text{ETT} = \text{ETX} \times t. \tag{5.2}$$

To calculate time t, a fixed packet size s is divided by the estimated bandwidth B of each link, $t = s/B$. The packet pair technique is used to calculate *B* per link. This technique consists of transmitting a sequence of two back-to-back packets to estimate bottleneck bandwidth. Each neighbor measures the interval period between the two packets and reports it back to the sender. The computed bandwidth is the size of the large packet of the sequence divided by the minimum delay received for that link.

In the second approach, the loss probability is estimated considering that the IEEE 802.11 uses data and ACK frames [25]. The idea is to periodically compute the loss rate of data and ACK frames to each neighbor. The former is estimated by broadcasting several packets of the same size as data frames, one packet for each data rate defined in the specified standard. The latter is estimated by broadcasting small packets of the same size as ACK frames and sent at the basic rate that is used for ACKs. ETT can be presented in the form

$$\text{ETT} = (r_t \times p_{\text{ACK}})^{-1}, \qquad (5.3)$$

where r_t represents best achievable throughput and p_{ACK} represents the delivery probability of ACK packets in the reverse direction. Similar to ETX, the chosen route is the one with the lowest sum of ETT values.

The fast link-quality variation is a critical problem of wireless media. Metrics based on average values computed on a time–window interval, such as ETX, may not follow the link-quality variations [6]. This problem is even more complex in an indoor environment. To cope with this, modified ETX (mETX) and the effective number of transmissions (ENT) have to be taken into account [27]. These metrics consider the standard deviation in addition to link-quality average values to project physical layer variations onto routing metrics. The difference between mETX and ETX is that mETX works at the bit level. The mETX metric computes the bit error probability using the position of corrupted bit in the probe and the dependence of these bit errors throughout successive transmissions. This is possible because probes are composed of a previously known sequence of bits. On the other side, ENT is an alternative approach that measures the number of successive transmissions per link considering variance. Also, ENT broadcasts probes and limits route computation to links that show an acceptable number of retransmissions according to upper-layer requirements. If a link shows several expected transmissions higher than the maximum tolerated by an upper layer protocol, ENT excludes this link from routing process, assigning to it an infinity metric [6].

The next metric that also considers link-quality variation is interference aware (iAWARE) [28]. This metric uses signal-to-noise ratio (SNR) and signal to interference and noise ratio (SINR) to continuously reproduce neighboring interference variation onto routing metrics. The iAWARE metric estimates the average time the medium is busy because of transmissions

TABLE 5.1

Main Routing Metric Characteristics

Metric	Quality Aware	Data Rate	Packet Size	Intraflow Interference	Interflow Interference	Medium Instability
Hop	×	×	×	×	×	×
ETX	+	×	×	×	×	×
ETT	+	+	+	×	×	×
ML	+	×	×	×	×	×
mETX	+	+	+	×	×	+
ENT	+	+	+	×	×	+
iAWARE	+	+	+	+	+	+

from each interfering neighbor, according to the rule that the higher the interference, the higher the iAWARE value. iAWARE considers intraflow and interflow interference, medium instability, and data transmission time. iAWARE for certain link l is defined as

$$iAWARE_l = \frac{ETT_l}{IR_l}, \tag{5.4}$$

where interference ratio IR_l from u to v is obtained as $\min\{IR_i(u), IR_i(v)\}$ and $IR_i(v) = SINR_i(v)/SNR_i(v)$. The iAWARE of a path is calculated in a similar way as ETT. However, most routing protocol implementations prefer metrics with simpler designs, such as ETX or ETT. The main characteristics of the discussed WMN routing metrics are summarized in Table 5.1.

5.5.2 ROUTING PROTOCOLS

Ad hoc routing protocols are usually proactive, reactive, or hybrid [6]. The proactive strategy operates like classic routing in wired networks. Routers keep at least one route to any destination in the network. On the other hand, reactive protocols request a route to a destination only when a node has a data packet to send. Many WMN routing protocols use similar strategies. Nevertheless, they are adapted to the WMN's characteristics, for example, by using a quality-aware routing metric.

WMN routing protocols can be classified into four classes [6]: ad hoc based, controlled flooding, traffic aware, and opportunistic. Each class mainly differs on route discovery and maintenance procedures. In WMNs, most routing protocols consider that the network is only composed by wireless backbone nodes. If a mobile device operates as a backbone node, it must run the same routing protocol. Because of fewer considerations regarding mobility, increasing traffic demand, and certain infrastructure-like design properties, routing protocols for WMNs have required exclusive focus from researchers. A more general classification of WMN routing protocols is presented by Pathak and Dutta [29]. Table 5.2 surveys WMN routing strategies in terms of their characteristics and objectives.

MANET-like routing protocols were designed for mobile nodes, intermittent links, and frequently changing topologies. These protocols often rely on flooding for route discovery and maintenance. Direct employment of such protocols is not suitable for relatively static mesh networks.

TABLE 5.2
Classification of WMN Routing Strategies

Routing Strategy	Characteristics/Objective	Representative Solutions
MANET-like routing	• Proactive or reactive • Adapt ad hoc routing protocols to relatively stable and high-bandwidth WMN environment • Incorporate a WMN routing metric in existing protocol	• OLSR • BATMAN • AODV-ST • AODV-MR • PROC
Opportunistic routing	• Hop-by-hop routing • Exploit fortunate long-distance receptions to make faster progress toward destination	• Ex-OR • ROMER • SOAR • MORE
Multipath routing	• Maintain redundant routes to destination • Determine divergent routes to mitigate the crowded center effect • Load balancing and fault tolerance	• MMESH
Geographic routing	• Use location information for forwarding in large mesh networks	• Efficient geographic routing based on NADV metric
Hierarchical routing	• Divide network into clusters and perform routing for better scalability	• IH-AODV
Multiradio/ channel routing	• Accommodate intrapath and interpath interference • Consider channel assignment constraints and switching cost	• MR-LQSR
Multicasting protocols	• Adapt existing multicast ad hoc mechanisms to WMNs	• High-throughput ODMRP
Broadcast routing	• Minimum latency broadcasting with the least number of retransmissions • Adapting to multichannel environment	• MRDT framework

Traditional MANET-like protocols can be largely classified in proactive and reactive routing protocols. Proactive routing protocols are table-driven protocols that require flooding in case of link failure and use hop count as a primary routing metric. They do not take link quality or any other dynamic wireless characteristics, such as intermediate packet losses, into consideration.

Many of the proactive routing protocols have been adopted or specifically designed for WMNs. As an example, optimized link state routing (OLSR) [30] has accommodated features for link-quality sensing, and it is being adapted for WMN implementations. OLSR was adapted to use ETX as a link metric. It uses the fraction of HELLO messages lost in a given interval of time to calculate ETX. Another interesting and advanced solution is the better approach to mobile ad hoc networking (BATMAN) protocol [31]. It uses a weighted and autoselective flooding concept in which every node maintains logical direction toward the destination and accordingly chooses next hop neighbor while routing. The flooding uses short control packets to make every mesh node aware of network topology. However, there is no topology information dissemination and no multipoint relaying node selection. Although routing loops are easily avoided by this modified flooding mechanism, there are still problems about self-interference.

Instead of developing new routing protocols for WMNs, many researchers have proposed modifications of ad hoc routing protocols. Most of these protocols try to adapt to the characteristics of WMNs such as lower mobility, stable routes, and so on. Also, a variety of such protocols uses previously discussed routing metrics. The following are a few examples of such protocols:

- AODV-spanning tree (AODV-ST) [32] is a routing protocol that improves on AODV in several ways to adapt to WMN characteristics. To avoid repetitive reactive route discovery with flooding, AODV-ST maintains spanning tree paths rooted at gateway from the nodes. The gateway periodically requests routes to every node in the network to update its routing table. It can incorporate high-throughput metrics such as ETT and ETX for high-performance spanning tree paths. AODV-ST also uses IP–IP encapsulation for avoiding large routing tables at relay nodes and can also perform load balancing for gateways.
- AODV-multiradio (AODV-MR) [33] presents extension for AODV protocol in which each node has multiple radios and channel assignment is performed with some predefined static technique. AODV-MR uses the iAWARE metric with the Bellman–Ford algorithm to find efficient low-interference paths. Links on such paths display low intraflow and interflow interference together with good link

quality. An extension of the AODV-MR that is adaptive to changes in link quality to detect deteriorated situations and to recover the path is proposed by Shin et al. [34].

- Progressive route calculation (PROC) protocol [35], which locates the optimal route around a preliminary route, is obtained from sketchy network scanning. The source node first establishes a preliminary route to the destination by broadcasting. The destination then initiates the building of a minimum cost-spanning tree to source with the nodes around the preliminary route. The source uses this optimal route for future data transfer.

As previously discussed, traditional shortest path and ad hoc routing protocols may not be suitable for mesh environments. Opportunistic routing protocols have been proposed to exploit the unpredictable nature of the wireless medium. They improve classic routing based on cooperative diversity schemes. Classic routing protocols compute a sequence of hops to the destination before sending a data packet, using either hop-by-hop or source routing. On the other hand, an opportunistic routing protocol defers to the next hop selection after the packet has been transmitted. If a packet fortunately makes it to a far distant node than expected, such useful transmissions should be fully exploited. Although there are many advantages to such mechanisms, for example, faster progress toward the destination, it requires complex coordination between the transmitters regarding the progress of the packets. Many protocols have been developed based on this concept, and some of the most representative are as follows:

- Extremely opportunistic routing (Ex-OR) protocol [36], which combines routing with MAC layer functionality. Routers send broadcast packets in batches (without previous route computation) to reduce protocol overhead. In addition, broadcasting the data packets improves reliability because only one intermediate router is required to overhear a transmission. However, it does not guarantee that packets are received because they are not acknowledged. Thus, an additional mechanism is required to indicate correct data reception. Among the intermediate routers that have heard the transmission, only one retransmits at a time. The source router defines a forwarding list and adds it to the header of the data packets. This list contains the addresses of neighbors, ordered by forwarding priority. Routers are classified in the forwarding list according to their proximity to the destination, computed by a metric similar to ETX. The metric used by Ex-OR considers only the loss rate in the forward direction. Upon reception of a data packet, the intermediate router checks the forwarding list. If its address is listed, it waits for the reception of the whole batch of packets. It is possible, however, that a router does not receive the entire batch. To cope with this problem, the highest-priority router that has received packets forwards them and indicates to the lower-priority routers the packets that were transmitted. Consequently, the lower-priority routers transmit the remaining packets, avoiding duplicates. This process continues until the batch of packets reaches the destination.

- Resilient opportunistic mesh routing (ROMER) protocol [37] combines long-term shortest-path or minimum-latency routes with on-the-fly opportunistic forwarding to provide resilient routes and to deal with short term variations on medium quality. A packet traverses through the nodes only around long-term and stable minimum cost path. These nodes build a dynamic forwarding mini-mesh of nodes on the fly. Each intermediate node opportunistically selects transient high-throughput links to take advantage of short-term channel variations. In this way, ROMER deals with node failures and link losses and also benefits from opportunistic high-throughput routing.

- Simple opportunistic adaptive routing (SOAR) protocol [38] improves on Ex-OR in certain ways to efficiently support multiple flows in WMNs. SOAR maximizes the progress each packet makes by using priority-based timers to ensure that the most preferred node forwards the packet with little coordination overhead. Moreover, it minimizes resource consumption and duplicates transmissions by reasonable selecting forwarding nodes to prevent routes from diverging. To further protect against packet losses, SOAR uses local recovery to retransmit a packet when an ACK is not received within a specified time.

- MAC-independent opportunistic routing and encoding (MORE) protocol [39] extends the Ex-OR with network coding. It randomly mixes packets before forwarding. This randomness ensures that routers that hear the same transmission do not forward the same packets. Thus, MORE needs no special scheduler to coordinate routers. Experimental results show that MORE's median unicast throughput is 22% higher than Ex-OR, and the gains increase to 45% when there is a chance of spatial reuse. For multicast, MORE's gains increase with the number of destinations and can be two times greater than Ex-OR.

Multipath routing protocols permits the establishment of multiple paths between a source and a destination. They can provide various benefits over single path protocols in MWNs, increasing reliability, enabling load balancing and bandwidth aggregation, and secure communications [40]. Using traditional routing approaches and metrics, many MRs may end up choosing already congested routing paths to reach the MGs. This can lead to low performance because of highly loaded routing paths. A multipath routing approach has the following advantages [41]:

1. Fault tolerance achieved by using redundant paths as alternative routes exist to deliver messages from a source to the destination.
2. When a link becomes a bottleneck because of heavy load at a specific time, multipath routing protocols can balance the load by diverting traffic through the available alternative paths.
3. Multipath routing protocols can split data to the same destination into multiple streams, each routed through a different path, to increase the aggregate bandwidth utilization of a network.
4. In multipath routing, the recovery delay in case of a fault can be reduced because backup routes are initiated during the route discovery phase. When a failure occurs, the predefined routes can be used instead of rediscovering a new route.

On the other hand, given a performance metric, the improvement of multipath routing protocols depends on the availability of disjoint routes between source and destination. Moreover, the complexity of the multipath routing (especially for the route discovery phase) can be high.

Multipath mesh (MMESH) routing protocol [42] proposes a mechanism in which every node derives multiple paths to reach gateway node using the source routing. It then performs load balancing by selecting one of the least loaded paths. A large set of multipath routing protocols are surveyed by Tsai and Moors [43].

Geographical routing (position-based routing) assumes that each node knows its own location and each source is aware of the location of its destination. The MANET routing protocols often assume the availability of location information at nodes to facilitate intelligent data forwarding. WMNs can benefit from such location information, and several routing protocols are presented for such geographic routing and related issues. An efficient geographic routing protocol in which packets are forwarded toward the neighbor closest to the destination is proposed by Lee et al. [44]. Forwarding decisions are made on a hop-by-hop basis using a normalized advance (NADV) link metric, which is defined as follows:

$$NADV(n) = \frac{ADV(n)}{Cost(n)} = \frac{D(S) - D(n)}{Cost(n)}. \qquad (5.5)$$

Here, $D(x)$ denotes the distance from node x to destination, and $\text{Cost}(n)$ can be any cost factor, such as packet error rate, delay, power consumption, and so on. This way, NADV reflects the amount of progress made toward the destination per unit cost. For example, suppose that $P^{\text{succ}}(n)$ is the fraction of data transmissions to neighbor n successfully. If $1/P^{\text{succ}}(n)$ is used as link cost, then

$$\text{NADV}(n) = \text{ADV}(n) \times P^{\text{succ}}(n), \tag{5.6}$$

which means the expected advance per transmission.

Hierarchical routing has importance especially in MANETs, but its applicability to mesh networks has been limited. One possible reason for this could lie in the fact that most of the hierarchical routing protocols [45] assume high mobility, which is rarely a case in mesh. Instead, WMNs show far static behavior (at least in MRs), and client mobility can usually be handled by traditional mobility management schemes. The efficient accommodation of clustering schemes together with channel assignment policies can explore the full capacity available, whereas designing such mechanisms with clustering is still an open issue. Hierarchical routing's advantage depends on the depth of nesting (the number of hierarchy levels) and the addressing scheme used [41]. In a high-density network, the performance of hierarchical routing is considered to be very good compared with other routing approaches. The reasons for this good performance are the low overhead, relatively short routing paths, easy adaptation to failures, and quick path-setup time. While designing the structure of the hierarchy, the cluster heads should be carefully selected to avoid a bottleneck and large power consumption. A WMN may experience implementation difficulties when the selected cluster head is not capable of handling the heavy traffic load. In addition, the complexity of maintaining the hierarchy may compromise the performance of the routing protocol.

An improved hierarchical AODV (IH-AODV) routing protocol is proposed by Tingrui et al. [46]. It exhibits better scalability and performance in the network and incurs less routing overhead for finding alternate routes when a route is lost. Furthermore, a novel technique for IH-AODV fresh route detection is presented. Simulations show that IH-AODV scales well for large networks, and other metrics are also better than or comparable with AODV in hybrid WMNs. With some modification, this protocol is also applicable in cognitive WMNs [47].

Multiradio/channel routing protocols mainly use corresponding metrics derived for multichannel environment in suitable well-known routing techniques, such as AODV, dynamic source routing (DSR), and so on. To fully exploit the availability of multiple channels in WMNs, routing algorithms should account for the existence of channel diversity on a path in the network [48].

Example 5.2

Consider the 10-node multichannel WMN shown in Figure 5.7. Each node is equipped with two radios. It means that each node can transmit or receive data on two nonoverlapping channels simultaneously. The label on each link indicates the operating channel.

There are three possible routes from node F to gateway node A in the network, including F-G-A (involving channel 2), F-G-J-A (involving channels 2 and 3), and F-E-B-A (involving channels 6, 5, and 1). It can be very difficult to determine optimal path from node F to node A because path F-G-A is the shortest, but paths F-G-J-A and F-E-B-A are more channel diverse.

The routing problem in multichannel WMNs is exacerbated by the fact that the network topology is determined by the channel assignment. For example, coming back to Figure 5.7, although nodes E and G are located within the transmission range of each other, they cannot communicate with each other directly without a radio tuned to a common channel. This implies that the routing paths between any two nodes in the network are also restricted by channel assignment. As a result, a well-designed routing protocol for multichannel WMNs can become useless with an improper channel assignment algorithm. In some types of multichannel WMNs, nodes have to dynamically negotiate the channels used for communication.

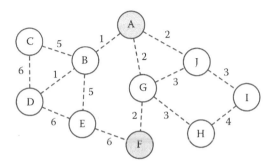

FIGURE 5.7   Multichannel WMN with nodes equipped with two radios.

Thus, it is difficult for the multichannel routing algorithm to predict the E2E performance of a path in such a dynamic environment [48].

Multiradio link-quality source routing (MR-LQSR) [49] presents an example of a widely accepted and deployed protocol for multiradio WMN environments. This protocol replaces the link-quality routing metric with the weighted cumulative ETT (WCETT) as an extension of ETX for multichannel wireless environments. The calculation of the WCETT metric can be divided into two parts:

1. The estimation of the E2E delay of the path and
2. The determination of the channel diversity of the path

As mentioned previously, ETT represents the expected total airtime spent in transmitting a packet successfully on a link. Therefore, ETT is obtained by multiplying the ETX value of a link by the transmission time of one packet. The calculation of WCETT then requires the sum of ETTs (SETT) for all links of the path, which corresponds to an estimation of the E2E delay experienced by the packet. To quantify the channel diversity, it needs to determine the bottleneck group ETT (BGETT). The group ETT (GETT) of a path for channel c is defined as the sum of ETTs for the path's links, which operate on channel c. The BGETT is then referred to as the largest GETT of the path. The rationale is that the total path throughput is dominated by the bottleneck channel (i.e., the busiest channel on the path). Thus, although low SETT implies short paths, low BGETT implies channel-diverse and high-bandwidth paths [48].

The WCETT metric can be defined as the weighted average of the sum of SETT and BGETT,

$$WCETT = (1 - \beta) \times SETT + \beta \times BGETT. \qquad (5.7)$$

Accordingly, the routing algorithm is to select the path whose WCETT is the lowest. The WCETT metric strikes a balance between channel diversity and path length (or between throughput and delay) by changing the weighting coefficient β.

Multicast routing provides an efficient solution, supporting collaborative applications such as video conferencing, distributed gaming, webcast, and distance learning among a group of users. Multicast is a bandwidth-conserving technology that reduces traffic by simultaneously delivering a single stream of packets to a group of recipients [48]. Many multicast routing protocols have been proposed for single-radio WMNs. A typical approach to supporting multicast in such an environment is to construct a multicast tree and let each parent node be responsible for multicasting data to its child nodes. This approach works under the assumption that a parent node and its child nodes share a common channel. However, in multichannel WMNs, this assumption may not hold. In addition, if the channel assignment is dynamic, extra overhead due to frequent tree reconstruction or retransmissions of multicast packets must be addressed. One possible solution is to use a common control channel or hybrid channel assignment strategy to coordinate the channels used by the parent and child nodes.

First insights about multicasting in WMNs [50] studied the performance improvement achieved by using different link-quality–based routing metrics via extensive simulation and experiments on a mesh network test bed using modified on-demand multicast routing protocol (ODMRP) as a representative approach. It has been shown that in case of broadcast, link quality in the backward direction should not be considered because there are no ACKs involved. Also, metric product over links of a path better reveals the overall quality of the path. For high-throughput ODMRP modification, adapted traditional unicast metrics such as ETX and ETT are used, as well as two derived metrics:

- Multicast ETX (METX) metric expressed as

$$\mathrm{METX} = \sum_{l=1}^{n} \frac{1}{\displaystyle\prod_{i=l}^{n}(1 - \mathrm{Perr}_i)}, \qquad (5.8)$$

  where l denotes lth link on n-hop path and Perr_l is the error rate of the link.
- Success probability product (SPP) metric, similar to METX, is defined as

$$\mathrm{SPP} = \prod_{l=1}^{n}(1 - \mathrm{Perr}_l). \qquad (5.9)$$

Considering link layer broadcast in multicast, SPP reflects the probability that the destination receives the packet without error. The routing protocol selects the path with the minimum 1/SPP (maximum SPP).

Example 5.3

Figure 5.8 presents a simple, but illustrative enough example showing why SPP is superior to a metric such as METX that tries

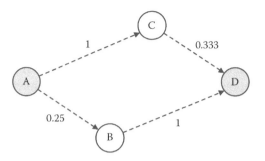

FIGURE 5.8   Example of high-throughput multicast routing.

to minimize the total number of transmissions. The labels over the links denote the forwarding probability $(1 - Perr_l)$ of each link.

Using Equations 5.8 and 5.9, respectively, it is possible to calculate METX and SPP for two possible paths from A to D:

$$METX_{ACD} = 6,\ SPP_{ACD} = 1/3,$$

$$METX_{ABD} = 5,\ SPP_{ABD} = 1/4.$$

SPP can choose higher-throughput paths than METX by minimizing the expected number of packet transmissions at the source.

Broadcast routing protocols are mainly related to the so-called minimum latency broadcasting problem in which the goal is to minimize latency, defined as the maximum delay between a packet's network-wide broadcast at the source and its eventual reception at all network nodes. Many broadcast-based applications of WMNs have strict latency requirements (e.g., audio conferencing, video feeds, and multiplayer gaming), and they can benefit from this routing approach. Efficient broadcasting in WMNs is especially challenging because of the desirability of exploiting the "wireless broadcast advantage," interface diversity, channel diversity, and rate diversity offered by networks. A distributed and localized heuristic framework called multiradio distributed tree (MRDT) is proposed by Qadir [51]. MRDT represents the first distributed solution to the minimum latency broadcasting problem for multiradio multirate multichannel (MR^2-MC) WMNs. This framework calculates a set of forwarding nodes and transmission rates at these forwarding nodes irrespective of the broadcast source. A forwarding tree is constructed, taking into consideration the source of the broadcast. This solution can greatly improve broadcast performance by exploiting the rate, interface, and channel diversity.

5.6   FAIR SCHEDULING IN WMNs

Scheduling is an active challenging issue, especially in commercial WMN applications. Many current deployments are optimized with respect to throughput, delay, or some other feature that gives little regard to fairness. Several representative scheduling and resource allocation techniques have been proposed for WMN in the open literature [52–54]. It is important to note that fairness can occur at different points in a WMN (e.g., router, link, or client).

Scheduling algorithms usually give preference to flows that are least expensive by some criteria. These criteria may be gateway distance, delay, small flows, and other similar metrics. However, this approach may allow for starvation or reduced QoS for flows that do not meet the criteria. Preference may be given to greedy flows, which on one extreme can be denoted as absolute or hard fairness. This side gives little priority to throughput and ensures that each client gets a fair share of the network resources. This may be achieved by using a time division mechanism or other similar approaches. The problem with this approach is that not all flows require the same amount of resources at all times, so the resources may remain occasionally unused, resulting in poor throughput. One approach that aims for a balance between the competing goals of fairness and throughput, denoted as max–min fairness [55], works by maximizing the minimum data rates for each flow. It results in higher throughput than hard fairness. However, the overall throughput is still much less than maximum throughput. The most interesting definition of fairness then is a compromise between hard fairness and maximum throughput [56].

5.6.1   CLASSIFICATION OF SCHEDULING

Scheduling algorithm can be observed as a centralized or distributed approach [57]. Usually, for situations in which the MRs are anticipated to be static, it may be easier and more beneficial to use centralized scheduling. On the other hand, when the MRs are mobile, it may be better to use a distributed approach in case the network becomes partitioned due to the movement of MRs.

Scheduling protocols for WMNs can be classified by degree of fairness into five categories [56]:

- Hard fairness (round-robin scheduling) [58] has been used in some of the earliest wireless networks and in simplistic network models because it is the least complex. It is the fairest scheme because each node is guaranteed exactly equal amounts of time in order. In networks where the nodes only require a small proportion of resources, hard fairness causes problems. Because each node is given time to transmit at regular intervals, if the node does not have any data to send, the time is wasted. This leads to very low overall throughput. At the same time, however, the problem of node starvation does not exist. Resources are assigned to each node inversely proportional to the number of flows through the node.

- Max–min fairness [55] allocates resources in order of increasing demand. The minimum amount of resources assigned to each node is maximized. So if there are more than enough resources for each node, every node is "contented." On the other hand, the resources are split evenly. This means that the nodes that require fewer resources get a higher proportion of their needs satisfied. The nodes that require more resources end up dropping many packets, and thus the network ends up with still quite low packet delivery ratio (PDR). This type of scheme works best in situations in which there are no large differences in resources requested at each node. This can be a problem in a WMN because intuitively, the nodes closer to the MG will experience much higher traffic than those on the outside of the network yet may end up dropping many of the packets anyway. This may be partially solved by increasing the resource capacity of the nodes closest to the MG.
- Proportional fairness [59] allocates resources proportional to some characteristic in the network. For example, one may choose to give priority to nodes that are close to the MGs. The amount of resources allocated then would be proportional to how close the node is to the MG. The strength of the proportionality can be controlled depending on the proportionality factor

$$R = \frac{1}{c^{\beta}}, \tag{5.10}$$

where R is the resources allocated to the node, c is the characteristic in which priority is given to ($c > 0$), and β is the proportionality factor ($\beta > 0$).

- Mixed-bias [59] scheduling allows for different levels of control over resources. Rather than just allowing for one bias, this scheme mixes two different biasing levels together. A certain proportion of the resources is assigned to one factor and the rest to another factor

$$R = \frac{\alpha}{c^{\beta_1}} + \frac{1-\alpha}{c^{\beta_2}}, \tag{5.11}$$

where β_1 and β_2 are the proportionality factors (β_1, $\beta_2 > 0$) and α is the fraction of resources assigned to each bias ($\alpha \geq 0$). This allows the scheduling algorithm to provide two different biasing levels (mixed biasing) against a certain characteristic. Rather than just strongly biasing against that characteristic, which may result in certain nodes to be starved, the mixed biasing allows for a combination of weak and strong biasing, meaning that a portion of the resources is reserved to provide a minimum service level, even for the nodes that are undesirable in terms of certain characteristics.

• Maximum throughput [60] scheduling has only one goal. As the name suggests, this goal is to maximize throughput. Whichever node requires the most resources or can transmit the fastest or most data gets access to the resources first. This ensures a very high throughput; however, there is a limitation with this approach. Nodes that have less priority, such as those far away from MG, those with fewer users, fewer flows, or less traffic demands, are essentially ignored. If enough time passes, all the packets waiting in the queues at these MRs are dropped. This should be avoided because of performance degradation problem.

5.6.2 FAIR SCHEDULING WITH MULTIPLE GATEWAYS

This perspective approach is related to the distributed requirement table mechanism [56], which is an enhancement of the original fair scheduling framework [57]. These requirements are necessary for generating the scheduling procedure because this information shows how busy each link is. Each MR keeps track of a local requirement table. In this table, the demand on each link between the router and a neighbor is kept. When a new schedule is requested, each MG asks for the partial requirement tables from each associated MR. The MG then combines these tables to form one complete requirement table, which it uses to generate cliques and eventually the scheduling.

One main difference from the approach used by Thomas [57] is that multiple MGs are assumed. This means that each MG is responsible for scheduling all the links that will forward packets toward it. The single gateway assumption is significant for two reasons:

1. The single MG causes an extreme bottleneck in the network. All traffic demands that flow in and out of the network must use this node and, hence, any scheduling procedure done in the network is limited by the single MG.
2. Similarly, the single MG causes a single point of failure in the network without recovery.

When multiple MGs are assumed, the bottleneck is eliminated. Not all of the traffic is destined to the same node in the network and is spread more evenly, especially with strategic MG placement. The single point of failure is eliminated as well. If one MG experiences an outage, the network has the ability to reconfigure itself to forward packets and perform scheduling from another MG. Once the requirement table is formed, the MG uses this information along with the clique information to form a scheduling plan. The clique information is all of the sets of links that may transmit at the same time without interference.

The type of fairness used in this solution is round-robin style with spatial reuse. Centralized schedule generation at the MG, which makes use of distributed routing tables located at the MG, is applied. The requirement propagation algorithm that allows each MG to distribute the requirements and routing table for scheduling into the network is proposed [56]. At each MR, the path to the MG is maintained in this table, together with the requirements for the links on this path. For each MC requesting to use corresponding MR, each link along the way to the gateway in the local table is given a requirement. When the MG signals the start time for new schedule generation, it requests the local requirement information from all of the MRs, which are currently using it as their primary MG. It then combines the requirements to help determine the scheduling as shown in Figure 5.9.

Here, each MR has a local requirement table. This requirement table keeps track of the requirement for itself and for all the nodes on the path.

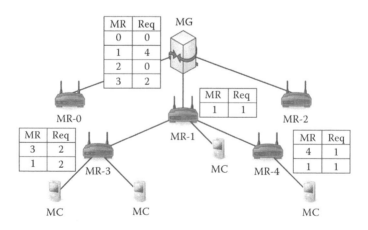

FIGURE 5.9   Distributed requirement tables combined at MG. (From J.B. Ernst. Scheduling Techniques in Wireless Mesh Networks. PhD Thesis. University of Guelph, Canada, 2009.)

A requirement is added when an MC sends data to an MR. At that particular MR, the requirement is incremented for itself and for all hops to the gateway in its local table because all these nodes will have to relay the packet. A single MG is responsible for generating the scheduling for all the nodes that route through it. When a new scheduling must be generated, the MGs request the requirement from each table. Each MG then combines the requirement information from each MR with the compatibility matrix [58]. The compatibility matrix represents the links that may transmit simultaneously without interference (Figure 5.10) and is computed or set up manually once the network is formed.

The MG then computes the scheduling. After the scheduling is computed, START packets are sent to the MRs when they are free to transmit, and END packets are sent to the same MRs when their transmission period has ended. This continues until the end of the current schedule.

Because of the positioning of the MRs and the communication ranges, if two MRs are not neighbors and do not share a common neighbor, they are not close enough to cause interference with each other, and they do not compete for the resources of a common neighbor. This way, both may communicate at the same time. The spatial time–division multiple access (TDMA) scheduling allows multiple links to be activated at the same time when they do not interfere. Hence, the network can be used more efficiently than it could if only one link in the entire network was active. Furthermore, because the algorithm uses the concept of compatibility, no two links are active that compete for resources, so collisions are avoided. This solution is different from many other TDMA solutions because it only allocates time for links that actually have requirements associated with them.

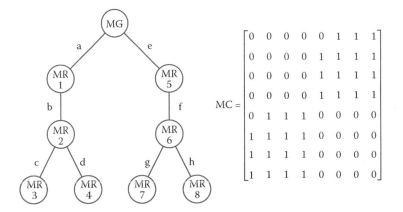

FIGURE 5.10 Example of compatibility matrix for corresponding network.

Example 5.4

One of the main uses for WMNs is to provide broadband access with expanded service areas from traditional WLANs. Having only one MG in this scenario is a major bottleneck, so the existing solutions should be extended to be able to support any number of MGs to make a truly scalable WMN. The performance evaluation of fair scheduling with multiple MGs can be carried out through simulation environment and parameters defined in Table 5.3 [56]. The simulation focuses on packet transmission from MRs to MGs. MCs are generated (using a uniform random distribution) at the start of the simulation and are randomly distributed within the simulation environment. Each MC is associated with the closest MR, and each MR routes its packets to the closest MG.

The simulation environment acts as an omniscient observer in that it performs the scheduling and distributes into the MGs. In a real environment implementation, this would need to be performed either through a centralized MG or via some kind of distributed MG solution. The interference model assumes that two nodes interfere if they are within range and transmitting at the same time or if there is a buffer collision. When interference occurs, retransmission is allowed until a threshold timeout is reached.

The first metric used in simulation is PDR, which is computed as follows:

TABLE 5.3
Simulation Parameters

Parameter	Value
Environment dimensions	1000 m × 1000 m
Node range	250 m
Number of MRs	10–55
Number of MCs	250
Number of MGs	1–6
Mean packet arrival	0.01 s
Mean hop delay	0.01 s
Retry threshold	0.01 s

Source: J.B. Ernst. Scheduling Techniques in Wireless Mesh Networks. PhD Thesis. University of Guelph, Canada (2009).

$$PDR = \frac{\sum_{j=1}^{n} PR_j}{\sum_{i=1}^{m} PS_i}, \tag{5.12}$$

where PR_j is the number of packets received at the destination MG, PS_i is the number of packets sent from source MR, m is the number of MGs, and n is the number of MRs. The second metric used in the simulation is delay. It measures the time taken by a packet to reach its destination. These metrics can help to gauge the performance of the protocol effectively. In the case of multiple MGs, the average values of these metrics are used.

The simulation results compare both fair scheduling against no scheduling and fair scheduling with multiple MGs against fair scheduling with a single MG.

Figure 5.11 shows the PDR as a function of the number of MRs for the case with a single MG and five MGs for both fair scheduling and no scheduling. As the network size increases, the difference between the techniques becomes more pronounced. The cases with multiple MGs have the greatest PDR. The results show that a single MG with no scheduling performs very poorly, delivering only 30% of the packets successfully to the Internet.

It is interesting to note that using multiple MGs without fair scheduling can actually perform better than fair scheduling with a single MG (Figure 5.12). This is because despite the use of fair

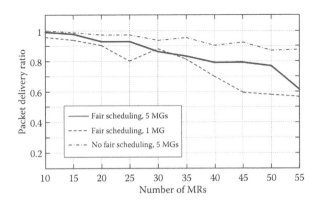

FIGURE 5.11 Packet delivery ratio versus number of MRs. (From J.B. Ernst. Scheduling Techniques in Wireless Mesh Networks. PhD Thesis. University of Guelph, Canada, 2009.)

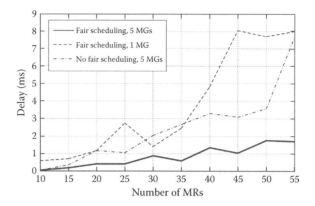

FIGURE 5.12 Delay versus number of MRs. (From J.B. Ernst. Scheduling Techniques in Wireless Mesh Networks. PhD Thesis. University of Guelph, Canada, 2009.)

scheduling, the single MG remains the major bottleneck in the network. This demonstrates how important it is to consider the architecture with multiple MGs.

5.7 VIDEO STREAMING IN WMNs

Research and development of WMNs was motivated by many applications that clearly demonstrated a promising market, such as broadband networking, telemedicine, transportation systems, military communications, security surveillance systems, and so on. With technological advances in video compression and networking, multisession video transmission over WMNs gains increasing research interest and enables a variety of multimedia services [61]. Video streaming is still a costly service because of the high-bandwidth requirements and the huge number of potential users. Moreover, video streaming over WMNs is a challenging issue because compressed video is very sensitive to transmission errors (packet loss), video transmission has stringent delay requirements, and video sessions compete with one another for limited network resources to maximize their QoS [62]. For overcoming these constraints, multiple disciplines, such as signal processing, wireless networking, and optimizations, must be involved.

The ultimate purpose of video compression and network transmission is providing the best presentation quality to the end users. E2E distortion is a commonly used metric for video presentation quality. The E2E distortion consists of two parts [63]: source coding and transmission distortion. The source coding distortion is caused by quantization errors during lossy video compression and is often measured by the mean squared error

(MSE) between the original source and the encoder reconstruction. On the other hand, the transmission distortion is caused by transmission errors such as packet loss, and it is measured by the MSE between encoder reconstruction and decoded material at the receiver [64]. The dominant type of transmission errors in video streaming over WMNs is packet loss due to network congestion and delay bounds violation, whereas bit errors are often very small especially within wired mesh environment [65]. In such an environment, video packets need to be transmitted over each link. At each link, the packets will be temporarily buffered, waiting to be scheduled for transmission. With multihop transmission, a packet may arrive at the destination after the scheduled playback time. In this case, the packet has violated its delay bound, is considered useless, and is thus dropped by the decoder.

5.7.1 MESH-BASED PEER-TO-PEER STREAMING SYSTEMS

Mesh-based systems, also known as swarm-based or data-driven P2P streaming systems, are widely deployed over the Internet. These systems have been proposed to distribute multimedia contents at low infrastructure costs [66]. In these systems, peers must share their resources to assist the server in delivering the video. The key idea is the same as for a P2P file-sharing system; file-sharing systems target elastic data transfers, whereas streaming systems focus on the efficient delivery of multimedia content under strict bandwidth and timing requirements. P2P streaming systems can be built in two ways, that is, tree-based systems in which one or more trees are constructed to connect peers for transferring contents and mesh-based systems in which peers connects to a few neighboring peers without an explicit topology before exchanging data with one another [67].

In mesh-based systems, a video sequence is divided into small segments that are transmitted from multiple senders to a receiver. The receiver coordinates the segment transmission from its senders. A receiver runs a scheduling algorithm to compose a transmission schedule for its senders, which specifies for each sender the assigned segments and their transmission times. When resources (and especially bandwidth) are enough to stream the videos in a P2P streaming system, almost all existing scheduling algorithms perform equally well. For higher video quality streams over the Internet, resources are never sufficient. Thus, to fully use the limited resources, more intelligent scheduling algorithms are needed.

As P2P streaming systems impose time constraints on segment transmission, composing segment transmission schedules is not easy. For example, segments arriving at the receiver after their decoding deadlines are not useful because they cannot be rendered for improving video quality.

Segment scheduling algorithms should strive to maximize the perceived video quality delivered by the on-time segments.

In P2P streaming systems, each swarm contains a subset of peers, and a peer may participate in multiple swarms. Data availability on peers is propagated by exchanging control messages, such as buffer maps that indicate video segment peers that are currently in their buffers. Using these maps, peers pull video segments from one another. More specifically, a receiver simultaneously requests segments from different senders. This is done by forming a segment transmission schedule by which the receiver specifies for each sender which segments and when to transmit.

5.7.2 PRINCIPLE OF VIDEO STREAMING OVER WMN

The principle of video streaming over WMN is illustrated in Figure 5.13. The kth precompressed video with variable source bit rate $R_k(t)$, $1 \leq k \leq K$, needs to be transmitted over an L-hop path to its destination. At each link, the video stream competes with other video sessions.

Here, in fact $R_k(t)$ represents the service rate of the kth session at time t, where

$$R_k^{\min} \leq R_k(t) \leq R_k^{\max}, \tag{5.13}$$

and R_k is the average source bit rate. The average available service rate for video stream is C_j ($1 \leq j \leq J$). In general, C_j is the capacity of link j available to the elastic sessions. The video session has a required delay bound of d_k. It can be seen that the packet delay bound violation probability p_k depends on input source bit rate $R_k(t)$, the network condition C_j of all links over the transmission path, and the characteristics of background traffic.

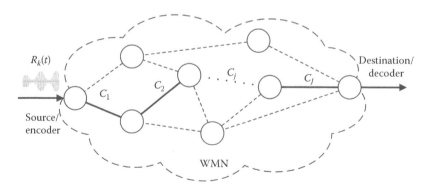

FIGURE 5.13   Video streaming over WMN.

As an extension of single-hop models to multihop models, a simple approximation is to assume that the loss process is independent on neighboring links (Kleinrock independence approximation) [68]. Here, the E2E packet loss probability for the kth session is given by

$$p_k = 1 - \prod_{j=0}^{H}(1 - p_{k,j}), \tag{5.14}$$

where $p_{k,j}$ represents the packet loss probability incurred to delay bound notation at a specific hop j, given the packet was not lost in previous hops.

In an article by Shaing and van der Schaar [69], the packet loss probability at the intermediate node m denoted by $p_{k,m}$ is computed as follows:

$$p_{k,m} = \Pr(W_{k,m} + E[W_{k,m}]) > d_k, \tag{5.15}$$

where $E[W_{k,m}]$ is the expected queuing delay of the packets at the source queue. In this case, the E2E packet loss probability p_k for stream k can be computed as follows:

$$p_k = 1 - (1 - p_{k,0})\left(1 - \sum_m p_{k,m}\right). \tag{5.16}$$

This is the two-parameter exponential model.

The next approach for modeling the multihop packet delay bound violation rate is related to a one-parameter exponential model, that is,

$$\Pr\{\text{Delay} > T\} \le e^{-(C' - R)T/L'}, \tag{5.17}$$

where T is the delay bound for the video stream and R is the total transmission rate. The parameter C' is related to the maximum video rate supported by the set of paths provided by the routing algorithm. It depends on the link capacity and the rate of background traffic over selected routes, whereas L' depends on the average packet size. These two parameters are empirically estimated from E2E delay statistics over the network [70].

5.7.3 Multimedia-Centric Routing for Multiple-Description Video Coding

The multiple-description (MD) coding technique is highly suitable for video communication in multihop wireless networks [71]. According to this coding technique, MDs are generated for a video source, each giving a low but acceptable video quality. Unlike traditional layered video coding, the descriptions are equally important in video decoding. At the destination, the video can be reconstructed from any subset of the received descriptions, with video quality commensurate with the number of received descriptions. Because of this, MD video coding represents a perspective framework for WMNs. Considering the routing problems, they can be addressed as follows [72]:

- How to characterize and model the multimedia-centric multipath routing problem
- The performance limits on video quality for a given mesh network
- How to design an efficient solution procedure for the formulated cross-layer routing problem
- How to implement the proposed routing solution in a distributed manner

Consider a WMN with N nodes (MR or MC) with existing connectivity between two nodes if they are within each other's transmission range. This network can be modeled as a directed graph $G(V, E)$, where V is the set of vertices representing N nodes and E is the set of links. Video session is considered for source node s to destination node t. The E2E path model can be derived from a concatenation of the link models along the path. The packet will be successfully delivered if and only if it survives the loss process at each link along the path. Available bandwidth can be measured by each node, whereas E2E path bandwidth represents the minimum of the available bandwidths of its constructive links. The quality of video with M descriptions is measured by the distortion of the received video, that is, the difference between a reconstructed video frame and the corresponding original video frame. The distortion metric D depends on the video codec (information loss due to quantization) and dynamics of all M paths (information loss due to transmission errors). For the double description video case (i.e., $M = 2$), the average video distortion can be characterized by the following expression [73]:

$$D = P_{00}d_{00} + P_{01}d_{01} + P_{10}d_{10} + P_{11}\sigma^2, \tag{5.18}$$

where P_{00} is the probability that both descriptions are received, P_{01} is the probability that description 1 is received and description 2 is lost, P_{10} is the probability that description 1 is lost and description 2 is received, and P_{11} is the probability that both descriptions are lost. Also, d_{00} is the distortion when both descriptions are received, d_{01} is the distortion when only description 1 is received, d_{10} is the distortion when only description 2 is received, and σ^2 is the distortion when both descriptions are lost (i.e., the variance of the source). The conditional distortions d_{00}, d_{01}, and d_{10} can be approximated using the Ozarow bound [74].

This is a cross-layer optimization problem that explicitly optimizes application performance via network layer operations as follows: minimizing D, subject to flow constraints and link stability constraints. The choice of paths as well as the dynamics on each path will determine the distortion value. The choice of a pair of paths can be expressed using a set of binary index variables for each link. When the index variable is 1, the corresponding link is used in a path; otherwise, it is not chosen. Once the set of paths are given, path performance metrics can be derived from the link metrics, which finally yield the distortion value D. Solving the previously mentioned problem will provide the values for the index variables (pair of paths) as well as the bit rate for each description. The multimedia-centric routing problem has a highly complex relation pertaining to the contribution of any link to the objective function, which depends, in general, on the other links that are included in the process.

5.8 CONCLUDING REMARKS

Wireless mesh networking is receiving a great deal of attention because it offers a promising solution to the challenges presented by the next generation network (NGN) environment. A WMN can assume different types of network architectures, and the type of architecture can affect wireless characteristics differently. Compared with mobile ad hoc networks, mesh networks have some appealing properties. First, the MRs can form the backbone of links, which are relatively more stable than those in mobile ad hoc networks. Second, MRs may be more powerful than classic mobile devices in terms of computational power and steady power supply. They can also be equipped with multiple interfaces for performance enhancement. Therefore, sophisticated algorithms can be used, whereas computational complexity and energy consumption are secondary issues.

There has been an impressive amount of research effort concentrating on the design of WMNs in the last few years. Both the research and commercial vendors' communities are attracted to the multihop paradigm because of its simplicity, robustness, ease of setup/maintenance, and self-organizing properties. Factors such as support for heterogeneity, opportunity for using affordable hardware, available community-driven infrastructure, and increasing open-source software development have resulted in a tremendous increase in the research and development of WMNs. From a provided survey, it seems clear that researchers recognize the importance of addressing theoretical issues in mesh design under realistic conditions of commodity hardware, protocols, and joint design.

However, the design of WMNs presents several open issues, starting from routing metrics and protocols to video streaming framework. Routing is an important process that has attracted researchers to enhance the performance of wireless networking solutions. Routing in WMN is much easier if every mesh node is in the range of at least one gateway, and thus only the last hop involves a human-operated device. Multihop WMN can be an extremely cost-effective means to provide wireless Internet access in cities. On the other hand, multihop packet loss modeling plays a critical role in video streaming. It enables the prediction of joint impact input video characteristics and network conditions on packet loss probability due to delay bound violation and the E2E video quality. One possible solution is a cross-layer design to improve routing efficiency. This is accomplished by better reflecting physical (PHY) layer variations onto routing metrics or by better using the available radio spectrum to directly improve the network throughput.

6 Wireless Multimedia Sensor Networks

The availability of low-cost micro-electro-mechanical systems technology enables the development of wireless multimedia sensor networks (WMSNs), that is, networks of resource-constrained wireless devices that can retrieve content such as video and audio streams as well as scalar sensor data. The design of these networks depends significantly on the applications. Features such as environmental monitoring, target tracking, as well as the application's design objectives, cost, and system constraints must be considered. In these types of applications, network nodes should ideally maximize the perceived quality of service and minimize energy expenditure. In recent years, extensive research has opened challenging issues for WMSNs deployment. Among numerous challenges faced while designing architectures and protocols, maintaining connectivity and maximizing the network lifetime stand out as critical considerations. As a fundamental objective, maintenance is a dependable operation of a network. However, the resource-constrained nature of sensor nodes and ad hoc network topology, often coupled with an unattended deployment, pose nonconventional challenges and motivate the need for special techniques with dependable design and management of wireless sensor networks (WSNs). Data gathered from sensor nodes in physical proximity tend to exhibit strong correlation. To minimize such redundancy and hence reduce the network load with a goal toward conserving energy, effective network schemes have been extensively proposed. To this end, routing techniques supporting data fusion are extremely important as they dictate where and when sensory data streams should interact with each other. Advances in WMSNs have led to many specifically designed new protocols in which a long-lived network is an essential consideration.

6.1 INTRODUCTION

WSNs feature a combination of many functionalities such as sensing, computation, and communication, which are potentially exercised over large number of battery-powered nodes, scattered densely in a geographical area of interest. They sense and gather data from the surrounding environment and transmit it to logically more potent nodes, called sinks, to perform more intricate processing. Sensor-based applications span a wide range of areas, including scientific research, military, disaster relief and rescue, health care, industrial, environmental, and household monitoring [1]. Recently, the availability of low-cost hardware components that are able to ubiquitously capture multimedia content from the environment has fostered the development of WMSNs [2], that is, networks of wirelessly interconnected devices that allow the retrieval of video and audio streams, still images, as well as scalar sensor data.

The phenomenon being monitored by a WSN is often time varying, so it is important to timestamp events accurately. Another motivation for having accurate knowledge of time is that many wireless distributed protocols require some sort of coordination, whereas clock synchronization is essential to achieve coordination among distributed entities [3]. Generally, sensor networks are cyber-physical systems, in which the time notation is a key issue for applications involving distributed control and tracking. In some applications, such as monitoring and localization, the accuracy of the time stamps is critical to the accuracy of the interference. Clock synchronization is also important to operate the network in an energy-efficient manner by synchronizing the active and idle states or scheduling other events.

Because the sensor nodes are expected to operate autonomously with small batteries for a number of months, energy efficiency is a fundamental criterion in WSN protocol design. The major power-consuming component of a sensor node is the radio interface, which is controlled by the medium access control (MAC) protocol. An efficient MAC protocol increases the WSN lifetime to a great extent [4]. An efficient MAC protocol can reduce collision and increase the achievable throughput, providing flexibility for various applications. In the early stages, efficient data delivery was not the first priority. However, to support multitasking and efficient delivery of bursty traffic, new protocols are being developed.

Because individual nodes have limited range and form ad hoc topology over a shared medium, the design and implementation of routing algorithms that are able to effectively and efficiently support information exchange and processing is a complex task in WSN. A number of issues and practical limitations must be taken into account. First of all, to maximize the network's lifetime, the mechanisms adopted for routing and information forwarding need to be energy efficient. Second, because the nodes usually operate in an unattended fashion, the network is expected to exhibit autonomic properties, that is, the implemented protocol must be self-organized and robust to failures and losses [5]. Last but not the least, the routing protocol must be able to handle large and dense networks, as well as the associated challenges resulting from interference and from the need to discover, maintain, and use potentially long multihop paths.

The requirements of routing protocols for WSNs are very similar to those of routing protocols for mobile ad hoc networks (MANETs). However, compared with MANETs, the restrictions on energy efficiency are more compelling, nodes are usually static, and the networks are assumed to be larger. Recently, the design of efficient routing protocols for duty cycled WSNs has attracted much attention [6].

To eliminate the limitations on system lifetime of the WSNs, wireless passive sensor networks (WPSNs) are introduced as a completely new sensor networking paradigm [7]. A WPSN is nondisposable, more functional and cost efficient, runs as power is delivered in, and remains idle but ready to operate when no power is incident on the network.

Today, the focus is shifting toward research aimed at revisiting the sensor networking paradigm to enable the delivery of multimedia content. The emergence of WMSNs has made it possible to realize multimedia delivery on tiny sensing devices. WMSNs enable the retrieval of multimedia streams and store, process in real-time, correlate, and fuse multimedia content captured by heterogeneous sources [8]. The volume and characteristics of multimedia traffic are quite different from the traffic generated in WSNs. That has raised the need to explore communications protocols for multimedia delivery in WMSNs. Due to the great success of WSNs in solving real-world problems (environmental monitoring, health monitoring, security surveillance, industrial process automation, etc.), WMSNs are rapidly gaining the interest of researchers. Recently, the research in WMSNs gained high momentum due to the introduction of new vision, known as the Internet of things (IoT), which is evolving very rapidly considering the tremendous progress in the field of embedded systems. This has given a new dimension to WSNs, and its derivatives (WMSNs) to explore the possibility of getting recognition of their segregated efforts by the rest of the networking community. Also, the implementation of IPv6 gives an opportunity to integrate WMSNs with the Internet, enabling remote surveillance and monitoring. Multimedia sensors have severe resource constraints on bandwidth, power supply, computation, etc. Therefore, complex encoding techniques cannot be feasibly applied. The communication paradigm in WMSNs is also many-to-one, in which a sink is the ultimate destination of many sources, and is capable of performing complex decoding operations. Moreover, data produced by sensors exhibit highly spatial correlation due to the high density of node deployment. For example, transporting video with the guaranteed QoS in WMSNs is of prime importance due to the higher rate requirements on limited and variable channel's capacity. A layered approach may achieve high performance in terms of the metrics related to each of these individual layers. However, they are not jointly optimized to maximize the overall network performance. Taking into account the resources in WMSNs, joint optimization of the networking layer, that is, cross-layer design [9], is the most promising alternative to the inefficient traditional layered protocol architectures. A number of cross-layer protocols are proposed. The main focus is on providing QoS to each individual stream according to the current network condition.

This chapter is organized as follows. First, a survey on WMSNs including the different types of nodes, architecture, and factors important for designing such networks is provided. The internal architecture of multi-media sensor devices is also included. Different perspective solutions for all layers of the communication protocol stack, together with some cross-layer approaches, are presented. Next, the convergence of mobile and sensor systems is analyzed and a brief overview of WMNSs applications is provided. Finally, research issues concerning problems related to the WSNs automated maintenance conclude the chapter.

6.2 WMSN ARCHITECTURE

A WMSN is a distributed system that interacts with the physical environment by observing it through multiple media [8]. Furthermore, it can perform online processing of the received information and react to it by combining technologies from communications and networking, signal processing, computer vision, control, and robotics.

To enable wireless sensor applications using sensor technologies, the range of tasks can be broadly classified into three groups [10]:

1. *System.* Each sensor node is an individual system. To support different application software on a sensor system, development of new platforms, operating systems, and storage schemes is needed
2. *Communication protocols.* Enable communication between the sensor nodes
3. *Services.* Developed to enhance the application and to improve system performance and network efficiency

From the application requirement and network management perspectives, it is important that sensor nodes are capable of self-organizing themselves into a network. Subsequently, they are able to control and manage themselves efficiently.

Implementation of protocols at different layers can significantly affect energy consumption, E2E delay, and system efficiency. It is important to optimize communication and minimize energy usage. Traditional networking protocols are not suitable for WSN environments because they are not designed to meet these requirements. The new energy-efficient protocols employ cross-layer optimization by supporting interactions across the protocol layers. Protocol state information at a particular layer is shared across all the layers to meet specific requirements.

WMSN lifetime depends on the number of active nodes and connectivity of the network. Thus, energy must be used efficiently to maximize the network lifetime. Energy conservation maximizes network lifetime and is addressed through efficient reliable wireless communication; through intelligent sensor placement to achieve adequate coverage, security, and efficient storage management; and through data aggregation and compression. For reliable communication, services such as congestion control, buffer monitoring, acknowledgements, and packet-loss recovery are necessary. Communication strength is dependent on the location of the sensor nodes. Sparse sensor placement may result in long-range transmission and higher energy usage whereas dense sensor placement may result in short-range transmission and lower energy consumption. Coverage is interrelated to sensor placement. The total number of sensors in the network and their placement determine the degree of network coverage [10]. Depending on the application, a higher degree of coverage may be required to increase the accuracy of the sensed data [11].

Challenges in WMSNs include high bandwidth demand, high energy consumption, QoS provisioning, data processing and compression techniques, as well as cross-layer design. Multimedia content requires high bandwidth in order to be delivered. As a result, high data rate leads to high energy consumption. QoS provisioning is a challenging task due to the variable delay and variable channel capacity. For reliable content delivery, a certain QoS level must be achieved. In network processing, filtering and compression can significantly improve network performance in terms of filtering and extracting redundant information and merging content. Similarly, cross-layer interaction among the layers can improve the processing and transmission.

Most proposals for scalar WSNs are based on a flat, homogenous architecture in which every sensor has the same physical capabilities and can only interact with neighboring sensors. Flat topologies may not always be suited to handle traffic intensity generated by multimedia applications including audio and video. Likewise, the processing power required for data processing and communications, and the power required to operate it, may not be available on each node. With the emergence of WMSNs, new types of sensor nodes (multimedia sensors, multimedia processing hubs, and storage hubs) with different capabilities and functionalities have been used. This raises the need to reconfigure the network into different architectures in such a way that the network can be more scalable and more efficient depending on its specific application and QoS requirements. Therefore, based on the designed network topology, the available resources in the network can be efficiently utilized and fairly distributed, and the desired operations of the multimedia content can be handled.

In general, concerning their capabilities and functionalities, various types of nodes exist in WMSN [8,12]:

- Video and audio sensors (VAS). These types of sensors capture sound and still or moving images of the sensed events. They can be arranged in a single-tier topology, or in a hierarchical manner.
- Scalar sensors (SS). These nodes sense scalar data and physical attributes like temperature, pressure, and humidity and report the measured values. They are typically resource-constrained devices in terms of energy supply, storage capacity, and processing capability.
- Multimedia processing hub (MPH). These devices have comparatively large computational resources and are suitable for aggregating multimedia streams from individual sensor nodes. They are integral in reducing both the dimensionality and the volume of data conveyed to the sink and storage devices.
- Storage hub (SH). Depending on the application, the multimedia stream may be desired in real-time or after further processing. SHs allow data-mining and feature extraction algorithms to identify the important characteristics of the event, even before the data is sent to the end user.
- Sink node (SN). This node is responsible for packaging high-level user queries to network-specific directives and return filtered portions of the multimedia stream back to the user. Multiple sinks may be needed in a large or heterogeneous networks.
- Gateway (GW). This element serves as the last mile connectivity by bridging the sink to the IP networks and is also the only IP-addressable component of the WMSN. It maintains a geographical estimate of the area covered under its sensing framework to allocate tasks to the appropriate sinks that forward sensed data through it.

Network architectures in WMSNs can be divided into three reference models [8,13]: single-tier flat architecture, single-tier clustered architecture, and multitier architecture with heterogeneous sensors.

In *single-tier flat architecture*, the WMSN is deployed with homogeneous sensor nodes of the same sensing, computational and communication capabilities, and functionalities (Figure 6.1). In this model, all the nodes can perform any function from image capturing through multimedia processing to data relaying toward the SN in multihop topology. Moreover, multimedia processing is distributed among the nodes, which prolongs network lifetime.

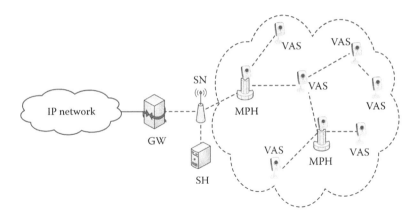

FIGURE 6.1 Single-tier flat WMSN architecture.

Single-tier clustered architecture is deployed with heterogeneous sensors in which multimedia and scalar sensors within each cluster relay data to a MPH (cluster head). The MPH has more resources and it is able to perform intensive data processing. It is wirelessly connected with the SN or the GW either directly or through other cluster heads in multihop fashion (Figure 6.2).

The third model, presented in Figure 6.3, is the *multitier architecture with heterogeneous sensors.* In this architecture, the first tier deployed with SSs performs simple tasks (e.g., motion detection and temperature monitoring), whereas the second tier of VASs can perform more complicated tasks such as object detection or object recognition. At the third tier, more powerful and high-resolution VASs are capable of performing more complex tasks, such as object tracking. Each tier can have a central hub to perform more data processing and communicate with the higher tiers. This

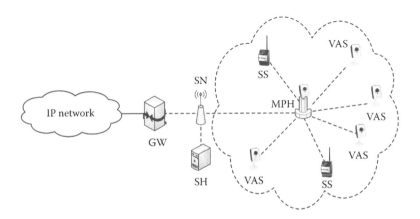

FIGURE 6.2 Single-tier clustered WMSN architecture.

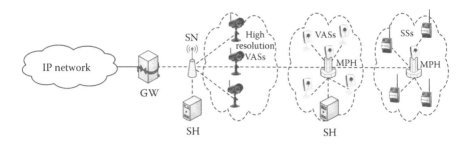

FIGURE 6.3   Multitier WMSN architecture.

TABLE 6.1
Comparative Characteristics of WMSN Architectures

Architecture	Single-Tier Flat	Single-Tier Clustered	Multitier
Types of sensors	Homogeneous	Heterogeneous	Heterogeneous
Processing	Distributed	Centralized	Distributed
Storage	Centralized	Centralized	Distributed

architecture can accomplish tasks with different needs with better balance among costs, coverage, functionality, and reliability requirements [14]. On the other hand, the use of just one node type in a homogeneous flat network is not scalable enough to enclose all complexities and dynamic range of applications offered over WMSNs. The main comparative characteristics of three reference architectures are presented in Table 6.1.

Example 6.1

A multitier, multimodal WMSN for environmental monitoring, using different sensing modalities and capabilities, is presented by Lopes et al. [15]. Multitier relates to the way the network elements are organized into hierarchical levels, according to their function-alities and capabilities. By multimodal, it is meant that different sensing modalities are employed to achieve a common goal.

Figure 6.4 shows a schematic representation of the proposed architecture. The network is self-organized into three tiers of dis-tinct sensor nodes, in which the first tier is equipped with pas-sive infrared (PIR) sensors that are capable of detecting objects in the sensing field as a first task in an environmental monitoring or tracking application. The second tier is equipped with smart video sensors that can identify objects in their field of view (FoV)

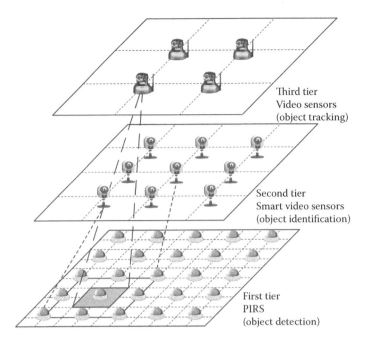

Third tier
Video sensors
(object tracking)

Second tier
Smart video sensors
(object identification)

First tier
PIRS
(object detection)

FIGURE 6.4   An example of a multitier, multimodal architecture.

by determining the object's predominant color. The third tier is equipped with video sensors that are capable of computing the location of the identified objects and their moving path, and is responsible for target tracking as they move through the network after being identified by the second tier sensors.

During network operation, the PIR sensors in the first tier are always active, as they are low-power sensors, and monitoring the network. In case they detect any object, they alert the camera sensors in the second tier and wake them up to start capturing images for the detected objects and try to identify them. In case objects are successfully identified according to their predominant color, the camera sensors in the second tier will wake up the powerful visual sensors in the third tier to track them.

A comparative simulation was done with the presented architecture and the multitier and single-tier WMSN approaches [15]. Experimental results have shown that the proposed architecture is, respectively, at least 2.2 and 11 times more efficient in terms of energy consumption than the classic multitier and single-tier WMSN approaches. The architecture faced target, for the first time, has seen a delay of 8 s greater than the best performance, even requiring the activation of three sensor tiers.

6.3 INTERNAL ARCHITECTURE OF A MULTIMEDIA SENSOR DEVICE

A multimedia sensor device is composed of several basic components [12]: a sensing unit, a central processing unit (CPU), a communication subsystem, a coordination subsystem, memory, and an optional mobility/actuation unit. The internal architecture of a multimedia sensor is shown in Figure 6.5.

Sensing unit is composed of two subunits: video, audio, and/or scalar sensors and analog to digital converter (ADC). The analog signals produced by the sensors based on the observed phenomenon are converted into digital signals by the ADC, and then fed into the CPU. The CPU executes the system software in charge of coordinating sensing and communication tasks and is interfaced with a memory. A communication subsystem interfaces the device to the network and is composed of a transceiver unit and of communication software. The latter includes communication protocol stack and system software such as middleware, operating systems, and virtual machines. A coordination subsystem is in charge of the operation of different network devices by performing operations such as network synchronization and location management. An optional mobility/actuation unit can enable the movement or manipulation of objects. The entire system is powered by a power supply unit (PSU), for example, batteries or solar cells.

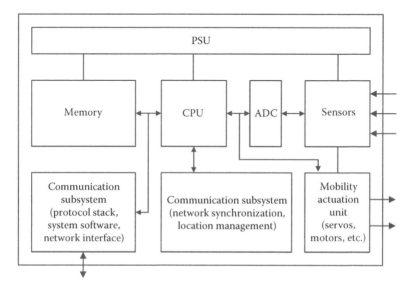

FIGURE 6.5 Internal architecture of a multimedia sensor.

6.4   PROTOCOL STACK FOR WMSN

The development of reliable and energy-efficient protocol stack is important for supporting various WSN and particular WMSN applications. Depending on the application, a network may consist of hundreds and thousands of nodes. Each sensor node uses the protocol stack to communicate with one another and with the SN. Hence, the protocol stack must be energy-efficient in terms of communication and must be able to work efficiently across multiple sensor nodes. The research challenges at different layers of the communication protocol stack are outlined in Table 6.2.

6.4.1   PHYSICAL LAYER FOR THE WMSN

The physical layer for the WMSN is related to the hardware transmission technologies defining the means of transmitting raw bits, rather than logical data packets, over the wireless links. It is also responsible for frequency selection, modulation, and channel encoding. The physical layer technology must work in a compatible way with higher layers in the protocol stack to support their application-specific requirements and to meet the design challenges of WMSNs [14]. This can be achieved with higher efficiency if a cross-layer approach is used between physical and MAC layers. The physical layer should utilize the available bandwidth and data rate in the best possible way, and should be more power efficient. The physical interface should have a good performance (gain) against noise and interference and provide enough flexibility for both different channel and multiple path selections.

Considering the WMSN projects from the open literature, physical layer technologies are mainly related to the families of IEEE 802.15 and IEEE

TABLE 6.2
Research Challenges Concerning Different Layers of the Protocol Stack

Layer	Challenging Issues
Application	High compression efficiency, low complexity encoder, error-resilient coding
Transport	Variations and modifications of traditional protocols or application of specific protocols related to congestion control and reliability
Network	Energy-efficient routing techniques with QoS assurances
Link	QoS support through channel access policies, scheduling and buffer management, and error correction techniques
Physical	Efficient bandwidth and energy utilization. Interference resistant and spectrum agile

802.11 standards (see Chapter 1). IEEE 802.15.4 (ZigBee) is the most common radio standard used in WSNs [16], because of its simplicity and its low-cost and low-power characteristics. However, the ZigBee standard is not suitable for high data rate applications such as multimedia streaming over WMSN and for guaranteeing application-specific QoS. On the other hand, other standards such as Bluetooth and Wi-Fi have higher data rates and code efficiency, but they consume more energy. Bluetooth transceivers have been used in earlier stages of WMSN test bed deployment [17], whereas Wi-Fi transceivers have been used in many WMSN practical applications and test beds projects [18].

With high coding efficiency (97.94%) and a data rate of up to 250 Mb/s, it was expected that ultra-wideband (UWB) would be a promising candidate for the physical layer standard of WMSNs. The physical layer of UWB is implemented by using either impulse radio (IR) of very short duration pulses (on the order of 10^{-10} s), or multiband orthogonal frequency division multiplexing (MB-OFDM) in which hybrid frequency hopping and OFDM are applied. IR-UWB has a simpler transmitter and rich resolvable multipath components but needs long channel acquisition times and requires high-speed ADCs. On the other hand, MB-OFDM-UWB offers robustness to narrowband interference, spectral flexibility, and efficiency but needs a slightly complex transmitter. Multiple access in IR-UWB can be realized by using direct sequence UWB (DS-UWB) or time-hopping UWB (TH-UWB). The low duty cycle of IR-UWB (<1%), caused by the short duration of the pulse, is a key advantage for low power consumption in WMSN. Also, spreading information over wide bandwidths decreases the power spectral density and in turn reduces interference with other systems and decreases the probability of interception.

Although UWB seems to be a promising physical layer technology and has many attractive features, many challenges must be solved to enable multihop networks of UWB devices. To further increase capacity and mitigate the impairment by fading and cochannel interference, multiantenna systems such as antenna diversity as well as smart antenna and MIMO systems can be combined with UWB. Further research has been carried out designing a cross-layer communication architecture based on UWB to support QoS in WMSNs as well as for guaranteeing provable latency and throughput bounds to multimedia flows in UWB [19].

6.4.2 Link Layer Quality of Service Support

In case of WMSNs that deliver various types of traffic, QoS support mechanisms are utilized to prioritize and manage the resource sharing according to the requirements of each traffic class. Because the link layer rules the sharing of the medium and all other upper-layer protocols are bound to that layer, it has the ability to severely affect the overall performance of the WMSNs. Therefore, this layer becomes a proper choice to implement QoS support [20]. Research efforts to provide link layer QoS can be classified mainly into channel access policies, scheduling and buffer management, and error correction.

Based on the nature of *channel access*, some MAC protocols are geared toward providing high link-level throughput, reduce delays, or guarantee QoS for a given packet type. They can be divided into two main categories [12]:

- *Contention-based protocols*—These protocols are also known as CSMA/CA-based MAC protocols and they are usually used in multihop wireless networking due to their simplicity and their adequacy for implementation in a decentralized environment like WSN. When a device is receiving data, transmissions from all the devices in its transmission range are impeded. However, this is achieved by the use of random timers and a carrier sense mechanism, which in turn, results in uncontrolled delay and idle energy consumption. To decrease the number of these collisions and to considerably reduce other sources of energy wastage, the wake-up/sleep mechanisms and the control messages RTS/CTS/ACK defined for IEEE 802.11 standards are used to design energy-efficient MAC protocols for WSN such as sensor MAC (S-MAC) [21] and time-out MAC (T-MAC) [22]. These protocols are not suitable for multimedia applications because they are designed to be energy efficient at the cost of increased latency and degraded network throughput.

- *Contention-free protocols*—Clustered on-demand multichannel MAC (COM-MAC) [23] is a representative contention-free MAC protocol. It exploits the fact that current sensor nodes already support multiple channels to effectively utilize the available channel capacity through cooperative work from the other sensor nodes. In this way, a better support for high data rates demanded by multimedia applications can be achieved. The operation of proposed protocol consists of three sessions: request session, scheduling session, and data transmission session. For COM-MAC to achieve high energy efficiency, first, a scheduled multichannel medium access is used within each cluster so that cluster members can operate in a contention-free manner within both time and frequency domains to avoid collision, idle listening, and overhearing. To maximize the network throughput, a traffic-adaptive and QoS-aware scheduling algorithm is executed to dynamically allocate time slots and channels for sensor nodes based on the current data traffic information and QoS requirements. Finally, to enhance transmission reliability, a spectrum-aware automatic repeat-request (ARQ) is incorporated to better exploit the unused spectrum for a balance between reliability and retransmission. Performed simulation studies indicate that COM-MAC can achieve increased network throughput at the cost of a small control and energy overhead.

Scheduling and buffer management in WMSNs is an open research issue that has attracted the research community in recent years, but is still not really solved. One of the approaches related to the scheduling and buffer management problem [24] is based on the fact that different network applications need different QoS requirements such as packet delay, packet loss, bandwidth, and availability. This can also be done not by increasing network capacity, such as in the COM-MAC approach, but by developing a network architecture that is able to guarantee QoS requirements for high-priority traffic. It argues that the sensor networks should be willing to spend more resources in disseminating packets that carry more important information by using a differentiated service model for WMSNs (Figure 6.6).

The proposed model can support two major different types of traffic classes: real-time class (expedited forwarding) and non–real-time traffic (assured forwarding), which is divided into three classes (high priority, medium priority, and low priority). Real-time traffic is buffered in a separate queue with low buffer size, whereas non–real-time traffic is managed by using random early detection (RED) [25]. The delay performance of different traffic classes is evaluated by using priority queuing and weighted round-robin (WRR) scheduling mechanisms [26]. It is shown that by using priority queuing for real-time traffic and WRR scheduler for non–real-time traffic classes, low delay bound and guaranteed network bandwidth for high-priority real-time traffic can be provided. This work, as it simply considers a scheduling system, does not provide any insight into the physical and MAC layers of WMSNs. The main drawback of this solution, however, is that it demonstrates that the proposed system can provide differentiated services, but it does not study signaling overheads or energy consumption.

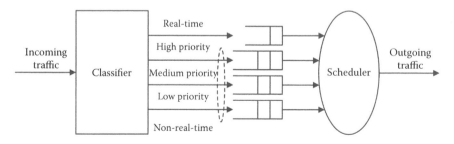

FIGURE 6.6   An example of scheduling and buffer management in WMSNs.

A similar approach is proposed by Yigitel et al. [20]. Here, a QoS-aware and a priority-based MAC protocol for WMSNs is introduced as Diff-MAC. It integrates different methods to meet the requirements of QoS provisioning to deliver heterogeneous traffic and provide a fair all-in-one QoS-aware MAC protocol. Diff-MAC aims to increase the utilization of the channel with effective service differentiation mechanisms while providing fair and fast delivery of the data. Performance evaluation results of Diff-MAC, obtained through extensive simulations, show significant improvements in terms of latency, data delivery, and energy efficiency compared with the nondifferentiated CSMA/CA approach.

To improve the perceptual quality of the received multimedia content, robust error correction and loss recovery techniques should be employed in WMSNs to overcome the unreliability of the wireless channel at the physical and link layers. Forward error correction (FEC) and ARQ are classic examples of these techniques. Applying different degrees of FEC to different parts of the video stream, depending on their relative importance (aka unequal protection) allows varying overhead on the transmitted packets. ARQ mechanisms, on the other hand, use bandwidth efficiently at the cost of additional latency involved with the retransmission process. Comparisons made between ARQ and FEC reveal that for certain FEC block codes, longer routes decrease both the energy consumption and the end-to-end latency, subject to a target packet error rate compared with ARQ [27]. Thus, FEC codes are an important candidate for delay-sensitive traffic in WMSNs. In recent research [28], a comprehensive performance evaluation of ARQ, FEC, hybrid FEC/ARQ, and cross-layer hybrid error control schemes for WMSNs are performed. Performance metrics such as energy efficiency, frame peak signal-to-noise ratio (PSNR), frame loss rate, cumulative jitter, and delay-constrained PSNR are investigated. The results of analysis show how wireless channel errors can affect the performance of WMSNs and how different error control scenarios can be effective for those networks. The results obtained also provide the required insights for efficient design of error control protocols in multimedia communications over WSNs.

Example 6.2

To capture bit-level errors in WMSN, a two-state discrete-time Markov chain called the Gilbert–Elliott channel model [29] can be applied. Figure 6.7 illustrates the state diagram for this channel model.

This model has memory, takes into account the correlation between consecutive errors, and abstracts bursty error distribution with a bad state (B) that represents a heavy error rate with a short interval and a good state (G) representing light error rate

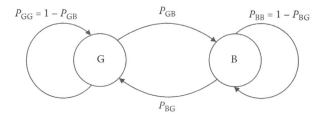

FIGURE 6.7   Gilbert–Elliott channel model.

with a longer interval. Each state has an associated bit error rate (BER) probability, that is, P_G for the good state and P_B for the bad state, and state transition probabilities could be derived from the experimental channel data. The stationary probabilities of the light and heavy error rate states can be obtained as

$$\pi_G = \frac{P_{BG}}{P_{BG} + P_{GB}} \text{ and } \pi_B = \frac{P_{GB}}{P_{BG} + P_{GB}}, \tag{6.1}$$

where P_{BG} is the probability of the channel state's transition from a bad state to a good state, and P_{GB} is the probability of transition from a good state to a bad state. According to the described model, every bit is erased with probability π_G at the light error rate and erased with probability π_B at the heavy error rate, independently of other bits. Furthermore, the average bit error probability of the WMSN channel can be expressed as

$$P = P_G \pi_G + P_B \pi_B. \tag{6.2}$$

Equation 6.2 shows how the probability of average BER depends on the transition probabilities between the light and heavy error states. It is noteworthy that even though the higher-order Markov chains can be used for characterizing the loss process in the wireless channel, this model provides good accuracy with less complexity and has been extensively used in the open literature to capture the erroneous nature of wireless channels from the bit-error process to the packet-loss process at different layers [28].

6.4.3 ENERGY-EFFICIENT ROUTING TECHNIQUES WITH QOS ASSURANCES

The network layer in WSNs handles the routing of sensed data from the sources to the SN, taking into account several design considerations such as energy efficiency, link quality, fault tolerance, and scalability. Although there are many routing protocols proposed for the traditional WSNs [6,30], the design of routing protocols for WMSN is still an active research issue. Specific characteristics and constraints due to multimedia content handling make the proposed routing protocols for WSNs not directly applicable for WMSNs.

According to the current research trends, routing techniques for WMSNs can be broadly classified into the following characteristic groups [1]:

1. *Latency-constrained routing protocols* based on the type of E2E delay guarantee (soft or hard real-time bounded latency). Sequential assignment routing (SAR) [31] is one of the first routing protocols considering QoS and energy efficiency for WSNs, which includes a table-driven multipath routing and path restoration technique to create trees routed from the one-hop neighbor of the sink. The objective of this routing algorithm is to minimize the average weighted QoS metric throughout the lifetime of the network. The multipath routing scheme ensures fault tolerance, whereas a path restoration technique eases the recovery in case of node failure. The limitation of the SAR approach is that it suffers from the overhead of maintaining the tables and status information for each sensor node when the number of nodes is huge.

   Energy-aware QoS (EAQoS) [32] is another previously proposed routing protocol that can run efficiently along with the best effort traffic. It aims to discover an optimal path in terms of energy consumption and error rate along which the E2E delay requirement can be satisfied. The proposed protocol finds the least cost-constrained and delay-constrained path for real-time data. The cost of the link is defined to capture communication parameters, such as the residual energy in the nodes, transmission energy, and error rate. Moreover, it maximizes the throughput for non–real-time data by adjusting the service rate for both real-time and non–real-time data coexisting at sensor nodes. EAQoS consistently performs well with respect to real-time and energy metrics, but the main weakness of this protocol is that it does not use any priority scheme to account for the different E2E delay requirements that multimedia traffic may have. Therefore, this protocol is not suitable for streaming applications. Also, the consideration of only propagation delay as well as average queuing delay in calculating E2E delay limits the ability of this protocol to satisfy the QoS requirements. In addition, the bandwidth ratio is initially set the same for all the nodes, which do not provide adaptive bandwidth sharing for different links. Moreover, the algorithm requires complete knowledge of the network topology at each node to calculate multiple paths, thus limiting the scalability of the approach.

Another QoS routing protocol that provides soft real-time E2E guarantees is SPEED [33]. This protocol requires that each node maintain localized information with minimal control overhead, and uses nondeterministic geographic forwarding to find paths. The main objective is to support a spatiotemporal communication service with a given maximum delivery speed across the network. SPEED performs better than classic dynamic source routing (DSR) and ad hoc on-demand distance vector (AODV) protocols in terms of E2E delays and miss ratios. Moreover, the total transmission energy is lower because of minimum control packet overhead, and a network-wide load balancing is achieved by the even distribution of traffic. On the other hand, energy consumption is not addressed directly in the protocol. In addition, SPEED does not adopt differentiated packet prioritization, and each forwarding node can only forward the packet at some speed less than or equal to the maximum achievable speed.

A power-efficient routing algorithm that ensures hard delay guarantee for WMSNs is proposed by Ergen and Varaiya [34]. It adopts a single sink model and aims at maximizing network lifetime by adjusting the number of traversing packets. To achieve this goal, first, the delay constraint is excluded and lifetime maximization is formulated as a linear programming problem, and solution is implemented in a distributed manner that uses an iterative algorithm to approximate the centralized optimal one. Then, delay guarantee is ensured into the energy efficient routing by limiting the length of paths from each node to the sink. This protocol can guarantee E2E delay requirement and prolong network lifetime.

2. A typical representative of *multiple QoS-constrained routing protocols* is multipath and multi-SPEED (MMSPEED) [35]. It supports probabilistic QoS guarantees by provisioning QoS in two domains, timeliness and reliability. These mechanisms for QoS provisioning are realized in a localized way without global network information by employing localized geographic packet-forwarding augmented by dynamic compensation, which compensates for local decision inaccuracies as a packet travels toward its destination. The main advantage of MMSPEED is that it guarantees E2E requirements in a localized way, which is desirable for scalability and adaptability to large-scale dynamic WMSNs. It can provide QoS differentiation in both domains and significantly improves the network capacity in terms of the number of flows that meet both reliability and timeliness requirements.

As an improvement, a timeliness and reliability QoS-aware, localized, multipath, and multichannel protocol is proposed by Hamid et al. [36]. Here, a routing decision is made according to the dynamic adjustment of the required bandwidth and path length–based proportional delay differentiation for real-time data. The proposed protocol works in a distributed manner to ensure bandwidth and E2E delay requirements of real-time traffic. At the same time, the throughput of non–real-time traffic is maximized by adjusting the service rate of real-time and non–real-time data. A differentiated priority-based classifier and scheduler that arranges packets according to delay and bandwidth requirement are the strengths of this approach. However, there are no alternative mechanisms to handle the delay whenever the buffer size increases and switching overhead affects the performance of the protocol.

Real-time routing protocol with load distribution (RTLD) [37] can provide efficient power consumption and high packet delivery ratio in WMSN. The main advantage of RTLD is that it can deliver packets within their E2E deadlines (packet lifetime) while minimizing the network loss ratio and power consumption. It combines the geocast forwarding with link quality, maximum velocity, and remaining power to achieve real-time routing in WMSN. The remaining power is used for mitigating the routing holes problem due to power expiration. The significant feature of RTLD is that it distributes the task of load forwarding to all forwarding candidates to avoid packet dropping caused by power termination and to prolong the network lifetime. A real-time routing in MWSN is an exciting area of research because messages in the network are delivered according to their packet lifetime in the case of mobile sensor nodes. An enhanced RTLD routing protocol for mobile MWSN is proposed by Ahmed [38].

3. *Routing for real-time streaming* in WMSNs is a challenging issue because video streaming data generally has a soft deadline with a tendency for shortest path routing approaches with the minimum E2E delay. Moreover, transmission requirements in terms of bandwidth can be several times higher than the maximum transmission capacity of sensor nodes, and thus needing multipath routing. Some protocols are proposed to explicitly handle real-time streaming by taking both E2E latency and bandwidth into consideration.

As an example, directional geographical routing (DGR) [39] protocol divides a single video stream into multiple substreams and exploits multiple disjoint paths to transmit them in parallel to make the best of limited bandwidth and energy in WMSNs and to achieve reliable delivery. Simulation results show that DGR exhibits low delay, substantially longer network lifetime, and a better received video quality. In particular, DGR improves the average video PSNR by up to 3 dB compared with a traditional geographic routing scheme. However, it tackles path failures with local repairs at the cost of additional overheads and transmission latency. In addition, DGR assumes that any node can send video packets to the sink at any instance, which limits the practicality of deploying it for high-density sensor environment.

4. Although most of the presented routing protocols are query-initiated, data delivery model–based routing is an alternative approach. Real-time and energy-aware routing (REAR) [40] is an event-driven protocol that uses metadata to establish multipath routing to reduce energy consumption. A cost function is constructed to evaluate the consumption of bandwidth on the transmission links, which trades off the relationship of energy and delay, and then QoS routing is chosen. The E2E delay of multihop routing not only depends on transmission distance but also relies on relay nodes processing and queuing delay. Because of the high bandwidth requirement of real-time multimedia data, a classified queue model is introduced at each node to deal with both types of data. This protocol saves energy by activating video sensors when monitoring events occur and using metadata instead of real data in the routing process. However, in case of streaming applications, the idea of metadata is not a good choice as the metadata for streaming data can itself be very huge and result in high energy and bandwidth consumption.

5. Recently, different adaptive classes of algorithms based on swarm intelligence have been considered to take *intelligent routing* decisions in WSNs. In general, this class of routing algorithm presents an interesting theoretical solution, but one that is difficult to implement in a real environment because of its complexity and long adaptation time. A comprehensive survey and comparison of routing protocols from classic solutions to swarm intelligence–based protocols are provided by Zungeru et al. [41].

Ant-based service-aware routing (ASAR) algorithm [42] chooses suitable paths to meet diverse QoS requirements from different kind of services and is mostly suited for WMSNs. ASAR periodically chooses three suitable paths to meet diverse QoS requirements from different kind of services (data query, stream query, and event driven) by positive feedback mechanism used in ant-based heuristics, thus maximizing utilization and improving network performance. It maintains an optimal path table and a pheromone path table at each cluster head. Routing selection for different data services is made based on delay, packet loss rate, bandwidth, and energy consumption required by the type of traffic. Besides the bottleneck problem of hierarchical models, new optimal path setup due to congestion requires extra calculation, which may decrease network performance by engaging extra energy for large networks.

AntSensNet [43] is an ant-based routing approach for WMSNs. This protocol uses an efficient multipath video packet scheduling to achieve minimum video distortion transmission. AntSensNet combines hierarchical structure with an ant colony optimization (ACO) heuristic to satisfy QoS requirements in WMSN. Besides its support for power-efficient multipath packet scheduling for minimum video distortion transmission, it is composed of both reactive and proactive components. It is reactive because routes are set up when needed, and proactive because, while a data session is in progress, paths are probed, maintained, and improved proactively using a set of special agents. The cluster network forms nodes into colonies, network route between clusters that meet the requirements of each application using ants, and forwarding of network traffic using the previously discovered route by the ants. In the clustering process, only the channel heads transmit information out of the cluster, which helps in preventing collisions between sensor nodes, hence promoting energy saving and latency.

6. *Hole-bypassing routing* can be described through a geographic energy-aware multipath stream (GEAMS)–based protocol [44]. GEAMS is a geographic, multipath-localized routing protocol designed to handle multimedia streaming data by maintaining both QoS restriction and energy efficiency. GEAMS routes information based on greedy perimeter stateless routing (GPSR) [45] while maintaining local knowledge for delivering this information on a multipath basis. In GPSR, the same nodes in close vicinity with the sinks are chosen repeatedly, which may cause early failure in most of the nodes. In contrast, using the GEAMS routing protocol, data streams are routed by different nodes and decisions are made locally at each hop. The decision policy at each node is based on the remaining energy at each neighbor, the number of hops made by the packet before it arrives at this node, the current distance between the node and its neighbors, as well as the history of the packets forwarded belonging to the same stream. GEAMS has two modes:

- Smart greedy forwarding, used when there is always a neighbor closer to the SN than the forwarder node, and
- Walking back, used to avoid and bypass holes

To meet the multimedia transmission constraints and to maximize the network lifetime, GEAMS exploits the network multipath capabilities to allow load balancing among nodes. GEAMS is more suitable for WMSNs compared with GPSR because it ensures uniform energy consumption and meets the delay and packet loss constraint.

Example 6.3

To carry out the optimal forwarding calculation for enhanced RTLD routing protocols [38], parameters such as packet velocity and received signal strength indicators (RSSI) such as link quality and remaining power for every one-hop neighbor needs to be determined. Then, the router management will forward a data packet to the one-hop neighbor that has an optimal forwarding. The optimal forwarding is computed according to

$$OF = \max\left[\lambda_1 \frac{RSSI_{max}}{RSSI} + \lambda_2 \frac{V_{batt}}{V_{mbatt}} + \lambda_3\left(1 - \frac{D}{D_{E2E}}\right)\right], \quad (6.3)$$

where $\lambda_1 + \lambda_2 + \lambda_3 = 1$.

Here, λ_1, λ_2, and λ_3 stand for weighting coefficients, which can be empirically estimated. $RSSI_{max}$ is the signal strength at reference point 1 m, which is equivalent to −45 dBm. V_{mbatt} is the maximum battery voltage for sensor nodes and is equal to 3.6 V. D is the average one-hop delay and D_{E2E} is E2E delay for real-time, which is commonly 250 ms.

The link quality is measured based on RSSI to reflect the diverse link qualities within the transmission range. According to the log-normal shadowing model [46], RSSI can be estimated as

$$RSSI(d) = P_t - PL(d_0) - 10\beta\log\left(\frac{d}{d_0}\right) + X_\sigma, \quad (6.4)$$

where P_t is the transmit power, $PL(d_0)$ is the path loss for a reference distance d_0, d is the transmitter–receiver distance, β is the path loss exponent (rate at which signal decays) depending on the specific propagation environment. X_σ is a zero-mean Gaussian distributed random variable with standard deviation σ.

The battery voltage can be computed as

$$V_{batt} = \begin{cases} P_{PT} \cdot Tx_{time} & \text{packet transmission} \\ P_{PR} \cdot Rx_{time} & \text{packet reception} \end{cases} \quad (6.5)$$

where P_{PT} is energy usage for packet transmission and P_{PR} is energy usage for packet reception.

The average delay D, to a one-hop neighbor from the source, can be calculated as half of the round-trip time (RTT). It is interesting to note that the average delay calculation does not require time synchronization because transmission time is inserted in the header of request to the route packet.

6.4.4   SPECIFICITY OF TRANSPORT LAYER IN WMSNs

Classic transport layer functionalities, such as providing E2E congestion control and guaranteeing reliability, become especially important in real-time delay-bounded applications such as multimedia streaming. Traditional transport protocols (e.g., TCP and UDP) cannot be directly implemented over WSN [47] because WSN in general, and WMSN in particular, have distinctive features that make them different from typical packet-switched networks, and they have a very wide range of applications with special requirements.

Some of the WMSN's features that affect the development of transport layer protocols are [14]

- *Network topology.* The WMSN environment is dynamic due to wireless link conditions and node status and, generally, it takes the shape of multihop many-to-one topology that is either flat or hierarchal. These variations in network topology should be taken into account in designing a transport protocol for WMSN.
- *Traffic characteristics.* Most of the traffic in WMSN is generated from the source nodes toward the sink and, depending on the application, this traffic can be continuous, event-driven, query-driven, or hybrid. Also in many cases, the source node can send its multimedia traffic using multipath route to the sink and this feature can be exploited to design a suitable transport protocol for ensuring the quality of multimedia streaming.
- *Resource constraints.* The sensor nodes have limited resources in terms of power supply, bandwidth, and memory, which require less expensive and more energy efficient solutions for congestion control and reliability.
- *Application-specific QoS.* WMSN has diverse applications from surveillance and target tracking to environmental and industrial applications. These applications may focus on different sensory data (scalar, snapshot, streaming) and, therefore, they need different QoS requirements in terms of reliability level, real-time delivery, certain data rate, fairness, etc.

- *Data redundancy.* Generally speaking, collected sensory data has relatively high redundancy and hence many WMSN applications use multimedia processing such as feature extraction, compression, data fusion, and aggregation to decrease the amount of data while keeping the important information. Therefore, reliability against packet loss becomes a challenging issue in WMSN especially if these packets contain important original data.

Obviously, no single transport layer solution exists that addresses the diverse concerns of WMSNs. As an example, defining reliability metrics, based on the packet content, and coupling application layer coding techniques to reduce congestion may be promising directions in this area [12]. Transport protocols for WSNs should have components that include congestion control and loss recovery because these have a direct effect on energy efficiency, reliability, and application QoS [47]. There are generally two approaches to performing this task.

The first would be to design separate protocols (algorithms), respectively, for congestion control and reliability. Most existing protocols use this approach and address congestion control or reliable transport separately. With this separate and usually modular design, applications that need reliability can invoke only a loss recovery algorithm, or invoke a congestion control algorithm if they need to control congestion. In the second approach, design considerations should be taken into account to achieve a full-fledged transport protocol that provides congestion and loss control in an integrated way. The first approach divides a problem into several subproblems and is more flexible to deal with. The second approach may optimize congestion control and reliability because loss recovery and congestion control in WMSNs are often correlated. Typical representatives of existing transport layer protocols for WMSNs are emphasized in Table 6.3.

6.4.5 SOURCE CODING TECHNIQUES FOR WMSNs

Multimedia processing techniques aim to reduce the amount of traffic transferred over the network by extracting useful information from the captured images and videos while at the same time maintaining the application-specific QoS requirements. Source coding, as one of the functionalities handled at the application layer, in the WMSN environment encompasses traditional communication problems as well as more general challenges. A real-time streaming application is more demanding than data-sensing applications in WSNs primarily due to its extensive requirements for multimedia encoding. The limitations of the sensor nodes requires video coding/compression that has low complexity, produces a low output bandwidth, tolerates loss, and consumes as little power as possible [51].

TABLE 6.3
Representative Transport Layer Solutions for WMSNs

Mechanism	Representative Algorithm	Main Characteristics
Congestion control	LRCC [48]	• Load repartition–based congestion control • Uses queue length as congestion detection indicator along with collision rate • Explicit congestion notification • Traffic redirection to the available paths
	QCCP-PS [49]	• Queue-based congestion control protocol with priority support • Hop-by-hop approach • Consists of congestion detection unit, congestion notification unit, and rate adjustment unit • Sending rate of each traffic source depends on congestion degree and priority
	DSCC [50]	• Differentiated services–based congestion control • Considers different types of applications and categorizes traffic into six different classes • Congested node optimally calculates the reduced share of the bandwidth for all one hop–away upstream nodes for which it is acting as a relaying node • Informs the upstream nodes about the modified data rate that they should use to alleviate the congestion
Reliability	M-DTSN [51]	• Multimedia distributed transport protocol for sensor networks • Improves flexibility by managing the trade-off between media quality and timely delivery for real-time data
Integrated (congestion control and reliability)	RSTP [52]	• Reliable synchronous transport protocol • Based on TCP • Reduces the large transmission delay of transferring multiple images between the source nodes and the sink due to transmission errors and limited bandwidth • Provides ordered delivery • Prioritizes different parts of the stream and schedule the transmission based on the information importance

In the traditional broadcasting paradigm, a video sequence is compressed once with a complex encoder and decoded several times with simpler decoders. Standard encoders used in widely spread MPEG and H.264 AVC rely on demanding processing algorithms [52–54] and may not be supported by sensor networks, which are characterized by nodes with low processing power and limited energy budgets. Hence, the aforementioned paradigm may be unfeasible for WMSNs and encoding techniques that move the processing burden to the decoder at the sink, such as distributed video coding (DVC) [55], seem to be promising.

DVC, used for low-complexity encoding by shifting the complexity to the sink side, incorporates concepts from source coding with decoder side information for creating an intracoded frame along with a side information frame. In this technique, multimedia content can be partitioned into multiple streams consisting of both intracoded and decoder side information frames by using simple and low-power encoder whereas the decoder at the destination side can be complex, exploiting the availability of resources such as energy and processing power capability.

Initial theoretical performance bounds for distributed source coding were established using an encoder–decoder from Slepian and Wolf [56]. It was proven that for lossless compression, separate encoding of information with combined decoding is as efficient as joint encoding. This result could not be practically implemented until determination of the relationship between distributed source coding and channel coding. Wyner and Ziv [57] suggested the application of linear channel codes for distributed source coding using the Slepian–Wolf approach.

Some of the applications of DVC in a WMSN have been previously proposed by Yaacoub et al. [58], Xue et al. [59], and Ahmad et al. [60]. In general, through practical implementation of DVC in WMSN, it can be concluded that there is a tradeoff between computation and transmission power consumption depending on the encoding schemes used [61]. Although a computationally intensive scheme such as discrete cosine transform (DCT) consumes more computational power, it achieves significant compression and hence, less transmission power is needed. On the other hand, a less computationally intensive scheme, such as a pixel-based codec, needs less computational power but more transmission power. Therefore, the choice of either scheme (DCT or pixel based) to implement the DVC in WMSNs depends on the tolerable distortion and power consumption.

6.5 CONVERGENCE OF MOBILE AND WIRELESS SENSOR SYSTEMS

Although the integration of mobile and wireless sensor systems is not an entirely new concept, forthcoming researches are dedicated to their convergence. Mobile networks and WMSN are evolving from heterogeneous to converged networks, to support emerging machine-to-machine (M2M) communication systems. M2M communications are highly related to the particular market scenarios having various optimization targets and can be realized separately within various environments. However, the terminals in M2M communications generally have less mobility.

The convergence of mobile systems and WMSNs can benefit both types of networks. Mobile networks can enable higher layer control and optimization to prolong WMSN lifetime, improve overall system performance, and provide QoS for sensor-oriented services. On the other hand, WMSN can enable the cognitive and intelligent aspects of the mobile systems. Furthermore, it is envisaged that the converged network architecture of mobile network and WSNs can enable better wireless services and more data-centric applications [62].

Currently, integration of mobile networks and WMSNs is realized through layered architecture. A group of sensor nodes construct the data detecting, whereas the GWs and BSs comprise the system control plane. WMSN is controlled indirectly by the BS through the dual-mode GW. The GW can just provide access for sensor nodes, and forward the detected data to the backhaul network. Communications between mobile and sensor networks use a GW data channel, which, however, decreases the system efficiency.

In the new network convergence approach proposed by Zhang et al. [63], the layered architecture is evolving into a flat architecture (Figure 6.8) to decrease the hierarchical signaling exchange between the two networks. In the converged architecture, the sensor nodes may have the ability to hear the downlink signaling from the BS. As a result, the mobile network can directly control and manage WMSN in a more efficient way. For example, the BS can help the sensor nodes to choose the optimal transmission path to route the traffic. Considering the uplink communications, due to the limited transmission range of sensor nodes, the data is still routed by the GW. Additional complexity is introduced into the sensor nodes to equip the downlink receiver, but the complexity will not be large because the device capabilities are much higher nowadays.

In the converged architecture, the effect on mobile networks and the complexity added to the sensor nodes should be evaluated to achieve an acceptable trade-off between the cost and performance gain. The authorization of the sensor nodes at the mobile cellular network needs to be considered. The information related to authorization can be relayed by the SN, which is already authorized in the mobile system. Moreover, new time coordination schemes need to be designed for the converged architecture because the multiaccess scheme in WMSN is contention based, whereas it is scheduling based in mobile systems. Hence, a jointly optimized coordination

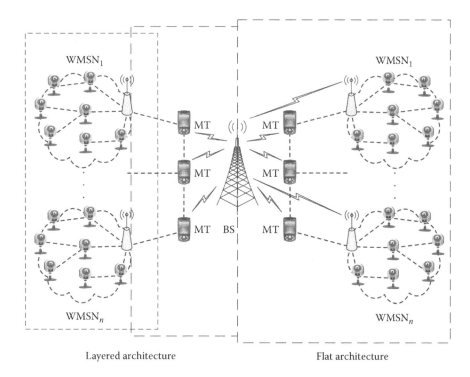

Layered architecture   Flat architecture

FIGURE 6.8 Layered and flat architecture for mobile system and WMSNs integration.

scheme should be designed to allow the sensor nodes to achieve a good trade-off between energy consumption and the system performance.

6.5.1 RADIO ACCESS TECHNOLOGIES CONVERGENCE

Currently, narrowband technologies or spread spectrum techniques are the main solutions for the radio interfaces of WSN (e.g., Bluetooth and ZigBee), whereas mobile systems use different radio access technologies (e.g., LTE and Mobile WiMAX). However, the design of a converged radio interface is the key challenge. To gain mutual benefits from the two types of networks, the following issues must be considered [63]:

- In mobile communications, OFDM and OFDMA techniques become the main solution for the radio interface. As higher data rate applications will be applied to WMSN in the future, the OFDM approach comes to an alternative for wireless sensor environments, and thus, the full convergence of the radio interface becomes possible. For example, noncontinuous OFDM [64] is one of the spectrum-pooling techniques for radio resource sharing between systems with different bandwidths. Each WMSN cluster shares a subset of OFDM subcarriers/carriers of the mobile system. Within the subset of subcarriers/carriers, the multiple access of sensor nodes is implemented.

- Because the coverage and channel conditions are quite different, the cyclic prefix of the two systems should be jointly designed. Besides, the two networks have different signal processing capabilities, and the bandwidth allocated for the two networks is different.
- If the two networks are working on different frequencies, some adaptive filters should be designed to aggregate a different number of subcarriers. Otherwise, if the two networks share the same frequency bandwidth, the converged radio frame needs to be designed to mitigate the multiaccess interference between the links from MT to sensor nodes and the links from MT to the BS.

6.5.2 PROTOCOL CONVERGENCE

In the converged scenario, the collected data from the sensor nodes can be routed to the BS by the GW. In a heterogeneous approach, the data channel between the two protocol stacks is usually implemented in the GW. In this case, data channels between the two independent stacks are implemented to exchange information. Because the network architecture and radio interface are highly converged for WMSN and mobile systems, the protocol and control signaling should also be tightly converged. In such a converged system, MAC and network layer protocols in the two stacks should be jointly optimized either to achieve some performance gains for WMSN or to extend the applications of mobile systems. However, the two protocol stacks are not independent while data and algorithms are shared between them.

In the converged networks, the downlink and uplink control signaling should be designed, and some cross-layer MAC solutions need to be implemented at the GW. The new signaling may affect the current WMSN and cellular standards. For the downlink, entry/exit of WMSN nodes and GWs can be managed by the BS, whereas for the uplink, the signaling from the WMSN nodes (e.g., on-demand periodic triggered requests for transmitting data) are also coordinated by the BS. Furthermore, in the PHY/MAC layer, the GW needs to convey sufficient/efficient control information to and from the BS for convergence optimization. The GW needs to request the resources from the BS for uplink and downlink transmission, and forward some system information to the sensor nodes. In the MAC layer, a two-level resource allocation scheme should be considered for the converged networks, especially for scenarios in which there are a large number of sensor nodes with heavy traffic. For example, the GW can map the data and resource requests of sensor nodes to the mobile system, and report to the BS. After that, the BS allocates a different WMSN channel group to each GW for intracommunications according to the requested information from different GWs. In the network layer, when a mobile GW enters the coverage area of WMSN, it may cause GW reselection or even reclustering of the sensor nodes. How to achieve a balanced trade-off among complexity, performance gain, and energy consumption via robust reselection and reclustering algorithms is an essential and challenging issue.

6.6 WMSN APPLICATIONS

WMSNs applications are rapidly gaining interest due to the great success of WSNs in solving real-world problems related to human interaction with the physical environment. They will not only enhance existing sensor network applications such as public safety, health care, building and industrial process automation, and others, but they will also enable some completely new applications. In addition to the ability to retrieve multimedia data, WMSNs will also be able to store, process in real-time, correlate, and fuse multimedia data originating from heterogeneous sources. Recently, the research in WMSNs gained high momentum due to the introduction of the IoT vision, which is evolving very rapidly considering the tremendous progress in the field of embedded systems (see Chapter 9). Significant results in this area over the last few years have ushered in a surge of civil and military applications [8]. Various fields of WMSNs applications together with some characteristic examples are presented in Figure 6.9.

As for many other communications systems, the development of WSNs was motivated by military applications. WMSNs can be used by military forces for a number of purposes such as monitoring or tracking the enemies and force protection. In military target tracking and surveillance, a WMSN can assist in intrusion detection and identification. Specific examples include spatially correlated and coordinated troop's movements. Unlike in commercial applications, a tactical military sensor network has different priority requirements for military usage [65]. Especially in the remote large-scale network, topology, self-configuration, network connectivity, maintenance, and energy consumption are the challenges.

Multimedia sensors can be used in public safety to enhance and complement existing surveillance systems against crime and terrorist attacks. Large-scale WMSNs can extend the ability of law enforcement agencies to monitor public areas and events, private properties, and borders. Multimedia sensor nodes could detect and record potentially relevant activities (thefts, car accidents, traffic violations) and make video/audio streams or reports available for future query. Multimedia content, along

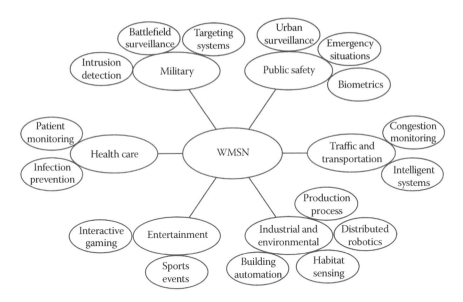

FIGURE 6.9 Characteristic examples of WMSN applications in various fields.

with advanced signal processing techniques, can be used to locate missing persons or to identify criminals or terrorists.

Conventional infrastructures have involved static sensors for surveillance in urban city areas, which are limited in spatial coverage. Recently, a new infrastructure for mobile surveillance was proposed, named vehicular sensor networks [66], which is a network of mobile sensors equipped on public transportation vehicles. This system facilitates the collection of surveillance data over a wider area than fixed infrastructure. At the same time, unlike traditional sensor nodes, vehicular sensors are typically not affected by strict energy constraints and the vehicles themselves can be equipped with powerful processing units and wireless transmitters. Vehicular multimedia sensors can continuously capture multimedia content from streets and maintain sensory data in their local storage. It may assist in the scene reconstruction of crimes, and more generally, the forensic investigations of monitored events, such as traffic accidents, terrorist attacks, and others.

As the transport and traffic networks are growing day-by-day, the question of how to obtain information about traffic conditions is becoming more and more challenging. In such an area, intelligent transportation systems has emerged as a key candidate that has benefited from the unique features and capabilities of WMSNs. Sensor nodes can be implemented in parallel with traffic infrastructure to provide monitoring function, or embedded in vehicles in which their main function is to avoid accidents and to assist drivers. Multimedia sensors are in a position to monitor vehicle traffic in big cities or highways and deploy services that offer traffic routing advice to avoid congestion. Also, multimedia sensors may monitor the traffic flows and retrieve aggregate information such as average speed and number of vehicles. Furthermore, smart parking systems based on WMSNs [67] allow the monitoring of available parking spaces and provide drivers with automated parking advice, thus improving mobility in densely populated areas.

Many industrial and environmental monitoring projects that use acoustic and video sensing, in which information has to be conveyed in a time-critical fashion, are being envisaged. Multimedia content such as imaging, temperature, or pressure, among others, may be used for time-critical industrial process controls. For example, in production process quality controls, the final industrial products are automatically inspected to find defects. In addition, machine vision systems can detect the position and orientation of parts of the product to be assembled by robotic systems. The integration of machine vision systems with WMSNs can simplify and add flexibility to systems for visual inspections and automated actions that require high speed, high magnification, and continuous operation. On the other hand, arrays of multimedia sensors are already used by biologists to determine the structure, functions, and evolution of ecosystems.

Virtual reality games and sports events are relatively new fields of WMSN applications. Virtual reality games that assimilate touch and sight inputs of the user as part of the player response need to return multimedia data under strict time constraints. Also, WMSN application in gaming systems are closely associated with sensor placement and the ease in which they can be carried on the person of the player. Recently, video sensor systems have started being used widely as support for judicial decision making.

Health care applications impose strict requirements on E2E system reliability and data delivery. WSNs in these applications are used to determine the patients' activities of daily living and provide data for studies [68]. To enable small device size with reasonable battery lifetimes, sensor nodes use low-power components. Another application domain for sensor networking in health care is high-resolution monitoring of moments and activity levels as a means to identify the causes of illness. Multimedia sensors can infer emergency situations and immediately connect patients with remote assistance services. Telemedicine sensor networks are integrated into mobile multimedia systems to provide ubiquitous health care services. Patients can carry medical sensors to monitor parameters such as body temperature, blood pressure, pulse oximetry, electrocardiogram, and breathing activity. Furthermore, remote medical centers perform advanced remote monitoring of their patients through video and audio sensors, location sensors, and motion or activity sensors.

6.7   WSN AUTOMATED MAINTENANCE

Although robotic wireless sensor networking is a well-proven concept [69], attempts at WMSN maintenance using mobile robots are recognized as new and challenging issues [70]. Robots are able to make complex decisions and take appropriate actions on themselves (controlled movement), sensor nodes (battery recharge and reposition), and environment (fire extinguishing). They can communicate directly or via multihop sensor paths. Robots can deploy and relocate sensors to improve network coverage, build routes and fix network partitions to ensure data communication, and change network topology to shape routing patterns and balance energy consumption. Wireless sensor and robotic networks represent the confluence of WSNs and multirobotic systems, and there is no limit to the benefits stemming from this mutual collaboration.

6.7.1   TASK ALLOCATION AND TASK FULFILLMENT PROBLEMS

The main challenges posed by mobile robots in WSNs are robot task allocation and robot task fulfillment [70]. A task is an action required upon the occurrence of an event originating in the sensor field (e.g., target detection or sensor failure), whereas its allocation represents the coordinated response of the robot team to such an event. Usually, sensor nodes transmit relevant event information such as location, intensity, and others, to one or more robots via multihop routes.

To optimize the overall system performance, task allocation (assignment) can be considered as a function of the individual utility of each robot. The utility measure is strictly task-dependent and robot-dependent, and takes into account the profit gained after task completion as well as the cost of carrying out the task. From the decision-making point of view, task allocation approaches are either centralized or distributed. In a centralized approach, decisions on robot task allocations are made by a single entity (e.g., coordinator robot or SN) once all the relevant information has been collected. Data flows from every robot to the coordinator, which runs a centralized algorithm and notifies each robot on its set of assigned tasks. On the other hand, a distributed approach brings a much higher degree of autonomy because multiple entities are involved in deciding on an appropriate response. Centralized implementations are generally more accurate than decentralized schemes, but they generate nonnegligible communication overhead, and have poor fault tolerance and scalability properties.

From the cooperation aspects, task allocation can be modeled as minimal or intentional. In the minimal model, a robot derives its individual action from the set of other agents learned through explicit interaction, but there is no negotiation among individuals for task assignments. Intentional cooperation exploits communication more heavily, as corporate actions are "negotiated" via interrobot message exchanging. This model of cooperation is achieving more attention in robotics research, probably due to its more intuitive formulation. Furthermore, it can be considered as an auction model in the field of game theory.

After being assigned a task, a robot needs to geographically relocate itself to fulfill the task. If the task is very specific or tied with a unique location, for example, repairing a particular sensor node, the robot will just need to move directly toward that location. If the task is otherwise described in a general form for a large geographic region, for example, fixing faulty sensors in the region, then the robot has to carefully plan its trajectory to guarantee service delivery as well as to satisfy task-specific requirements, mostly concerning service delivery latency.

6.7.2 TOPOLOGY CONTROL AND SENSOR LOCALIZATION

Robots can transport the sensors, while moving over the region of service, and place them at proper positions to establish a network with a desired topology in terms of coverage or connectivity. In the case of confined regions, the sensor deployment problem deduces to a graph traversal problem, in which a virtual geographic graph is precomputed according to the topology requirement. The goal is to find an efficient traversal algorithm that minimizes the total moving distance and thus the deployment latency. As robots possibly start at different locations, their deployed sensors may not finally interconnect unless they carry sufficient sensors to cover the entire region.

Example 6.4

As an example of topology control, the focused coverage problem is addressed by Falcon et al. [71]. Here, sensors are required to surround a coverage focus, called the point of interest (PoI), and maximize the coverage radius, that is, the minimal distance from the PoI to the uncovered areas. A localized carrier-based coverage augmentation (CBCA) protocol was proposed to incrementally construct a biconnected network with optimal focused coverage. Biconnectivity implies that there are at least two disjointed paths between every pair of nodes. The protocol relies on a triangle tessellation graph that is locally computable provided the PoI coordinates are given and nodes agree on a common orientation.

In CBCA, robots enter the region of service from fixed locations (base points), and advance straight to the PoI. As soon as they get in touch with previously deployed sensors, they search by multihop communication along the network border for the optimal sensor placement points with respect to focused coverage optimization. Border nodes store the locations of failed sensors inside the network as well as adjacent available deployment spots outside the network, and recommend them to robots during the search stage. This process is repeated until robots run out of sensors. Robots then return to base points for sensor reloading and re-enter the environment afterward to augment existing focused coverage, as shown in Figure 6.10. Because robots move at different speeds, they can exchange their targets when in contact so as to minimize coverage augmentation delay. Various techniques were developed to prevent sensor deployment point contention (sensors recommending the same target to multiple robots) and resolve robot collision (multiple robots serving the same node).

In addition, robots may also be used for improving the topology of an existing WSN. Due to random deployment, some sensors are very likely to appear structurally redundant from a local perspective, although there are topologically vulnerable points (in terms of coverage or communication) in other parts of the network. The mobile robots can employ the redundant sensors for topology control subject to the limitation of cargo capacity on individual robots. The problem is then to find the optimal route for robotic sensor pickup and delivery such that total travel distance (for energy efficiency) or maximum per robot travel distance (for delay efficiency) is minimized. This problem can be modeled as a variant of the vehicle routing

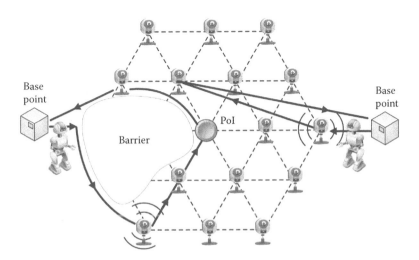

FIGURE 6.10   Example of WSN topology control using a CBCA protocol.

problem. Efficient population-based metaheuristics like ACO algorithms can be applied due to their concurrent exploration of the search space. The preferred approach is the one that returns the highest-quality solutions and remains robust as the network size grows.

Localization is a problem dealing with how a sensor determines its spatial coordinates (position). A simple, but not cost-effective, approach is to equip each sensor with a GPS receiver that provides an accurate-enough location estimation. On the other hand, non–GPS-based localization often requires certain location-aware devices (localization beacons), which periodically emit beacon signals containing their spatial coordinates. The number of beacons and their distribution have a direct effect on localization performance. A large number of uniformly distributed beacons will lead to better performance than a few crowdedly or linearly deployed ones. Localized sensors may become new beacons and help other sensors to self-localize. This iterative method reduces the initial number of beacons required but brings about aggregated localization error. In this case, mobile robots are introduced as an economical alternative in delay-tolerant scenarios. Robots are aware of their own location while transmitting beacon signals conveying their up-to-date coordinates. On an individual sensor, a localization procedure is engaged during the robot visit. The advantages of this approach lie in reduced deployment cost and communication overhead because a few beacons are required and only local communication is involved. It should be noted that sensors can be localized when the robots are within their communication range and after they receive sufficient beacon signals from the robots. These advantages come at the expense of increased localization delay. Robot trajectory has to be properly envisioned as optimal in length yet ensuring a quick, full, and accurate localization to every sensor.

6.8 CONCLUDING REMARKS

In recent years, WMSNs have received much attention from the networking research community as well as from other disciplinary fields. There exists a need for a great deal of research efforts on developing efficient protocols, algorithms, topology, and services to realize WMSN applications.

In the design of WMSN, the platform must deal with challenges in energy efficiency, cost, and application requirements. Application requirements vary in terms of computation, storage, and user interface. Consequently, there is no a single platform that can be applied in a real environment.

Considering the link layers, a cross-layer approach is essential for efficient MAC designs in WMSNs, together with queue management and traffic classification/prioritization, as long as high QoS level is required for multimedia traffic. QoS is clearly a needed feature in many WMSN applications, especially if there is some sort of real-time streaming. Nevertheless, many proposed scenarios, in which the use of different message priorities, are going to give more or less the same performance with less complexity. While analyzing routing approaches from the open literature, it is possible to conclude that multipath routing represents optimistic and perspective solutions because WMSNs need to exploit the resources to its limits and sometimes in short bursts. Although most of the proposed application-specific transport protocols do not take into consideration the multimedia requirements, designing an algorithm with appropriate performance metrics for both reliability and congestion control, and based on the application layer source coding techniques, will be a promising direction in this research area.

Because WSNs in general and WMSNs in particular are considered as resource-constrained environments, the correlation characteristics and functionality interdependencies among the layers cannot be neglected and should be exploited for better performance and efficient communication. Consequently, cross-layer design stands as the most promising alternative to inefficient traditional layered protocol architectures. Recent works in WMSNs show that cross-layer integration and design techniques result in significant improvement in terms of QoS, energy conservation, and exchanging information between different layers of the protocol stack.

WMSNs and mobile systems are evolving from heterogeneous to converged architectures in order to meet the increasing requirement for M2M communications. Although there are many technical challenges in the convergence process, such as new network architecture, radio interface, and protocols, the convergence will bring about notable advantages.

Wireless multimedia sensor and robot networks have emerged as a paradigmatic class of cyber-physical systems with cooperating objects. Because of the robots' potential to unleash a wider set of networking means and thus augment network performance, automated maintenance of WSNs by mobile robots has become a rapidly growing research area.

7 Security in Wireless Multimedia Communications

Security is a crucial part of wireless multimedia systems. Generally, recognizing the significance of security in wireless networks for current and future users is very important. The increasing use in application areas requires the provision of several properties at the network or application layer, which are typically considered security properties: privacy of communications, nonrepudiation of communication members' actions, authentication of parties in an application, and high availability of links and services among several others. Security technologies enable the provision of several of these properties at the required level. The increasing demand of users for various wireless multimedia communication services has led to the development and to the coexistence of different and often incompatible technologies with unique applications and characteristics. To integrate several wireless networks into a single architecture, there are a number of challenges that must be addressed. One of the most important challenges is secure interoperability. Cognitive radio networks (CRNs) are more flexible and exposed to wireless networks compared with other traditional radio networks. Hence, there are many security threats to cognitive communications, more so than other traditional radio environments. The unique characteristics of CRNs make security a more challenging task. A typical public key infrastructure scheme, which achieves secure routing and other purposes in typical ad hoc networks, is not enough to guarantee security under limited communication and computational resources. However, there has been increasing research attention on security threats caused specifically by cognitive radio characteristics. Wireless mesh network (WMN) operates as an access network to other communications technologies and, because of that, it is exposed to numerous security challenges not only in the specific transmission operations but also in the overall security against foreign attacks. Also, security is critical for many wireless sensor networks applications due to the limited capabilities of sensor nodes.

7.1   INTRODUCTION

Wireless networking is transitioning from conventional infrastructure-based last-hop wireless networks to more dynamic, self-organizing peer-to-peer (P2P) networks. This transition has significant implications for both security and privacy. In traditional multimedia wireless systems, the high cost of setting up the infrastructure limits the network operators to sufficiently large entities that presumably care about their brand and overall reputation. Thus, they are expected to be trusted by the end users. Accordingly, security in these networks is centered on protecting end-user data from being exposed to outsiders and preventing the infrastructure from being accessed by unauthorized parties. In emerging wireless multimedia systems, most communications are realized via peers. Even if there is some form of central authority, it may be that there is no central authority at all, at least not one that is present all the time. Consequently, nodes can be compromised, removed, or destroyed without immediate or rapid detection. Moreover, without a centralized authority, devices have to work together to accomplish tasks such as network formation, routing, and adjusting to network dynamics. They might also need to take part in mitigating attacks. All these changes require different security and privacy schemes.

Future wireless multimedia systems will constitute heterogeneous radio access technologies and networks underneath a common Internet protocol (IP) layer, where security presents a very critical part and very complex and broad topic. IP security (IPsec) [1] is a suite of protocols for securing communications by authenticating or encrypting each packet (or both) in a data stream. IPsec also includes protocols for cryptographic key establishment.

If networks do not have any inherent security, as a result, there is no guarantee that a received message is from the channel sender, contains the original data that the sender put in it, or was not sniffed during the transit. Multimedia itself may not bring to mind any new considerations for security if we simply regard it as the compilation of the voice, data, image, and video contents, areas that have at least some current security mechanisms. Protection of the networks is an important security issue for service providers, as is the ability to offer customers value-added security services as needed or desired. Service or network providers planning for implementation of security mechanisms need to consider the nature of the security threat, the strength of security needed, the location of security solutions, the cost of available mechanisms, the speed and the practicality of mechanisms, and interoperability.

CRNs are vulnerable to various attacks because they are usually deployed in unattended environments and use unreliable wireless communication [2]. However, it is not simple to implement security defenses in

CRNs. As application-specific networks, CRNs have some unique features and, correspondingly, some unique security requirements. CRNs face new security threats and challenges that have arisen due to their unique cognitive characteristics [3]. One of the major obstacles in deploying security on current CRNs is that they have limited computational and communication capabilities. With this in mind, many researchers have begun to ensure security for CRNs with different security mechanisms. Security mechanisms, including trust management, have the ability to secure CRNs against attackers.

The routing operation over WMNs creates a vulnerable security system because of the multihop traffic transmission and loose P2P data exchange during the internode authentication mechanism while routing neighborhood nodes information and exchanging new node updates. In addition, the multihop behavioral characteristics of the WMN creates challenges in the security of traffic operations while in transmission through the gateway to the wireless cloud [4]. The dynamic topology updates further expose the whole network security to persistent and corruptible attacks. The data reliability and authentication in WMNs during the neighborhood node exchanges through link state and routing are quite loose and very insecure. The ease in WMN integration with other wireless networks, like in broadband and multimedia, has also established the necessity for an unyielding privacy protection and security mechanism [5,6].

The distributed-sequenced mechanism in the medium access control (MAC) channel frames also creates susceptibility to attacks whereas the mobile mesh client nodes and its consequent dynamic topology also establish the need for a more effective, resilient, and comprehensive security system in WMNs. The constraint in WMN security creates the challenge of possible attacks by invasive malicious codes when an attack becomes distributed in the architecture through the simple dynamism of mesh. These attacks compromise confidentiality and integrity and violate the privacy of the WMN users. The nodes can also be compromised by the operation of traffic transmission, unverified router information exchange, and network notification infiltration.

In the context of wireless sensor networking, security means to protect sensed data against unauthorized access and modification, as well as to ensure the availability of network communication and services despite malicious activities. If collected data are private and sensitive, such as user location information, then privacy issues are also of concern. On the other hand, security in wireless multimedia sensor networks (WMSNs) is a very young research field. From a security point of view, there is no strict border between WSNs and WMSNs. Therefore, some of the traditional security solutions for WSNs can be easily adapted for WMSNs. However, WMSN also has some novel features that stem from the fact that some of the nodes will have multimedia sensors and higher computational capabilities. This brings new security challenges as well as new protection opportunities [7]. Everything seems to indicate that video surveillance and monitoring will be one of the killer applications for WMSNs. Because the security issue in WMSN is so complex, different solutions are going to be application-dependent and environment-dependent. In the case of surveillance and monitoring, most of the time, it can be assumed that all the nodes are trusted initially but later on can be compromised. Concerning this scenario, denial of service (DoS) is going to be one of the most common attacks. Other problems like resilience to traffic analysis and compromised nodes are also going to be of the utmost importance.

This chapter starts with general security issues including security attacks in wireless networks. Then, security requirements are presented. Next, the security aspects in emerging mobile networks are analyzed. Security of cognitive radio and WMNs are also outlined. Some security aspects characteristic of WMSNs conclude the chapter.

7.2 GENERAL SECURITY ISSUES IN WIRELESS COMMUNICATIONS

Dramatic advances in technologies enable and support multimedia services. These advances provide both benefits and challenges when it comes to security. Of particular interest is the level of interactivity experienced with multimedia service. Two of the key multimedia network characteristics are internetworking and interactivity as well as their relation to security [8]. Wireless networks are more vulnerable than their wired counterparts. Complications arise in the presence of node mobility and dynamic network topology. At the same time, node resource constraints, due to power limitations, bandwidth and memory requirements, make the direct application of existing security solutions difficult. Finally, in some environments, network size and physical inaccessibility of nodes further exacerbates the security problems.

Factors that contribute to security problems include [9]: channel, mobility, resources, accessibility. Air interface usually involves broadcast communications, which make eavesdropping and jamming easier. Here, a physical connection is replaced by a logical association. The latter can be interrupted and must be renewed whenever a wireless device moves beyond the transmission range. Establishing secure association in the presence of mobility is challenging, especially in high-mobility environments such as vehicular ad hoc networks (VANETs). If a wireless device is affiliated with a person, tracking the device reveals that person's location and mobility patterns. In that way, privacy becomes an important concern. Most users' terminals are still resource-constrained. One of the fundamental reasons is the need to keep physical size small to enable mobility and embeddability. High-end terminals are battery-powered, which limits computation and communication. Such limitations invoke DoS attacks aimed at battery depletion. Some devices are personal and are usually attended by their owners, whereas others (sensors nodes or robots) are generally left unattended, and are placed in remote locations. This greatly increases their vulnerability to physical attacks.

The fundamental mechanism used to protect information in secure systems is termed access control. With adequate access controls in place, both the confidentiality and the integrity of an information object can be assured to some defined confidence level. Confidentiality is at risk whenever information flows into an object. When adequate physical controls exist, information can be stored and moved from one place to another without taking additional steps to hide it. To ensure the confidentiality of information flows sent via insecure paths, it is necessary to encrypt the information at the source and to decrypt it at the destination. To ensure integrity, validation methods must be used to verify that stored or transferred information is not corrupted. Security criteria establish discrete reference levels for evaluating how well a candidate product or system element meets the requirements of a security policy [10]. The criteria do not establish what protection needs to be provided in a given context: the information is provided by an applicable security policy.

Table 7.1 represents a set of various security criteria and the corresponding definitions. To qualify as compliant with respect to a particular criterion, a candidate product or system must satisfy all of the requirements. Security criteria establish a finite, discrete set of terms and conditions for evaluating the ability of a product, system block, or a complete system to protect designated types of resources.

TABLE 7.1

Various Security Criteria and the Corresponding Definition

Security Criteria	Definition
General	Define general compliance requirements
Confidentiality	Define levels of protection against unauthorized disclosure
Integrity	Define levels of protection against unauthorized modification
Accountability	Define levels of ability to correlate events and users
Reliability	Establish levels to measure how well a product/service performs
Assurance	Define overall trust levels for product/service or system

Source: B. Bakmaz et al., *International Journal of Applied Mathematics and Informatics* 1, 2: 70–75, 2007.

The complexity of a system and the tasks it needs to perform determine how granular an appropriate set of security criteria needs to be. A security policy specifies which security criteria must be satisfied by a qualifying product to ensure that an adequate level of protection is provided to resources. The overall rating of a product is based on the set of criteria that the product can satisfy. If the capabilities of a product match the requirements of a security policy, then the product can be certified as compliant at a particular level. To evaluate a system that provides more than one level of protection, different subcategories of security criteria requirements need to be defined. Security policies define which security criteria requirements apply to each defined protected information object and user class. A product is evaluated against each required security criterion subcategory for compliance one by one.

7.2.1 SECURITY ATTACKS

Communications over wireless channels are, by nature, insecure and easily susceptible to various kinds of threats. Typical security attacks in wireless networks can be categorized based on their effects, including data integrity and confidentiality, power consumption, routing, identity, and service availability [11]. Classification of typical security threats in wireless multimedia communications and their effects/targets is presented in Table 7.2.

DoS attack presents an adversary's attempt to exhaust the resources available to its legitimate users. Also, jamming can be used to launch DoS attacks at the physical layer [12]. An adversary can utilize jamming signals to make the attacked nodes suffer from DoS in a specific region.

Node capture represents a characteristic threat in WSNs through which an intruder can perform various operations on the network and easily compromise the entire system [13]. An attacker physically captures sensor nodes and compromises them such that sensor readings are inaccurate or manipulated. In addition, the attacker may attempt to extract essential cryptographic keys (e.g., a group key) from wireless nodes that are used to protect communications in most wireless networks.

In *eavesdropping attacks* a malicious user secretly monitors on ongoing communication between targeted nodes to collect connection-related information (e.g., MAC address) and cryptography (e.g., session key materials) [11]. Eavesdropping can be performed even if the messages are encrypted. Moreover, information collected by eavesdropping is usually used for other forms of attack.

Denial of sleep attacks are the most common type of power consumption threats. In general, this type of attack attempts to exhaust the device's limited power supply, which is one of the most valuable assets in wireless networks. The worst case would cause a collapse of network communications. During a sleep period, the MAC layer protocol reduces the node's

TABLE 7.2

Classification of Typical Security Threats in Wireless Multimedia Networks

Class of Threats	Threat	Characteristic Effects/Targets
Data integrity and confidentiality	DoS	Deny legitimate users access to a particular resource
	Node capture	Cryptographic keys from nodes
	Eavesdropping	Connection information (MAC address), cryptography (session key)
Power consumption	Denial of sleep	MAC layer protocol, decreasing or disabling the sleep period
Service availability and bandwidth consumption	Flooding	Large number of useless packets sent to the victims
	Jamming	Breaking connectivity with jamming radio signals
	Replaying	Repeatedly and continuously sending of packets copies
Routing	Unauthorized routing update	Changing of routing information maintained by routing hosts
	Wormhole	Intercepting sessions originated by the sender
	Sinkhole	Attracts all nodes to send packets through colluding nodes
Identity and privacy	Impersonate	Impersonates the nodes' identity (MAC or IP address)
	Sybil	Single node presents itself to other nodes with multiple spoofed identifications
	Traffic analysis	Depredation of the network, traffic, and nodes related information

power consumption to extend its lifetime. Thus, the attacker affects the MAC layer protocol to shorten or disable the sleep period [14]. Denial of sleep attack specifically targets the energy-efficient protocols unique to WSN deployments.

Flooding presents the process in which the attacker sends a large number of packets to the point of attachment (PoA) to prevent the victim or the whole network from establishing or continuing communications. Most of the flooding attacks launched to date have tried to make the victims' services unavailable, leading to revenue losses and increased costs of mitigating the attacks and restoring the services [15]. This type of threats typically target upper layers protocols, for example, ICMP, TCP, UDP, DNS, and SIP. If these attacks result in a denial of service to legitimate users, they can also be referred as a variant of distributed DoS attacks.

Jamming attacks can disrupt wireless connectivity among nodes by transmitting continuous radio signals such that other authorized users are denied from accessing a particular frequency channel. The attacker can also transmit jamming radio signals to intentionally collide with legitimate signals originated by target nodes [11].

Replay attacks occur when an attacker copies a forwarded packet and later sends out the copies repeatedly and continuously to the victim to exhaust the buffers or power supplies. In addition, the replayed packets can crash weak applications or exploit vulnerable holes in poor system designs. These attacks are usually performed during authorization or key agreement protocols to perform, for example, masquerade attacks [16].

Unauthorized routing update attack attempts to update routing information maintained by routing hosts (PoAs or data aggregation nodes), in an attempt exploit the routing protocols, fabricate the routing update messages, and falsely update the routing table [11]. This attack can lead to several incidents including:

- Isolation of some nodes from PoAs
- Network partitioning
- Routing in a loop
- Forwarding messages to unauthorized attackers
- Misrouting or packet delay by false routings, etc.

In a *wormhole attack*, an attacker records packets (or bits) from one location in the network, tunnels them to another location, and retransmits them into the network. The wormhole attack poses many threats, especially to routing protocols and location-based wireless security systems [17]. Many subsequent attacks (e.g., selectively forwarding and sinkholes) can be performed after the wormhole's path has attracted a large amount of traversing packets.

Sinkhole attacks attract nodes to send all packets through one or several colluding (sinkhole) nodes, so that the attacker (and its colluding group) has access to all traversing packets. To attract the victimized nodes, the sinkhole node is usually presented as an attractive forwarding node such as having a higher trust level, being advertised as a node in the shortest distance or short delay path to a base station, or a nearest data aggregating node (in WSNs) [11].

Impersonate and Sybil attacks expose the node's identity (MAC or IP address) to establish a connection with or launch other attacks on a victim. The attacker may also use the victim's identity to establish a connection with other nodes or launch other attacks on behalf of the victim. As a consequence, packets traversed on a route consisting of fake identities are selectively dropped or modified. Also, a threshold-based signature mechanism that relies on a specified number of nodes can be corrupted. In general, identity-related attacks cooperate with eavesdropping attacks or other network-sniffing software to obtain vulnerable MAC and IP addresses [11].

Traffic analysis attacks may include examining the message length, message pattern or coding, and duration the message stayed in the router. In addition, the attacker can correlate all incoming and outgoing packets at any router. Such an attack violates privacy and can harm entities that are linked with these messages. Also, the attacker can also perversely link any two entities with any unrelated connections. Traffic analysis can also be used to determine the locations and identities of the communicating parties by intercepting and examining the transmitted messages [12].

7.2.2 SECURITY REQUIREMENTS IN WIRELESS MULTIMEDIA COMMUNICATIONS

Secured services in wireless networks should satisfy certain requirements such as the following [18]: authentication, nonrepudiation, confidentiality, access control, integrity and availability, and resistance to jamming and eavesdropping.

Entity or data origin *authentication* is used to confirm that a communication request comes from a legitimate user. The entity authentication is frequently used to justify the identities of the parties in the communication sessions. The data origin authentication focuses on confirming the identity of a data creator. On the other hand, *nonrepudiation* guarantees that the transmitter of a message cannot deny having sent the message, and the recipient cannot deny having received the message. Digital signatures, which function as unique identifiers for individual users, much like fingerprints, are widely used for nonrepudiation purposes.

Confidentiality is the protection of transmitted data from passive attacks preventing access by, or disclosure to, unauthorized users. It is closely related to data privacy, such as encryption and key management. The other aspect of confidentiality is the protection of traffic flows from any attacker's analysis. It requires that an attacker is not able to determine any traffic-related information, such as the source/destination location, transmission frequency, session length, and others.

As an alternative confidentiality mechanism, *access control* limits and governs the devices that have access to the communication links. Thus, each entity must be authenticated or identified beforehand to gain access to the communication channels. Because of the broadcast nature of the wireless medium, it is difficult to control access, and hence, it is vulnerable to eavesdropping.

Integrity and availability are the trustworthiness and reliability of information. *Integrity* means that the data from the source node should reach the destination node without being altered. It is possible for a malicious node to alter the message during transmission. Integrity can even involve whether a person or an entity entered the right information. *Availability* can be observed in the case that communication should remain fully operational when a legitimate user is communicating. Also, it must be robust enough to tolerate various attacks during any authorized transmission, and should be able to provide guaranteed transmissions whenever authorized users require them.

Making the wireless channel *resistant to jamming* is a challenging security issue. A jammer may broadcast an interference signal on a broad spectral band to disrupt legitimate signal reception. The reaction of jammers can be active or passive. Active jammers send out random bits or a radio signal continuously into the channel and therefore block the communications of users. On the other hand, a passive jammer is idle until it senses transmission activities occurring in the channel. Then it transmits jamming signals to interrupt an ongoing transmission. Because the jammer must detect transmission activities before issuing its jamming signal, the transceiver may improve its own low probability of detection to avoid jamming attacks.

A typical secrecy problem in a wireless communication system involves three entities: the transmitter, the receiver, and the eavesdropper. The eavesdropper is assumed to be passive, and therefore its location is unknown to the transmitter and receiver. Perfect secrecy is achieved when the transmitter delivers a positive information rate to the legitimate receiver and ensures that the eavesdropper cannot obtain any information [12].

The way to avoid *eavesdropping* is to use a cipher to encrypt each transmitted data stream, which can only be decrypted at the intended receiver using a privately shared key. Another widely used approach is to force the transmitter and receiver to adopt some information-hiding measures to prevent unauthorized detection of any transmission activities.

Example 7.1

The symmetric data encryption/decryption algorithm, which is widely used in networks, is shown in Figure 7.1. The data encryption standard (DES) with a common private key is normally shared

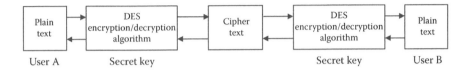

FIGURE 7.1 Symmetric cryptographic technique.

by two users. Secret key cryptology operates in both transmission directions.

User A sends an encrypted message to user B with a secret key. User B can use the secret key to decipher the message. Because this message has been encrypted, even if the message is intercepted, the eavesdropper between user A and user B will not have the secret key to decipher the message. If these two users do not have this private key, a secure channel is required for the key exchange. Instead of using an additional channel, the physical layer methods can be applied to distribute secret keys, to supply location privacy, and to supplement upper-layer security algorithms.

7.3 PHYSICAL LAYER SECURITY

The application of physical layer security schemes makes it more difficult for attackers to decipher transmitted information. The existing physical layer security techniques can be classified into major categories [12]: theoretically secure capacity, channel, coding, power and signal detection approaches.

7.3.1 THEORETICALLY SECURE CAPACITY

The secrecy capacity, defined as the maximum transmission rate at which the eavesdropper is unable to decode any information [19], is equal to the difference between the two channel capacities:

$$C_s = C_M - C_W = \frac{1}{2} \log_2 \left(1 + \frac{P}{N_M} \right) - \frac{1}{2} \log_2 \left(1 + \frac{P}{N_W} \right), \quad (7.1)$$

where C_M is the capacity of the main channel and C_W denotes the capacity of the eavesdropper's channel. Here, P corresponds to the average transmit signal power, whereas N_M and N_W are power of the noise in the main channel and the eavesdropper's channel, respectively.

Information-theoretic security is an average-information measure. The system can be designed and tuned for a specific level of security. On the other hand, it may not be able to guarantee security with probability. Furthermore, it requires knowledge about the communication channel that may not be accurate in practice. A few systems (e.g., quantum cryptograph) have been deployed, but the technology is not widely available due to its implementation costs.

7.3.2 CHANNEL

The perspective techniques that have been proposed to increase security based on the exploitation of channel characteristics include the following:

- Algebraic channel decomposition multiplexing (ACDM) precoding scheme [20], in which the transmitted code vectors are generated by singular value decomposition of the correlation matrix, which describes the channel's characteristic features between the transmitter and the intended receiver. Because any potential transmitter–eavesdropper channel is going to have a different multipath structure, the eavesdropper's ability to detect and decode the transmissions can be severely reduced.
- Randomization of multiple-input, multiple-output (MIMO) transmission coefficients [21] is a procedure in which the transmitter generates a diagonal matrix dependent on the impulse response matrix of the transmitter–receiver channel. This diagonal matrix has a unique property that makes the matrix undetectable to the attackers but easily detectable to the intended receiver. This method reduces the signal interception capability of the intruder and leads to a blind deconvolution problem due to the redundancy of MIMO transmissions. The proposed scheme indicates that the physical layer technique can assist upper layer security designs by providing secret key agreement with information-theoretic secrecy.
- Radiofrequency (RF) fingerprinting system [22] consists of multiple sensor systems that capture and extract RF features from each received signal. An intrusion detector processes the feature sets and generates a dynamic fingerprint for each internal source identifier derived from individual packets. This RF system monitors the temporal evolution of each fingerprint and issues an intrusion alert when a strange fingerprint is detected, thus helping distinguish an intruder from a legitimate user.

7.3.3 CODING

The objective of coding approaches is to improve resilience against jamming and eavesdropping [12]. These approaches mainly include the use of error correction coding and spread spectrum coding.

In a conventional cryptographic method, a single error in the received cipher text will cause a large number of errors in the plain text after channel decoding. A combination of turbo coding and advanced encryption standard cryptosystem can be used to set up a secure communication session. The main advantages of secure turbo code include higher-speed encryption and decryption with higher security, smaller encoder/decoder size, and greater efficiency.

On the other hand, spread spectrum is a signaling technique in which a signal is spread by a pseudo-noise sequence over a wide frequency band with frequency bandwidth much wider than that contained in the frequency ambit of the original information. The main difference between conventional cryptographic systems and spread-spectrum systems lies in their key sizes. Traditional cryptographic systems can have very large key spaces. However, in a spread spectrum system, the key space is limited by the range of carrier frequencies and the number of different sequences.

7.3.4 POWER AND SIGNAL DETECTION

Data protection can also be facilitated using power and signal detection approaches. The usual schemes in these approaches involves the use of directional antennas and the injection of artificial noise.

Directional transmission can improve spatial reuse and enlarge geographical coverage, as beam width is inversely proportional to peak gain in a directional antenna. If an omnidirectional antenna is used, a node in the coverage range of a jammer would not be able to receive data securely. On the other hand, if a directional antenna is used, the node would still be able to receive data from the directions not covered by the jamming signals. Hence, the deployment of directional antennas can improve wireless network capacity, avoid physical jamming attacks, and enhance data availability.

Artificial noise is generated using multiple antennas or the coordination of relaying nodes [23]. This noise is utilized to impair the intruder's channel, but it does not affect the intended receiver's channel because the noise is generated in the null-subspace of the legitimate channel. Relying on artificial noise, secret communications can be achieved even if the intruder enjoys a much better channel condition than the intended receiver.

Example 7.2

Figure 7.2 shows the BER performance of a legitimate receiver and the eavesdropper with respect to the ratio of the variance of the artificial noise to that of the legitimate receiver's channel noise (α) for different ratio of the energy per bit to the artificial noise (β). A random MIMO 4×4 system with BPSK modulation is adopted.

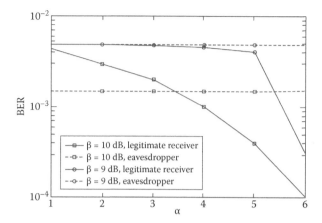

FIGURE 7.2 BER performance of a legitimate receiver and an eavesdropper when artificial noise is added at the transmitter. (From Y.-S. Shiu et al., *Physical Layer Security in Wireless Networks: A Tutorial. IEEE Wireless Communications* 18, 2: 66–74, 2011.)

The BER performance of both receivers improves as α increases, but the eavesdropper's performance is kept almost constant with respect to ratio β, whereas the BER for legitimate receiver improves as α increases. If the legitimate receiver's channel noise is given, parameter α can be increased by increasing the variance of the artificial noise while simultaneously increasing bit energy such that β stays unchanged. The artificial noise can, thus, be adjusted with the aid of experimental data to choose an operating point that maximizes the performance gain between the legitimate and eavesdropper receivers.

7.4 SECURITY ASPECTS FOR MOBILE SYSTEMS

Because of the characteristics of wireless medium, limited protection of mobile nodes, nature of connectivity, and lack of centralized managing point, mobile networks are too vulnerable and often subject to attacks. If a network does not have any inherent security, as a result, there is no guarantee that a received message (a) is from the channel sender, (b) contains the original data that the sender put in it, or (c) was not sniffed during transit [24].

To distinguish from other aspects of security, the goals of infrastructure security of mobile communication networks are identified as the following [25]:

- Physical security of the network infrastructure including the nodes and the cables
- Access control to the network infrastructure nodes

- Availability of the network infrastructure
- Security of network management signaling
- Amenable to security management
- Support secure coupling to foreign networks

The mobile networks are open to different services and service providers. This means that the mobile network operator has trust relations with different networks and services it provides to its users. Based on the security goals, there is a nonexhaustive list of major requirements on infrastructure security. At first, all the elements of a communication network should be hardened so that it is resistant to various security attacks.

Maintaining latest update levels of network elements and their software is an essential part of any operations and management concept applied in a network domain. Critical infrastructure elements must be identified and well protected. The necessity of redundant elements should be carefully evaluated, for instance, to still allow management during exceptional situations. Information on the network's internal structure, including the topology of the platform types, the distribution of fundamental elements, capacities, or customer data concerning location, service usage, usage pattern, account information, and so on, should be made available only to authorized parties and only to the extent that is actually required. The limited availability of this information reduces the knowledge about potential targets.

Also, an intrusion detection system should be deployed in the network to detect, monitor, and report security attacks such as DoS and attacks that utilize system flows, and any compromise of system security. Fast response to security attacks and automatic recovery of security compromise must be provided to increase the probability of business operation and to mitigate the effects of attacks. On the other hand, the network should be easy to manage and realistic security policies must be developed. Also, rapid couplings of the networks of different administrative domains must be secured (authentication, integrity, confidentiality, availability, and antireplay protection) against external attackers. Mutual dependencies between the infrastructure security and new scenario to be created must be minimized. Finally, authentication, confidentiality, integrity, antireplay protections for network management signaling will be provided.

7.4.1 Network Operators' Security Requirements

A successful mobile network operator will play a central role in future mobile networks while having a "partnership" with various participants of the value chain. Because the network operator is the main contact point of the user, their security requirements on mobile systems should be carefully considered by the operator with support from other parities. The user's security requirements, such as protecting its terminal and data from possible network attack, are regarded as part of the operators requirements because of the direct influence on operator's business when the requirements are not fulfilled. Mobile network operators have the responsibility to take care of party-specific (security) concerns while balancing this with suitable security solutions to protect the mobile network infrastructure.

Because user satisfaction is crucial for the network operators' and users' security, requirements must be regarded as one of the most important issues for them. Any compromise of security that has an effect on the user's assets may finally turn out to be a serious problem for the network operator's business. The users' requirements of security in future mobile networks can be categorized as

- Terminal security (includes access control, virus-proof, and theft prevention)
- Communication and data privacy (includes security of voice and data communication, privacy of location, call setup information, user ID, call pattern, etc.)
- Service provision security (service availability should be ensured to prevent DoS attack as well as to provide e-commerce security)
- Security against fraudulent service providers

Because of competition, a mobile network operator will usually enhance its service provision (capacity, quality, variety of services, etc.) to satisfy its customers, possibly by cooperating with other partners, which could be service providers or other network operators.

A heterogeneous network means the combination of different network technologies and possible opening of the operators' managed/controlled network to the Internet, which is not under the control of anyone. This, besides the general security requirements for a network, brings several new security issues:

- Secure attachment and detachment to/from network must be provided
- Access control to various services or network elements must be provided by the operator
- Trust relationship should be built between heterogeneous systems
- Changes in wireless medium require adaptation in physical and MAC layers should not compromise security
- To prevent rogue BSs, the mobile network operator should have mechanisms that will identify such BSs and thus protect the users
- Network must be installation and repair resistant

- Operations and management of security solutions must be possible and relatively simple
- Extension of network should not lead to weakness in security
- Network architecture should not compromise the extensibility of security services

Services and contents for the user can be provided by the mobile network operator, service providers, content providers, and other business or suite owners. There are some security requirements that arise from the mobile network operator's point of view. First of all, service should be provided to the specified set of users (authentication), according to the contractual obligation agreed (authorization), and the usage should be accountable. Rogue service or content providers can appear and methods must be developed to deter them. Then, secure access to service from any partner should be provided. Also, the operator should take care that the service providers are correctly charged, or if the service provider is paying the operator, then the operator should take care that bills the service provider correctly. Because cooperation with many other service providers is expected in future communication and service provision systems, nonrepudiation will become very important between operators and service providers to prevent and combat fraud. However, appropriate business models may be more efficient than technology means. Finally, in the future, the number of network operators will increase and, thus, a service that can be provided is openness toward each other, which will create a positive perception for users.

7.4.2 Security Architecture for Mobile Systems

Because the emerging mobile architectures (WiMAX, LTE, LTE-A, etc.) are designed to support flat IP connectivity and full interworking with heterogeneous wireless access networks, the novel unique features bring some new challenges in the design of the security mechanisms. The LTE network has been specified by the 3GPP on the way toward next generation mobile systems to ensure the dominance of the cellular communication technologies (see Chapter 1).

Security architecture for mobile systems (Figure 7.3), with five feature groups, is defined by the 3GPP committee [26]. Each of these feature security groups meets certain threats and accomplishes certain security objectives:

- Network access security (I)—provides the user equipment with secure access to the evolved packet core (EPC) and protects against various attacks on the radio access link. At this level, security

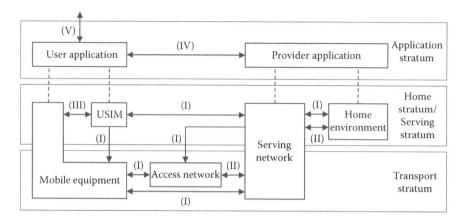

FIGURE 7.3   3GPP security architecture.

mechanisms such as integrity protection and ciphering between the universal subscriber identity module (USIM), mobile equipment (ME), the evolved-universal terrestrial radio access network (E-UTRAN), and the entities in the EPC, are implemented

- Network domain security (II)—enables nodes to securely exchange signaling data and user data, and protect against attacks on the wire line network
- User domain security (III)—provides a mutual authentication between the USIM and the ME before the USIM accesses the ME
- Application domain security (IV)—enables applications in the user equipment (UE) and in the service provider domain to securely exchange messages
- Visibility and configurability of security (V)—enables the user to be informed whether a security feature is in operation or not and whether the use and provision of services should depend on the security feature

7.4.3   Security in LTE Systems

Concerning researches on the LTE/LTE-A features, the following security aspects can be identified [27]: cellular security, handover security, IMS security, HeNB security, and machine type communication (MTC) security.

7.4.3.1   Cellular Security

A mutual authentication between the UE and the EPC is the most important issue in the LTE security framework. LTE systems utilize the authentication and key agreement (AKA) procedure to achieve mutual authentication between the UE and the EPC and generate a ciphering key (CK) and an integrity key (IK), which are used to derive different session keys for encryption and for integrity protection. Owing to the support of non-3GPP access, several different AKA procedures are implemented in the LTE security architecture when the UEs are connecting to the EPC over distinct access networks.

In the case when an UE connects to the EPC over the E-UTRAN, the mobility management entity (MME) represents the EPC to perform a mutual authentication by the evolved packet system (EPS) AKA protocol [26], as shown in Figure 7.4. On the other hand, when a UE connects to the EPC via non-3GPP access networks, the non-3GPP access authentication will be executed between the UE and the AAA server. The authentication signaling will pass through the proxy AAA server in the roaming scenarios. The trusted non-3GPP access networks can be preconfigured at the UE. If there is no preconfigured information at the UE, the non-3GPP network will be considered as an untrusted access. For a trusted non-3GPP access network, the UE and the AAA server will implement the extensible authentication protocol-AKA (EAP-AKA) or improved EAP-AKA to accomplish access authentication. In the case of untrusted non-3GP access network, the UE and the evolved packet data gateway (ePDG) need to perform the IPsec tunnel establishment. To establish the IPSec security associations, the UE and the ePDG shall use the Internet key exchange version 2 (IKEv2) protocol with EAP-AKA or improved EAP-AKA.

The EPS AKA protocol has some improvements over the UMTS AKA so that it can prevent some security attacks such as redirection attacks, rogue

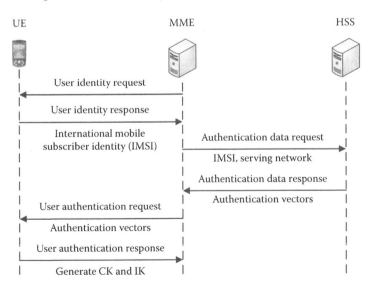

FIGURE 7.4 EPS AKA authentication.

base station attacks, and man-in-the-middle (MitM) attacks. However, there are still some weaknesses in the current LTE security mechanism [27], for example, lack of privacy protection and DoS attacks prevention in the EPS AKA scheme. As an improvement, a security-enhanced AKA (SE-EPS AKA) scheme based on wireless public key infrastructure (WPKI) has been proposed [28]. The scheme ensures the security of user identity and the exchanged message with limited energy consumption by using elliptic curve cipher encryption.

7.4.3.2 Handover Security

The 3GPP committee has specified the security features and procedures regarding mobility within E-UTRAN as well as between the E-UTRAN and the others RANs. To mitigate the security threats posed by malicious BSs, the LTE security mechanism provides a new handover key management scheme to refresh the key materials between an UE and an eNB whenever the UE moves from one eNB to another. However, a lot of vulnerabilities can still been found in the LTE mobility management procedure and the handover key mechanism (lack of backward security, vulnerability to desynchronization and replay attacks, etc.).

As roaming users expect a seamless handover experience (see Chapter 4) when switching from one wireless network to another, fast and secure handover operations need to be supported by the networks. As perspective solutions, fast and secure reauthentication protocols for the LTE subscribers to perform handovers between the WiMAX and the Wi-Fi systems have been proposed by Al Shidhani and Leung [29], which avoids contacting authentication servers in the LTE networks during the handovers. By these schemes, the EAP-AKA protocol for the handovers from a WiMAX to a Wi-Fi and initial network entry authentication (INEA) protocol for the handovers from a Wi-Fi to a WiMAX can be improved by including extra security parameters and keys to speed up the reauthentications in future vertical handovers. The modified version of the EAP-AKA and INEA has the same messaging sequences as that in the standard EAP-AKA and INEA, which can avoid interoperability problems with other services without a loss of capabilities due to the modifications. The proposed scheme can achieve an outstanding performance in terms of the signaling traffic reauthentication and reauthentication delay compared with the current 3GPP standard protocols. Also, this approach can provide several security features including forward and backward secrecy. However, the scheme can only support single-hop communications between UE and PoA.

7.4.3.3 IMS Security

IMS is an overlay access-independent architecture that provides the 3GPP networks with multimedia services such as IP telephony, video conferencing, etc. [30]. To access multimedia services, the UE needs a new IMS subscriber identity module (ISIM) with stored authentication keys and functions. Because SIP is used for control and signaling, the main architectural component in the IMS is the SIP proxy, known as the call session control function (CSCF). The CSCF represents the home subscriber server (HSS), providing session authentication and control of the multimedia services.

Currently, in the IMS security issues, most research is focused on registration authentication procedures. Multimedia services will not be provided until the UE has successfully established a security association with the network. In addition, a separate security association is required between the UE and the IMS before access is granted to multimedia services. UE has to be authenticated in both the LTE network layer and the IMS service layer [31]. A UE first needs to accomplish mutual authentication with the LTE network through the EPS-AKA before being given access to multimedia services. Then, an IMS AKA procedure is executed between the ISIM and the home network for AKA for the IMS. Once the network authentication and the IMS authentication are successful, the subscriber will be granted access to the multimedia services.

Because the IMS AKA works based on the EAP AKA scheme, it has several shortcomings such as vulnerability to DoS and MitM attacks, lack of sequence number synchronization, and high bandwidth and energy consumption. As mentioned previously, an IMS UE needs to execute two AKA protocols, the EPS AKA in the LTE access authentication and the IMS AKA in the IMS authentication, which brings high energy consumption for the energy-limited UE. In addition, these two AKA procedures share many similar operations, which increase the overall system complexity and results in QoS degradation.

7.4.3.4 HeNB Security

A HeNB is a low-power AP (aka femtocell) typically installed by a subscriber in residences or small offices to increase indoor coverage for high-speed multimedia services with the advantages of low cost and high QoS. This node is connected to the EPC over the Internet via the broadband

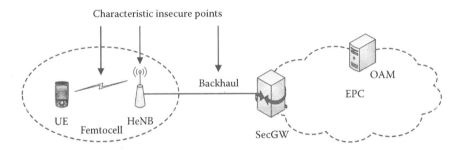

FIGURE 7.5   Characteristic insecure points in HeNB architecture.

backhaul (Figure 7.5). The backhaul between the HeNB and security gateway (SecGW) can usually be insecure. The SecGW represents the EPC to perform a mutual authentication with the HeNB by the IKEv2 with the EAP-AKA or certificates-based scheme. A HeNB needs to be configured and authorized by the operation, administration, and maintenance (OAM) entity. When a UE accesses the network via a HeNB, the MME will first check whether the UE is allowed to access the target HeNB based on the allowed closed subscriber group (CSG) list. Then, a secure access authentication between the UE and the MME will be performed by the EPS AKA.

Most of the HeNB security vulnerabilities arise from the insecure wireless links between the UE and the HeNB and the backhaul between the HeNB and the EPC. To overcome these vulnerabilities, the corresponding countermeasures have been discussed by the 3GPP committee [32]. However, current 3GPP specification has still not addressed some requirements of the HeNB security [27]:

- Lack of a mutual authentication between the UE and the HeNB. Current HeNB security mechanisms cannot prevent various protocol attacks including eavesdropping attacks, MitM attacks, masquerade attacks, and compromising subscriber access list because it does not provide robust mutual authentications between the UE and the HeNB. In addition, the HeNB is not a sufficiently trusted party if the core network and OAM authenticate it independently because the fairness between them is not valid in the IP-based network.
- Vulnerability to DoS attacks. Because of the exposure of the entrance points of the core network to the public Internet, the HeNB architecture in the LTE network is subject to several Internet-based attacks, especially DoS attacks.

The issues of the authentication and access control of the HeNB users have been addressed by Golaup et al. [33]. When a UE wants to access the network via a HeNB, the core network is additionally responsible for performing access controls. To perform access control, the core network must maintain and update a CSG list. The information contained in the CSG list is stored as subscription data for the UE in the HSS and is provided to the MME for access control. An improved AKA mechanism for HeNB, which adapts proxy signature and proxy-signed proxy signature, was proposed by Han et al. [34]. By this scheme, the OAM and the core network have a contractual agreement on the installation, operation, and management of the HeNB by issuing a proxy signature for each other. Then, the OAM redelegates its proxy-signing capability to a HeNB. The core network also redelegates its proxy-signing capability to the HeNB and issues its own signature to a UE. Finally, the mutual authentication between the UE and the HeNB can be achieved with the proxy signature on behalf of the OAM and the core network. The scheme can prevent various attacks such as masquerading HeNB, MitM attacks, and DoS attacks. However, it incurs a large amount of computational and storage costs and requires several changes to the existing architecture, which makes the system more infeasible in a real environment.

Significant threats to the security and the privacy of the femtocell-enabled LTE networks have been reviewed in an article by Bilogrevic et al. [35], with novel research directions tackling some of these threats. A solution to the issue of identity and location tracking by assigning and changing identifiers based on the context was introduced. This approach provides a new user-triggered identifier change strategy instead of a network-controlled strategy. Here, the UE can dynamically decide when to change identifiers based on their own observation of the surroundings, such as node density, device speed, and mobility pattern. In addition, a protection mechanism against DoS attacks with a HeNB deployment in the LTE architecture has also been suggested. It has been pointed out that solutions relying on cooperation among several participating entities, such as ISPs, can provide efficient protection against DoS attacks.

7.4.3.5 MTC Security

The MTC, also called M2M communication, is viewed as one of the next sophisticated techniques for future wireless networking. Different from the traditional wireless networks, the MTC is defined as a form of data communication between entities that does not necessarily need human interaction [36]. It is mainly used for automatically collecting and delivering information for measurements. MTC security architecture includes three different security areas:

1. Security between the MTC device and 3GPP network
2. Security between the 3GPP network and the MTC server/user/application
3. Security between the MTC server/user/application and the MTC device

The introduction of the MTC into the LTE systems remains in its infancy. This involves a lot of unique features such as massive number of devices, small and infrequent data transmissions, distinct service scenarios, and fewer opportunities for recharging the devices, which brings unprecedented challenges for the 3GPP to achieving its standardization. Thus, the current LTE network needs to overcome many technical obstacles in the system architecture, radio interface, resource, and QoS management to promote the rapid development of the MTC. Security issues inherent in the MTC have not been well explored by the 3GPP committee and other researchers. Even more, there is no adequate security mechanism between the MTC device and RAN. In addition, there is no specific mechanism to ensure the security of communications among MTC devices. Furthermore, to support diverse MTC features, the LTE system architecture will be improved, which can incur some new security issues. Thus, the 3GPP committee needs to further investigate the security aspects of the MTC.

MTC devices are extremely vulnerable to several types of attacks such as physical attacks, compromise of credentials, protocol attacks, and attacks to the core network because they are typically required to have low capabilities in terms of both energy and computing resources, deployed without human supervision for a long time.

A group-based access authentication scheme [37], through which a good deal of MTC devices can be simultaneously authenticated by the network and establish an independent session key with the network, can achieve robust security with desirable efficiency and avoid the signaling overhead in the LTE networks. When multiple MTC devices request access to the network simultaneously, the MME authenticates the MTC group by verifying the aggregate signature generated by the group leader on behalf of all the group members. Then, each trusts the MME by verifying the elliptic curve digital signature via the group leader. Finally, a distinct session key between each MTC device and the MME will be established according to the different key agreement parameters. The scheme can not only achieve a mutual authentication and a key agreement between each MTC device in a group and the MME at the same time but can also greatly reduce the signaling traffic and thus avoid network congestion. However, it may bring a lot of computational costs due to the use of the elliptic curve digital signature and the aggregate signature. In addition, the scheme requires the devices to support both LTE and Wi-Fi interfaces.

There are still many open challenges for the practical enforcement of the MTC security [27]:

- Security mechanisms to ensure reliable high-speed connectivity for sensitive data are required
- A balance between the encryption and the small amount of information transmission needs to be achieved
- Novel access authentication schemes for congestion avoidance for the simultaneous authentication of multiple devices are required
- E2E secure mechanisms for MTC are mandatory
- Secure mechanisms for supporting restricted and high-speed mobility of the MTC devices are required

7.5 SECURITY IN CRNs

CRNs (see Chapter 2) are vulnerable to various attacks because they are usually deployed in unattended environments and use unreliable wireless medium. However, it is not a simple task to organize the implementation of security defenses in CRNs. One of the major obstacles in deploying security in CRNs is that they have limited computational and communication capabilities. Security mechanisms, including trust management, have the ability to secure CRNs against attackers. Specific CRN applications have some unique features and correspondingly, some unique security requirements.

7.5.1 GENERAL SECURITY REQUIREMENTS IN CRNs

CR technology is more susceptible to attack compared with traditional wireless networks due to its intrinsic nature. Although security requirements may vary in different application environments, there are in fact some general requirements that provide basic safety controls such as [38] access control, confidentiality, authentication, identification, integrity, nonrepudiation, and availability.

Access control is a security requirement for the physical layer that restricts the network's resources to authorized users. Because different secondary users (SUs) coexist in CRNs, collisions can happen if they simultaneously move to and use the same spectrum band according to their spectrum sensing results. Thus, the access control property should coordinate the spectrum access to avoid collisions.

Confidentiality is closely related to integrity. Although integrity ensures that data is not maliciously modified in transit, confidentiality ensures that the data is transformed in such a way that it is unintelligible to an unauthorized entity. This issue is even more pronounced in CRNs, in which the SU's access to the network is opportunistic and spectrum availability is not guaranteed.

Authentication has the primary objective of preventing unauthorized users from gaining access to protected systems. It can be considered as one of the basic requirements for CRNs because there is an inherent requirement to distinguish between primary users (PUs) and SUs. An authentication problem occurs in CRNs when a receiver detects a signal at a particular spectrum, that is, how can a receiver be sure that the signal was indeed sent by the primary owner of the spectrum? According to Tan et al. [39], it is practically impossible to conduct authentication in CRNs other than in the physical layer. For example, a CR receiver is able to receive signals from TV stations and process them at the physical layer, but it may lack the component to understand the data in the signals. Therefore, if the authentication depends on the correct understanding of the data, at upper layers, the CR receiver will be unable to authenticate the PU. One way is to allow PUs to add a cryptographic link signature to its signal, so the spectrum usage by PUs can be authenticated.

Identification is one of the basic security requirements for any communication device, whereby a user is associated with his or her unique identifier (e.g., IMSI in 3GPP networks). A tamper-proof identification mechanism is built into the SU unlicensed devices. It would be advantageous for a CR to know how many networks exist, how many users are associated with each network, and even certain properties about the devices themselves. To achieve this level of information, it is essential for a CR to gather an accurate notion of the RF environment. Service discovery and device identification provide the necessary building blocks for constructing efficient and trustworthy CRNs.

Integrity is of importance in a wireless environment because, unlike their wired counterparts, the medium is easily accessible to intruders. It is related to the detection of any intentional or unintentional changes to the data occurring during transmission. Data integrity in CRNs can be achieved by applying higher cryptographic techniques.

Nonrepudiation techniques prevent either the sender or receiver from denying a transmitted message. In a CR ad hoc network setting, if malicious SUs violating the protocol are identified, nonrepudiation techniques can be used to prove the misbehavior and disassociate/ban the malicious users from the secondary network. The proof of an activity that has already happened should be available in CRNs.

Availability refers to the ability of PUs and SUs to access the spectrum in CRNs. For PUs, availability refers to being able to transmit in the licensed band without harmful interference from SUs. On the other hand, for SUs, availability refers to the existence of chunks of spectrum, in which the SU can transmit without causing harmful interference to the PUs. In CR, one of the important functions of this service is to prevent energy starvation and DoS attacks, as well as misbehavior.

7.5.2 CHARACTERISTIC ATTACKS IN CRNs

Depending on their target in security requirements, attacks in CRNs can be broadly categorized into two types [38]:

- *Selfish attacks* occur when an intruder wants to use the spectrum with higher priority. This attack meets its target by misleading other unlicensed users to believe he or she is a licensed user. In that way, the adversarial user can occupy the spectrum resource as long as that user wants. This selfish behavior does not obey the spectrum sharing scheme [40]. Selfish SUs increase their accessing probability by changing the transmission parameters to enhance their own utilities by degrading the performance of other users. Hence, the CRN performance is degraded.

- *Malicious attacks* are related to the cases when the adversary prevents other unlicensed users from using the spectrum and causes DoS. These types of attacks drastically decrease the available bandwidth and break down the whole traffic.

There are other different types of attacks; for example, attacks on spectrum managers [41]. If the spectrum manager is not available, communication between CR nodes is not possible. The spectrum availability should be distributed and replicated in CRNs, whereas the attack can be prevented by specific pilot channels in the licensed band. As for eavesdropping on the transmission range of CR, it is not limited to a short distance because it is using unlicensed bands.

In Table 7.3, characteristic attacks in CRNs are emphasized and classified depending on various protocol stack layers. Also, these attacks can adversely affect the final layer of the communication protocol stack because protocols that run at the application layer rest on the services provided by lower layers. Most of these attacks are comprehensively analyzed by Mathur and Subbalakshmi [42].

However, adversaries can launch attacks targeting multiple layers. These are also known as cross-layer attacks and can affect the entire cognitive cycle because attacks at all layers become feasible [3].

7.5.3 SECURE SPECTRUM MANAGEMENT

One of CR's functions is to detect spectrum holes by spectrum sensing. It keeps monitoring a given spectrum band and captures the information. CR users can temporarily use the spectrum holes without creating any harmful interference to the PUs. CR must periodically sense the spectrum to detect the presence of incumbents and quit the band once detected. The detection techniques that are often used in local sensing are energy detection, matched filter, and cylcostationary feature detection [2].

The main benefit of introducing security in the spectrum decision process is a stronger guarantee that the service of PUs will not be significantly disrupted. The resilience of the spectrum decision against malicious attackers protects the secondary network at no additional cost. Many existing dynamic spectrum access protocols make spectrum decisions based under the assumption that all involved parties are honest and there is no malicious outsider that can manipulate the decision process. Jakimoski and

TABLE 7.3

Characteristic Attacks in CRNs

Layer	Security Attack	Description
Transport layer	Key depletion attack	Because of a high number of sessions, the number of key establishments can increase the probability of the same key being used twice
	Jellyfish attack	An attacker causes the victim's cognitive node to switch from one to another frequency band, thereby causing considerable delay
	Lion attack	A malicious node actually causes the jamming to slow down the throughput of the TCP by forcing handover's frequency
Network layer	Network endoparasite attack	The malicious nodes attempt to increase the interference at a heavily loaded high-priority channel
	Channel endoparasite attack	A compromised node launches an attack by switching all its interfaces to the channel that is being used by the highest priority link
	Low-cost Ripple effect attack	Compromised node can transmit the misleading channel information and forces other nodes to adjust their channel assignments
Link layer	Biased utility attack	A malicious SU can intentionally tweak parameters of the utility function to increase its bandwidth
	Asynchronous sensing attack	A malicious SU can transmit asynchronously instead of synchronizing the sensing activity with other SUs in the network during the sensing process
	False feedback attack	False feedback from one user or a group of malicious users can make other SUs take inappropriate action and violate the protocol terms
Physical layer	Intentional jamming attack	The malicious SU jams PUs and other SUs by intentionally and continuously transmitting in a licensed band
	Primary receiver jamming attack	A lack of knowledge about the location of primary receivers can be used by a malicious entity to intentionally cause harmful interference to a victim primary receiver
	Sensitivity amplifying attack	Some PU detection techniques have higher sensitivity toward primary transmissions with a view to preventing interference to the primary network
	Overlapping attack	Transmissions from malicious entities can cause harm to PUs and SUs, not only in one network but also in other CRNs belonging to the same geographical region

Subbalakshmi [43] assumed that there is some synchronization among the nodes in the cluster in the network. The time is divided into equal length intervals, whereas the nodes know when each cycle begins and ends. They are also aware of the schedule of the events during a cycle, that is, which node sends its channel availability data, which channels it uses, etc. Three main events are handled in a given cycle: (1) one or more nodes can join the spectrum decision process in a given cluster, (2) the nodes of the cluster send their spectrum sensing data, and (3) the cluster head sends the final channel assignment to the other nodes.

The protocols of different layers of CRNs must be able to adapt to the channel parameters of the operating frequency. Also, they must be apparent to the spectrum handover and related latency. When implementing an algorithm, the best available spectrum should be chosen depending on the channel characteristics of the available spectrum and the QoS requirements of the CR user.

7.6 SECURITY IN WMNs

The multihop behavioral characteristics of WMNs (see Chapter 5) creates challenges in the security of traffic operations while in transmission through the WG to the wireless cloud. The dynamic topology updates further expose the whole network security to persistent and corruptible attacks. The reliability and authentication of data traffic in WMN during neighborhood node exchanges through link state and in routing operations are loose and very insecure. The ease in WMN integration with other wireless nodes and communication networks, like in broadband and multimedia, has also established the necessity for an unyielding privacy protection and security mechanisms.

The distributed-sequenced mechanism in the MAC channel frames also creates susceptibility to attacks whereas the mobile mesh client nodes and its consequent dynamic topology in the wireless mesh infrastructure also establish the need for a more effective, resilient, and comprehensive security system in WMNs [4]. The constraint in WMN security creates the challenge of possible attacks by invasive malicious codes when an attack becomes distributed in the architecture through the simple dynamism of mesh. These attacks compromise confidentiality and integrity, and violate the privacy of the users. Furthermore, the nodes can also be compromised by the operation of traffic transmission, unverified router information exchange traffic, and network notification infiltration. Security in WMNs explores key security challenges set in diverse scenarios, as well as emerging standards that include authentication, access control and authorization, attacks, privacy and trust, encryption, key management, identity management, DoS attacks, intrusion detection and prevention, secure routing, and security policies [44].

7.6.1 Typical Security Attacks in WMNs

Distributed-sequenced network architecture, vulnerability of the shared wireless medium in the access channel, and the neighborhood clients' information exchanges expose the WMN to various security attacks. The WMN is a dynamic multihop environment and frequent changes in the topology require security authentication for notification updates. Security attacks can occur at the routing layers, during the client node updates, and as notification message infection. In addition, there are communication security risks during routing operation and packet transmission. Typical attacks and security impairments in WMN protocol stacks are presented in Table 7.4. These attacks create negative impairments, message corruption, network infiltration, and depreciate network resources.

In summary, these attacks damage and compromise data traffic by affecting essential network elements and functionalities such as APs, MTs power supplies, node mobility, routing tables, and cache [4]. Security challenges can be resolved by counterattacked mechanisms, intrusion detection concepts, and network resolution or diffusion of affected threats in the network. In multihop wireless network architecture, the security mechanism uses a comprehensive security mesh key and encryption. A secure MAC layer also ensures that traffic from WMNs is authorized, and doing this safeguards access into neighborhood nodes. WMN security attacks can occur on a protocol layer or on a multiprotocol communication. Therefore, a multiprotocol mesh solution approach remains the most promising technique to resolving these security issues.

DoS is a major security attack that frequently occurs in WMNs. In distributed-sequenced traffic processing, it results in a worse scenario called distributed denial of service (DDoS) and it is an even more severe threat to WMNs. This security threat occurs when requesting nodes are not offered with requested services and updates in a maximum given

TABLE 7.4
Security Attacks in WMN Protocol Stack

Application layer	Resources depletion and application authentication errors
Transport layer	Domain name system (DNS) spoofing and traffic attacks spurious message
Network layer	Black hole, wormhole, gray hole, rushing attacks, location disclosure, and DoS
Link layer	Jamming and flooding attacks
Physical layer	Tampering, collision jamming, physical attacks and battery exhaustion

time. It affects network availability and reduces communication between network devices in data transmission and reception. Also, it violates the access security and authentication of traffic in ubiquitous WMN, thus preventing communication of transmissions by stopping a sending/receiving device and possibly any links as well. DDoS attack on mesh routers (MR) can paralyze the entire transmission of data packets across the WMN. The DDoS threat can appear as a wormhole, black hole, or gray hole, and as distributed flooding DDoS attacks in the network layer.

DDoS is always very difficult to manage in large WWANs. They consume large network resources to the extent of rendering the WMN ineffective. DDoS are spread by the naturally distributed processing architecture of the network. It normally quickly floods AP and backbone MR using those hierarchal control points to congest the WMN traffic. Antivirus software is usually employed to neutralize the attacks and to destroy the DDoS zombies.

The DDoS can also be perceived as traffic flooding attack in the WMN, where the attack compromises a large number of mesh clients in a campus network. It overflows the network and causes flooding. It congests the network resource and occupies the available bandwidth. The WMN routing mechanism in multihop transmission keeps changing from disjointed traffic, to slow and quiet traffic in the backbone, to active exchanges in the peripherals. The attackers can infiltrate the routing mechanism and its performance by altering, tampering, or even forging of false messages or router notifications. The attacker can modify packet data messages using replicate nodes to carry out DDoS attacks. In spoofing of wireless infrastructure, the attacker uses a replicate or silent MitM attack to execute information disclosure threat in an enterprise deployment such as WMN. These attacks can be alleviated using the complete EAP method [45], which allows authentication between clients and the WMN infrastructure.

Furthermore, wormholes and black holes are attacks that cause security weaknesses in the WMN routing. The black hole is a security attack situation in which a malicious node uses the routing protocol to advertise itself as having the shortest path to the node. In this situation, the malicious node advertises itself to a node that it wants to intercept the packet. In a wormhole security attack, an intruder receives packets at one location and tunnels them selectively to another location in the network. Once the wormhole link (tunnel) is established, it can be used by the attacker to compromise the integrity of the WMN security. In a wormhole attack, the attacker uses DDoS techniques to threaten the WMN security.

In the physical layer, tampering with security threats involves the attacker modifying the traffic information on routed packets. The WMN works on the principle of mutual meshing and supportive hierarchal network. Because of this fact, the attacker can distort the sequence numbering, number of hops, or other frame fields while transmitting data. Consequently, this causes network resources to reroute, redirect, and reconfigure routes, taking up bandwidth and processing time and, consequently, degrading the eventual performance of the entire WMN [4]. It can sometimes result in a looping problem and high overheads in the network. Tampering generally occurs when routing information is lacking in reliability checks and accuracy.

In addition, the physical layer is susceptible to signal jamming. This is the security threat that occurs when an attacker jams the interface of a transmitting or routing node on the physical channel. They usually employ frequency hopping on the communicating node or may even use the defense techniques of spread frequency spectrum matching over the communicating signal range. This security threat is very difficult to prevent using detection intrusion because the attacker blocks the sensing devices. In some level of the hierarchal model, the attack is flooding and causes a frequency jamming attack in the physical layer.

Other security threats involve forging. In this scenario, the attacker forges network notification messages and broadcasts wrong information to the network and other surrounding nodes. Wrong routing information such as link availability and hop counts are usually forged by the attacker. The poor verification and authentication of the packet data contributes more to this threat in WMN. In addition, other threats in the WMN routing include resource depletion attacks and rushing attacks. These security threats attack bandwidth, the node header, routing table, cache, and the battery of the node routers. In the rushing and resource depletion attacks, the attacker continuously sends a lot of route request packets over the WMN in a short time, creating a bottleneck and congestion for nodes processing these network route notifications. These insidious attacks deplete or slow down the network resources.

7.6.2 SELECTIVE JAMMING

A wireless environment is invariably vulnerable to external and internal attacks. External attacks take the form of random channel jamming, packet replay, and packet fabrication, and are launched by foreign devices that are unaware of network secrets (e.g., cryptographic credentials and pseudorandom spreading codes). They are relatively easier to counter through a combination of cryptography-based and robust communication techniques. In contrast, internal (insider) attacks, which are launched from compromised nodes, are much more sophisticated. These attacks exploit knowledge of network secrets and protocol semantics to selectively and adaptively target critical network functions.

Although all wireless networks are susceptible to insider attacks, WMNs are particularly vulnerable to them for a number of reasons [46]:

- Mesh nodes are relatively cheap devices with poor physical security, which makes them potential targets for node capture and compromise
- Given their relatively advanced hardware, WMNs often adopt a multichannel design, with one or more channels dedicated to control/broadcast purposes. Such static design makes it easier for an attacker to selectively target control/broadcast information
- The reliance on multihop routes further accentuates the WMN's vulnerability to compromised relays, which can drop control messages to enforce a certain routing behavior (e.g., force packets to follow long or inconsistent routes)

Attack selectivity can be achieved, for example, by overhearing the first few bits of a packet or by classification of transmissions based on protocol semantics. Internal attacks cannot be mitigated using only proactive methods that rely on network secrets because the attacker already has access to such secrets. They additionally require protocols with built-in security measures, through which the attacker can be detected and its selective nature neutralized.

Jamming in wireless networks has been primarily analyzed under an external adversarial model as a severe form of DoS against the physical layer. Existing antijamming strategies employ some form of spread spectrum communication, in which the signal is spread across a large bandwidth according to a pseudo-noise code. However, the spread spectrum technique can protect wireless communications only to the extent that the pseudo-noise codes remain secret. Insiders with knowledge of the commonly shared pseudo-noise codes can still launch jamming attacks. Using their knowledge of the protocol specifics, they can selectively target particular channels/layers/protocols/packets.

7.6.2.1 Channel-Selective Jamming

Control channels in a WMN are reserved for control information broadcasting. These channels facilitate operations such as network discovery, time synchronization, coordination of shared medium access, and routing path discovery without interfering with the communications of stations with mesh APs (MAPs) [46]. An adversary who selectively targets the control channels can efficiently launch a DoS attack with a fairly limited amount of resources. To launch a channel-selective jamming attack, an intruder must be aware of the location of the targeted channel, whether defined by a separate frequency band, time slot, or pseudo-noise code. It should be noted that the control channels are inherently broadcast. Thus, every intended receiver must be aware of the secrets used to protect the transmission of control packets.

Example 7.3

Consider the effect of channel-selective jamming on MAC protocols for multichannel WMNs. A multichannel MAC (MMAC) protocol with split-phase design [47] is employed to coordinate access for multiple nodes residing in the same collision domain to the common set of channels. In this design, time is split into alternating control and data transmission phases. During the control phase, every node converges to a default channel to negotiate the channel assignment. In the data transmission phase, devices switch to the agreed upon channels to perform data transmissions.

By employing a channel-selective strategy, an insider can jam only the default channel and only during the control phase. Any node that is unable to access the default channel during the control phase must defer the channel negotiation process to the next control phase, thus remaining inactive during the following data transmission phase. This attack is illustrated in Figure 7.6. It should be noted that the effect of this channel-selective jamming attack is propagated to all frequency bands at a low energy overhead because only a single channel is targeted and only for a fraction of time.

Several antijamming methods have been identified to address channel-selective attacks from insider nodes [46]:

- Replication of control information represents an intuitive approach to counter channel-selective jamming using multiple broadcast channels. In this case, an insider with limited hardware resources cannot jam all broadcasts simultaneously.

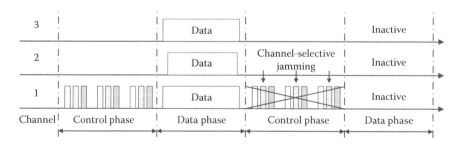

FIGURE 7.6 Channel-selective jamming attack on MMAC with split-phase design.

- Assignment of unique pseudo-noise codes is an alternative method for neutralizing channel-selective attacks by dynamically varying the location of the broadcast channel based on the physical location of the communicating nodes. Broadcast communication can be repaired locally if a jammer appears, without the need to reestablish a global broadcast channel.
- Elimination of secrets is designed to counter selective insider jamming attacks in the first place. A transmitter randomly selects a public pseudo-noise code. To recover a transmitted packet, receivers must record the transmitted signal and attempt to decode it using every code in the public codebook.

7.6.2.2 Data-Selective Jamming

To improve the energy efficiency of selective jamming and reduce the risk of detection, an inside attacker can exercise a greater degree of selectivity by targeting specific packets of high importance. An example of this attack is shown in Figure 7.7, where MP_a transmits a packet to MP_b. Jammer (compromised) MAP_j classifies the transmitted packet after overhearing its first few bytes. MAP_j then interferes with the reception of the rest of the packet at MP_b.

Packet classification can also be achieved by observing headers of various layers or implicit packet identifiers such as packet length, or precise protocol timing information. For example, control packets are usually much smaller than data packets. The packet length of an eminent transmission can be inferred by decoding the network allocation vector field of request-to-send (RTS) and clear-to-send (CTS) messages.

An intuitive solution for preventing packet classification is to encrypt transmitted packets with a secret key. In this case, the entire packet,

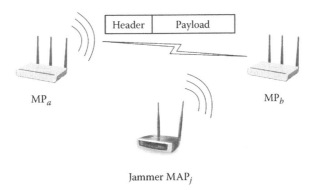

| Header | Payload |

MP_a MP_b

Jammer MAP_j

FIGURE 7.7 Data-selective jamming attack.

including its headers, has to be encrypted. Although a shared key suffices to protect P2P communications, for broadcast packets, this key must be shared by all intended receivers. In symmetric block encryption schemes, reception of one cipher text block is sufficient to obtain the corresponding plaintext block if the decryption key is known. Hence, encryption alone does not prevent insiders from classifying broadcast packets. To prevent classification, a packet must remain hidden until it is transmitted in its entirety.

7.7 SECURITY ASPECTS IN WMSNs

WMSNs have many unique features that differ from other wireless networks (see Chapter 6). However, similar to other wireless networks, one of the most significant challenges of sensor networks is the provision of secure communication. The lessons learned from protocol design for mobile ad hoc networks and scalar WSNs leads researchers to the conclusion that security is critical and should be taken into account from the start. The WMSN research community might be tempted to hastily assemble protocols without properly provisioned security features. Nevertheless, attempts have been made to retrofit security measures to those protocols. Most of the time, these will not be feasible and the original design decisions will no longer be valid.

From the general security point of view, there is no strict border between scalar WSNs and WMSNs. Therefore, some of the traditional security solutions for WSNs can be easily adopted for WMSNs. However, WMSNs have some novel features that stem from the fact that some of the sensor nodes will have multimedia and higher computational capabilities. This brings new security challenges as well as new protection opportunities [7].

7.7.1 POTENTIAL OF ASYMMETRIC CRYPTOGRAPHY

The area where multimedia streaming applications are different from other applications in wireless sensor networks is in the usability of encryption techniques to ensure confidentiality [48]. In a WSN, the public key cryptography schemes are not suitable because of their high power and computational requirements. Standard and advanced symmetric encryption schemes are commonly used. However, these schemes are unsuitable considering multimedia content. Multimedia data is generally larger in size and the use of these symmetric encryption schemes has memory and computational requirements that are unsupportable by the sensor nodes.

Using continuously developing technology, multimedia sensor nodes will be increasingly more powerful in the following years, even for the sake of properly handling the multimedia content. Moreover, in-node processing is going to be necessary to minimize bandwidth utilization. Abstractions of the sensed data will be computed inside the node and the node itself will decide whether the information should be transmitted and whether it should first send as a part of the abstracted data or as a part of the sensed data. Therefore, asymmetric cryptography will be feasible for wireless multimedia sensor nodes. Moreover, it is a perspective approach in many cases, from authentication to routing.

Furthermore, it has been shown that even nodes with highly constrained computing power limitations can use asymmetric cryptography when efficient asymmetric algorithms are used. Probably, elliptic curve cryptography algorithms are going to be the most attractive ones because of their unique characteristics such as the small key size, fast computations, and bandwidth savings [49]. Even relatively lightweight WMSN nodes will choose asymmetric cryptography as the best approach in most cases. In addition, time synchronization is generally not going to be needed because it is only required to provide authentication with symmetric cryptography.

Example 7.4

An elliptic curve over a finite field associated with a prime number $p > 3$ can be written as

$$y^2 = x^3 + ax + b, \tag{7.2}$$

where a and b are two integers that satisfy $4a^3 + 27b^2 \neq 0$ (mod p). Then, the elliptic group, $E_p(a, b)$, is the set of pairs (x, y), where $0 \leq x, y < p$, satisfying Equation 7.2.

Let $A = (x_1, y_1)$ and $B = (x_2, y_2)$ be in $E_p(a, b)$, then $A \odot B = (x_3, y_3)$ is defined as

$$x_3 = \lambda^2 - x_1 - x_2 \text{ (mod } p),$$
$$\tag{7.3}$$
$$y_3 = \lambda(x_1 - x_3) - y_1 \text{ (mod } p),$$

where

$$\lambda = \begin{cases} \dfrac{y_2 - y_1}{x_2 - x_1} & \text{if } A \neq B \\ \\ \dfrac{3x_1^2 + a}{2y_1} & \text{if } A = B \end{cases}. \tag{7.4}$$

As an illustrative example, points on the elliptic curve $E_{31}(1, 5)$ are given in Table 7.5.

TABLE 7.5

Points of the Elliptic Curve $E_{31}(1, 5)$

(0, 6)	(0, 25)	(1, 10)	(1, 21)	(3, 2)	(3, 29)	(6, 14)	(6, 17)	(7, 13)
(7, 18)	(11, 13)	(11, 18)	(12, 3)	(12, 28)	(13, 13)	(13, 18)	(14, 2)	(14, 29)
(15, 4)	(15, 27)	(16, 5)	(16, 26)	(19, 1)	(19, 30)	(21, 7)	(21, 24)	(25, 0)

7.7.2 VULNERABILITIES OF WMSNs

Due to the characteristics of restricted resources and operation in a hostile environment, WMSNs are subjected to numerous threats and vulnerabilities mainly related to DoS and traffic analysis aspects. One advantage of WMSN nodes is that they are more powerful than scalar nodes, which translates into more computational power for resisting DoS-related attacks. Nevertheless, attackers can always bring even more powerful computing resources to perform their attacks. DoS attacks are very complex because they can be performed in so many different ways, and against any of the different communication layers. Moreover, protecting a subset of WMSNs is obviously useless.

It is a well-known fact that trusting tamper resistance is problematic. Probably the best approach is to assume that tamper resistance is going to have limited effectiveness. Limiting the scope and the amount of sensible information, including keys, stored in each sensor is also of importance. In the case of networks that use asymmetric cryptography, in every node, there only needs to be its own private key and several public keys. So, when one node is compromised and tampered, the damage is more controlled. This is another point that supports the use of asymmetric cryptography over symmetric cryptography solutions.

Reactive jamming attacks are a specific sort of jamming attack in which the intruder only jams channels with detected activity. These types of attacks are characteristic for surveillance applications where it can be performed to block the sink node to prevent it from receiving reports of an ongoing intrusion.

Collision attacks represent a robust and efficient way of producing the same end result as the jamming attacks. In a collision attack, an intruder only needs to send a byte when it overhears a packet to force the receiver of the packet to discard the entire message. The amount of energy to perform such attacks is much lower when compared with jamming attacks. Thus far, a proposed solution to this attack has been the use of strong error-correcting codes. Nevertheless, for a given encoding, the attacker can just send more bytes than the error correction codes can correct. In addition, error-correcting codes add processing and traffic overhead.

Typically, attacks against the MAC protocols may be very hard to prevent but their effects are mostly local and can be isolated using redundant routing techniques. On the other hand, attacks at the networking layer, specifically those against the routing protocols, can disrupt the entire network operation. Existing WMSN routing protocols are extremely vulnerable to black hole and wormhole attacks.

Traffic analysis is the process of intercepting and examining messages to deduce information from patterns in communication. It is a very effective way to determine the geographic location of a sink node. For example, if a sink node is well-concealed visually, an adversary cannot determine its location by visually scanning the area where the WMSN is deployed. That person needs to analyze the network traffic to determine the location of a sink node in such cases. Furthermore, if the WMSN covers a large area, it is very difficult for the adversary to scan every location to find a sink node. However, by analyzing network traffic, an intruder can quickly track its location. In some other cases, it is impractical for the adversary to freely move from place to place to visually search a sink node. For example, an adversary monitoring a WMSN needs to hide from sensor nodes. Traffic analysis provides that person an efficient way to find the location of the sink node.

Even with hop-by-hop re-encrypted packets, an adversary can still deduce significant information that can reveal the sink node location by monitoring traffic volume, or by looking at time correlations [50]. The act of transmitting itself reveals information to the attacker, regardless of whether packet contents can be inspected. In the case of rate monitoring, the volume of transmissions can be exploited, whereas in the case of time correlation, an adversary can listen to a transmission and also the next-hop forwarding along a relay path. The adversary infers some path relationship between two neighboring nodes regardless of whether the packet is redisguised at each hop.

In a rate monitoring attack, an adversary monitors the packet sending rate of nodes near the adversary, and moves closer to the nodes that have a higher packet sending rate. On the other hand, in a time correlation attack, an adversary observes the correlation in sending time between a node and its neighboring node that is assumed to be forwarding the same packet. The adversary infers the path by following each forwarding operation as the packet propagates toward the sink node.

Besides basic countermeasures (e.g., hop-by-hop re-encryption, imposition of a uniform packet rate, and removal of correlation between a packet's receipt time and its forwarding time), more sophisticated countermeasures are proposed in a study by Deng et al. [50]. This technique introduces randomness into the path taken by a packet. Packets may also fork into multiple fake paths to further confuse an adversary. A technique is introduced to create multiple random areas of high communication activity to deceive an adversary about the true location of the sink node. The effectiveness of these countermeasures against traffic analysis attacks is demonstrated analytically and via simulation using three evaluation criteria: total entropy of the network, total overhead/energy consumed, and the ability to frustrate heuristic-based search techniques to locate a sink node.

7.8   CONCLUDING REMARKS

Wireless security has been an active and very broad research area since the last decade. Communications over wireless channels is, by nature, insecure and easily susceptible to various kinds of attacks. Regardless of how complex any wireless system becomes, the issue of security should always be approached and managed in a structured and uniform way. Theoretically, security can be viewed through different levels (e.g., content, communication, and network) as well as from a lot of different requirement perspectives (e.g., user, network, service). Generally, it is critical to ensure that confidential data are accessible only to the intended users rather than intruders.

Although physical layers are mainly different considering various radio access technologies, security aspects and challenges at this layer are practically common for all wireless multimedia systems. Numerous physical layer security approaches have been introduced and evaluated in terms of their abilities and computational complexity. The implementation of physical layer security in a real environment is part of a layered approach, and the design of protocols that combine traditional cryptographic techniques with physical layer techniques is an interesting research direction.

Mobile networks are known as a reliable and secure platform for voice communication. With the implementation of next generation mobile networks comes the need to open up these closely guarded networks to a much more diverse set of users and multimedia services. A lot of security issues for the LTE networks are still open research issues without comprehensive resolutions. Future mobile networks need to alleviate the overheads of the cryptographic operations to achieve a trade-off between the security functionality and the overheads. Design of efficient and secure group-based access authentication schemes for mass device connection is still a key challenging issue.

Along with the realization of cognitive radios, new security threats have been raised. Intruders can exploit several vulnerabilities of this new technology and cause severe performance degradation. Security threats are mainly related to two fundamental characteristics of cognitive radios: cognitive capability and reconfigurability. Threats related to cognitive capability include attacks launched by intruders that mimic primary transmitters and transmission of false observations related to spectrum sensing. On the other hand, reconfiguration can be exploited by attackers through the use of malicious code installed in cognitive radios. Furthermore, as CRNs are wireless in nature, they face all the classic threats present in traditional wireless networks.

Wireless mesh topology is exposed to numerous security challenges not only in the transmission operation but also in the overall security against foreign attacks. The vulnerabilities in broadband domain, challenges in the routing layer, as well as new concepts to solving security challenges in WMNs using traffic engineering resolution techniques, can be identified.

DoS, which is a main challenging problem in traditional sensor systems, will continue to be one of the main security issues in WMSNs. Because visual surveillance is expected to be the most important application, traffic analysis attacks are the next challenging area for WMSN security. Here, secure high-level data aggregation, intruder nodes and multiple identity attacks, detection of compromised nodes, and privacy concerns constitute some of the most important security challenges that need to be addressed.

8 Wireless Communication Systems in the Smart Grid

With the increasing interest from both academic and industrial communities, this chapter describes the developments in communication technology for the smart grid (SG). The smart grid can be defined as an electrical system that uses information, two-way secure communication technologies, and computational intelligence in an integrated fashion across the entire spectrum of the energy system from the generation to the end points of the power consumption. Communication networks play a critical role in the smart grid, as the intelligence of this complex system is built based on information exchange across the power grid. In particular, wireless networks will be deployed widely in the smart grid for data collection and remote control purposes. The design of the communication network associated with the smart grid involves a detailed analysis of requirements, including the choice of the most suitable technologies for each case study and the architecture for the resultant heterogeneous system. Fundamental control technologies for communications, data management diagnostic analysis, and work management are also required. The smart grid uses communication and information technologies to provide utilities regarding the state of the grid. Using intelligent wireless communications, load scheduling can be implemented so that peak demand can be flattened. This reduces the need to bring additional expensive generation plants online.

8.1 INTRODUCTION

Affordable and reliable electric power is fundamental to modern society and economy. With the application of microprocessors, reliable and quality electric power is becoming increasingly important. On the other hand, the electric grid was cited by the National Academy of Engineering as the supreme engineering achievement of the twentieth century [1]. Efforts to modernize the grid are motivated by several goals:

a. To make the production and delivery of electricity more cost-effective
b. To provide consumers with electronically available information and automated tools to help them make more informed decisions about their energy consumption and cost control
c. To help reduce the production of greenhouse gas emissions in generating electricity by permitting greater use of renewable sources
d. To improve the reliability service
e. To prepare the grid to support a growing fleet of electric vehicles to reduce dependence on oil

There exists several ways to make the transmission and distribution systems more efficient. In most cases, there are no sensors in distribution grids that communicate the actual voltages delivered to customers, so systems must be operated based on estimates of losses along the line. The deployment of automated sensors and control that permit dynamic voltage and reactive power optimization may permit the reduction of voltage levels by a few percent and reduce power consumption. Another source of inefficiency arises from the peaked nature of electricity demands. Investment reduction that will be required to replace old generation and transmission facilities can be achieved also by using electricity more efficiently. Electricity is consumed by three categories of users: residential, commercial, and industrial. Each category accounts for roughly one-third of overall consumption. Automation and information technologies have been used for many years to increase efficiency when necessary in commercial and industrial environments. As for the consumers, they generally pay little attention to how they use electricity because they do not receive any timely or actionable information about consumption. Smart meters that electronically record interval data can provide near real-time information about electricity usage and cost to an in-home display or software application that allows consumers to make reasonable choices about energy use and cost control. A computer-based energy management system can receive dynamic pricing information over the Internet and adjust lighting, air conditioning, and application operation in accordance with the consumer's preferences to minimize cost.

The electricity production and consumption, being under rules for almost a century, is undergoing a revolution supported by information and communication technologies (ICTs). Modern society critically depends on a reliable supply of high-quality electrical energy. The essential concept of the SG is the integration of advanced information technology, digital communication, sensing, measurement, and control technologies into the power system [2]. Using ICT in a smart way can help reduce energy consumption in different energy-human sectors, such as telecommunications, transport, logistics, and so on. The development of the SG presents one global priority with numerous benefits. In the United States, the transition to the SG is under way. Significant SG efforts are also under way in other countries, including China, Japan, Korea, Australia, and EU, among others [2,3]. The worldwide use of energy is growing, generating greater demand, but the supply is limited.

This chapter starts by enumerating the key requirements and by establishing the standards for the SG. Then, it deals with the main component of the SG together with the corresponding communication architecture. A brief description of the effective demand load control is presented. The importance of wireless mesh networking as well as the heterogeneous network integration to coordinate the SG functions is invoked too. Next, it deals with SG and sensor networks together with smart microgrid network. Finally, an outline of the SG demand response (DR) concludes the chapter followed by a discussion.

8.2 KEY REQUIREMENTS OF THE SMART GRID

The SG evolves the architecture of legacy grid, which can be characterized as providing a one-way flow of centrally generated power to end users into a more distributed, dynamic system characterized by a two-way flow of power and information. The SG will involve networking vast numbers of sensors in transmission and distribution facilities, smart meters, back-office systems, and home devices that will interact with the grid [2]. Large amounts of data traffic will be generated by meters, sensors, and synchrophasors.

Although networking technologies and systems have been greatly enhanced, the SG faces challenges in terms of reliability and security in both wired and wireless communication environments. Some of the key requirements of the SG from the aspects of global multimedia communications include the following [2]:

- Integration of renewable energy resources
- Active end user perception to enable energy conservation
- Secure communications
- Better utilization of existing assets to address long-term sustainability
- Management of distributed generation and information storage
- Integration, communication, and control across the information system to promote open system interoperability and to increase safety and operational flexibility

From a technical component's perspective, the SG is a highly complex combination and integration of multiple analog and digital technologies and systems. The automation being applied to the electric grid is similar in concept to the network management and operations support systems that were applied to the telecommunication networks in the 1970s and 1980s. Applying ICT to the grid is not straightforward because it must account for constraints that did not exist in automating the telecommunication network. For example, unlike the communication network, which routes packets of information, the electric grid routes power flows that are constrained by the laws of physics. Some SG applications that control generators or sub-stations have latency requirements measured in milliseconds, although the consequences of failing to deliver a control packet on time can be harmful. On the other hand, applications such as the communication of smart meter interval data have much less stringent requirements. Characterizing the performance requirements is crucial to understanding which communication technology should be used for various SG applications.

Advanced and new grid components allow for more efficient energy supply, better reliability, and availability of power. Components include, for example, advanced conductors and superconductors, improved electric storage components, advanced power electronics, and distributed energy generation. Smart devices and smart metering include sensors and sensor networks. Information provided by sensors need to be transmitted over a communication backbone. This backbone is characterized by a high-speed and two-way communication. Different multimedia communication applications and technologies form the communication backbone. Utilities have the choice between multiple and diverse technologies in the area of multimedia communication networking.

8.2.1 ESTABLISHING STANDARDS FOR THE SMART GRID

Control systems operated by different electric utilities whose networks interconnect will need to exchange information. User-owned smart appliances, energy managements systems, and electric vehicles need to communicate with the SG. Standards defining the meaning, representation, and protocols for data transport are essential for this complex system to interoperate seamlessly and securely. The development of standards for the SG requires efforts at regional, national, and international levels. Although electric utilities typically operate within national boundaries, there are interconnections across borders. In addition, many of the equipment and systems suppliers are global companies that seek to address markets worldwide. Unnecessary variations in equipment and systems to meet different national standards add costs, which get passed on to consumers. International standards promote supplier competition and expend the range of options available to utilities, resulting in cost reduction.

The technical standards for SG are being development by more than 20 organizations, most of them international in scope, including the International Electrotechnical Commission (IEC), the International Organization for Standardization (ISO), the Institute of Electrical and Electronics Engineers (IEEE), the International Telecommunication Union (ITU), and the Internet Engineering Task Force (IETF), among others. During 2009, the National Institute of Standards and Technology (NIST) engaged more than 1500 stakeholders representing hundreds of organizations in a series of public workshops over a 9-month period to create a high-level architectural model for the SG, to analyze the cases used, and to identify applicable standards, gaps in currently available standards, and priorities for new standardization activities [3]. The result of the study "Release 1.0 NIST Framework and Roadmap for Smart Grid Interoperability" was published in January 2010 [4]. To evolve the initial framework, late in 2009, NIST established a new partnership, the Smart Grid Interoperability Panel (SGIP), which has international participation. This body, with more than 640 organizations and companies, is also guiding the testing and certification framework for the SG. NIST and SGIP are establishing cooperative relationships with the corresponding SG standardization initiatives that are under way in other parts of the world [5]. For example, in Europe, the Joint Working Group on Standards for Smart Grids has been established together with the European Telecommunications Standards Institute (ETSI) and the European Committee for Standardization [6]. Japan has developed an initial standard for the SG and has established a Smart Community Alliance [7], which has extended the concept of the SG beyond the metric system to encompass energy efficiency and for the intelligent management of other resources such as water, gas, and transportation. The government of South Korea has announced a plan for a national SG network implementation and is beginning to work on a standards roadmap [8]. The State Grid Corporation of China has developed a draft "Framework and Roadmap for Strong and Smart Grid Standards" [9]. International cooperation among these efforts is under way through a recently established International Smart Grid Action Network (ISGAN) [5]. ISGAN's vision is to accelerate progress on the key aspects of SG policy, technology, and related standards through voluntary participation by governments in specific projects and programs.

8.2.2 COMPONENTS OF THE SMART GRID

The main components of the SG are new and advanced grid components, smart devices and smart metering, integrated communication technologies, programs for decision support and human interfaces, and advanced control systems. An overview of the main components of an SG is shown in Figure 8.1.

New and advanced grid components allow for a more efficient energy supply and better reliability and availability of power, as it was mentioned. Smart devices and smart metering include sensor networks. Sensors are used at multiple places along the grid, at transformers and substations, or at users' homes [10]. They play an outstanding role in the area of remote monitoring, and they enable new business processes such as real-time pricing (RTP). Control centers can immediately receive accurate information about the actual grid condition. Smart meters allow for real-time determination and information storage of energy consumption and provide the possibility to read consumption both locally and remotely [11].

Information provided by sensors and smart meters needs to be transmitted over a communication backbone, which is composed of heterogeneous technologies and applications. These can be classified into the communication services group shown in Table 8.1, together with brief descriptions and examples. Usually, WANs and LANs are deployed within a SG [12].

The data volume in the SG will increase tremendously compared with traditional grids. Tools and applications include systems based on artificial intelligence and semiautonomous agent software, visualization technologies, alerting tools, advanced control, and performance review applications as well as data and simulation applications and geographic information system (GIS). Artificial intelligence methods as well as semiautonomous agent software, for example, contribute to minimize data volume and to create a format most effective for user comprehension in which the software has features that learn from input and adapts. As for GIS, it provides

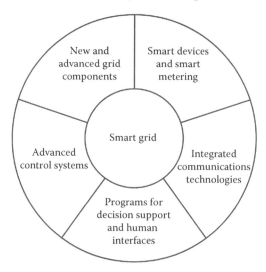

FIGURE 8.1   Main components of the smart grid.

TABLE 8.1

Smart Grid Communication Applications and Technologies

Core networking	Protocols needed to provide interoperability in a network that may significantly vary in topology and bandwidth
Security	Security measures for consumer portals because they directly deal with information and billing process
Network management	Standard technologies for collecting statistics, alarms, and status information on the communication network itself
Data structuring and application	"Metadata" for family describing and exchanging how devices are configured and how they report data
Power system operations	Several of the key applications for portals involve integration with distribution system operations, such as outage detection and power quality monitoring
Consumer applications	Electrical metering and various aspects of building (home) automation

location and spatial information and tailors them to the specific requirements for decision support systems along the SG.

Advanced control systems monitor and control the essential elements of the SG. Computer-based algorithms allow efficient data collection and analysis, provide solutions to human operators, and are also able to act autonomously. Faults can be detected much faster compared with traditional grids, and outage times can be significantly reduced. To fulfill these requirements, the SG communication infrastructure has to integrate enabling networking technologies.

8.3 COMMUNICATION ARCHITECTURE FOR THE SMART GRID

The SG is usually deployed in a considerably wide geographical area. Consequently, the communication infrastructure of the SG has to cover the entire region with the intention to connect a large set of nodes. The communication infrastructure is envisioned to be a multilayer structure that extends across the entire SG, as shown in Figure 8.2.

In particular, home area networks (HANs) communicate with various smart devices (meters, sensors, and actuators) to provide energy efficiency management and DR. Neighborhood area networks (NANs) connect multiple HANs to local access points. WANs provide communication links between the NANs and the utility systems to transfer information. The three-layered structure of the communication networks provides a potential operation of the SG to operate economically, efficiently, reliably, and

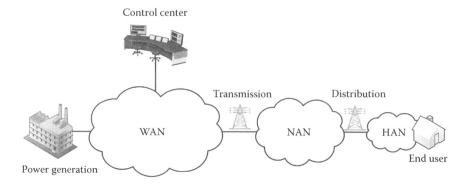

FIGURE 8.2   Multilayer structure across the smart grid.

securely. On the other hand, there are many challenges imposed on the design of the SG communication architecture: energy services, interoperability, tremendous data amount, highly varying traffic, QoS, and security.

A unique characteristic of the SG is the integration of distributed renewable energy sources (e.g., solar and wind power). For an NAN, there are two main power sources: the power from the utility and the distributed renewable energy. These power sources have two essential differences: price and availability. Balancing the usage of different energy sources will be very important for power grid stability, availability, and operation cost.

Data will flow over generation, transmission, distribution, and user networks in the SG. A variety of technologies will be used to set up the communication architecture to provide enough information to the control centers. One of the major problems of the multitier topology of communication networks is interoperability among so many subnetworks [13].

The amount of data generated by smart devices will experience explosive growth in the future. These tremendous data place considerable load on the communication infrastructure of the SG. There are large amounts of real-time and archival operational data in the SG. The amount of data varies tremendously during a day, so the traffic conditions change rapidly. During peak hours, the data communications require higher data rate and more reliable services.

Different categories of data have different QoS priorities in terms of transmission latency, bandwidth, reliability, and security. Information including device's state, load, and power pricing should flow over the communication network accurately, effectively, and reliably. A higher priority and guaranteed QoS should be provided to the meter data, and the power price data used for summarizing the monthly bill for electric usage have normal priority and QoS.

The SG will use computer networks for controlling and monitoring the power infrastructures. This in turn exposes the SG to outside attacks.

There are many potential threats within utilities, such as indiscretions by employees and authorization violation [14]. SG networks have to carry reliable and real-time information to the control centers of the utilities.

Cognitive radio-based communication architectures for the SG have been proposed by Yu et al. [15] and Wang et al. [16]. This architecture uses cognitive radio technology to enable the communication infrastructure more economically, flexibly, efficiently, and reliably. Cognitive radio refers to the potentiality that wireless systems are context aware and capable of reconfiguration based on the surrounding environments and their own properties (see Chapter 2).

The communication architecture has the hierarchical tree structure, including HAN, NAN, and WAN (Figure 8.3). Cognitive communications that operate in the unlicensed spectrum are applied in the HAN to coordinate the heterogeneous wireless technologies. On the other side, cognitive communications that operate in the licensed spectrum are applied in the NAN and WAN to dynamically access the unoccupied spectrum. Table 8.2 summarizes the features and technologies in the three subareas.

The HAN is a fundamental component that enables two-way communications to provide DR management services in the SG. Commissioning and control are two necessary functionalities in the HAN. Commissioning is specified to identify new devices and to manage the formation of the self-organizing network. Control is used to maintain the communication link between devices and to ensure interoperability within the SG. The HAN consists of a cognitive home gateway (CHGW), smart meters, sensors, actuators, and other intelligent devices. It usually uses a star topology with either wired technologies (e.g., power line communication [PLC]) or different wireless technologies (e.g., ZigBee and Wi-Fi) [15].

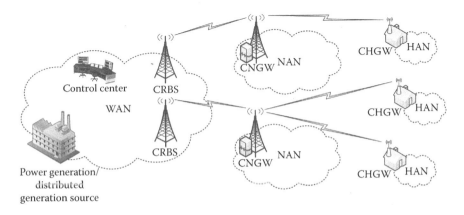

FIGURE 8.3 Cognitive radio-based communication architecture in the smart grid.

TABLE 8.2

Features of the Cognitive Radio-Based Hierarchical Communication Infrastructure for the Smart Grid

Cognitive Area Networks	Home Area Network	Neighborhood Area Network	Wide Area Network
Spectrum band	Unlicensed band	Licensed band	Licensed band
Network topology	Decentralized/ centralized	Centralized	Centralized
Network elements	Smart meters/ sensors/actuators, CHGWs	CHGWs, CNGWs	CNGWs, spectrum broker
Featured techniques	Cross-layer spectrum sharing, access control, power coordination	Hybrid dynamic spectrum access, guard channel, spectrum handover	Optimal spectrum leasing, joint spectrum management

In SG applications, the NANs will collect energy consumption information from households in a neighborhood and distribute them to a utility company through WAN. Cognitive NAN gateway (CNGW) connects several HGWs from multiple HANs. The HGWs are the data access points of the HANs to the outside NAN, and they also act as the cognitive nodes located in the NAN. From the technology point of view, the CNGW can be considered as the cognitive radio access point to provide a single-hop connection with CHGWs in a hybrid access manner. The CNGW manages the access of the CHGWs and distributes spectrum bands to them.

In the WAN, each CNGW is no longer an access point but a cognitive node with the capability to communicate with the control center through unused frequency space. This center is connected with cognitive radio base stations (CRBS) that are dispersed over a wide area (e.g., a city). In conjunction with the control center, there is a spectrum broker that plays an important role in sharing the spectrum resources among different NANs to enable the coexistence of multiple NANs. In a large geographical distribution of NANs, several CNGWs may not be within the geographic area covered by base stations. These CNGWs have to communicate in an ad hoc mode to share unoccupied spectrum by themselves.

Example 8.1

The problem of transmitting power coordination among wireless nodes in a HAN can be formulated as follows. This is a non-cooperative game in which the main participants are (a) player,

(b) action, and (c) utility. Each wireless node is an individual player in the game. On the other side, each player will take an action to adjust its transmitting power level according to that it offers. Finally, to define the utility for each player, we have to consider a HAN where, for example, I wireless nodes are intending to transmit data [15]. The available rate of the ith node is given by the well-known Shannon's formula, that is,

$$R_i = \int_0^{B_i} \log_2 \left(1 + \frac{p_i(f)}{N_0 + \sum_{j \neq i} p_j(f)} \right) df \qquad (8.1)$$

where B_i is the channel bandwidth, while $p_i(f)$ and N_0 are the power spectral density function of the system i and noise in receivers, respectively. Then the utility function of the ith node can be calculated from

$$U_i = R_i - \alpha_i \int_{B_i} p_i(f) df. \qquad (8.2)$$

Here, α_i is the constant for the corresponding node. This function defines the utility for each player, which reflects the player's demand to find equilibrium between transmitting rate and energy consumption. It should be mentioned that this is one of the crucial assignments in a cognitive-based SG communication architecture realization.

8.4 ROLE OF EFFECTIVE DEMAND LOAD CONTROL IN THE SMART GRID

Smart metering and two-way communication infrastructures realize the real-time interconnection of IP-addressable components at the consumer and operator premises over the Internet. These systems allow automated power consumption monitoring and control together with real-time electricity price signaling and fault diagnosis. These technologies together with electric energy storage entities, backup devices, and plug-in hybrid electric vehicles provide unique opportunities to address the challenge of managing consumer power demands in real time [17].

The main objective of demand load control in the SG is to alleviate peak loads, which cause major expenditures in power utility. This can be achieved by forwarding nonemergency power demands from peak load to off-peak load times or by using energy collected and stored during off-peak load times.

In principle, demand load control does not reduce total power consumption because it either shifts demands in time or uses previously stored energy. Most important, demand load management reduces or eliminates the need for activating the supplementary means of power generation so as to satisfy increased demand during peak time. Effective demand load management reduces the cost of operating the grid, and consequently, it lowers electricity price.

The ground to address challenges in demand load control for the SG is now very high because of advances in communication infrastructures and the creation of a two-way channel for real-time communication between utility operators and consumers. Using this channel, information about power consumption is distributed to the operator control, which in turn takes automated demand load control decisions and makes them available to consumers through appropriate signaling.

Demand task scheduling means scheduling power demand tasks for activation at appropriate times. It can be performed by the operator or the user. In RTP, the operator can distribute price signals to encourage consumers to shift their power demands during periods of low-cost energy, thus providing them to reduce their power consumption at peak-load times.

Example 8.2

Consider a scenario with real-time bidirectional connection between a controller located at the premises of the grid operator center and the IP-addressable smart metering devices (SMDs), each of which is connected to SG-enabled consumer appliances. The connection can be realized over a wireless network that interconnects the SMDs and the ZigBee-enabled plugs of appliances. ZigBee is often discussed as a standard for smart energy in HANs, with the intention of low-cost devices and low-energy use [18]. The basic components of the architecture are presented in Figure 8.4.

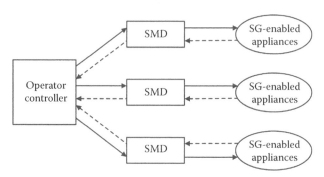

FIGURE 8.4 System architecture for power demand scheduling.

The SG-enabled consumer appliances send power demand requests (dashed arrows) to the SMD, which further dispatches them to the controller located at the operator premises. The controller returns a schedule for each task (solid arrows), which is passed to the appliances through SMD. The received requests have different power requirements, durations (which may be even unknown), and time flexibility into their satisfaction. Flexibility could be modeled as a deadline by which a demand is to be activated. The controller decides on the time of the activation of different power demand tasks. Then it sends the corresponding command for activation to the SMD from which the task emanated. The SMD transfers the command to the corresponding appliance, whereas the power demand is activated at the time defined by the operator controller. The communication takes place through a high-speed connection with near zero delay. The operator has full control over the individual consumer appliances, which in turn comply to the dictated schedule and start the task at the prescribed time.

Example 8.3

Each of n power demand task is characterized by the following parameters:

- A time of generation a_i, which is the time instant at which the request arises at the controller
- A time duration of s_i time units, which denotes the required amount of time for which the task need to be active

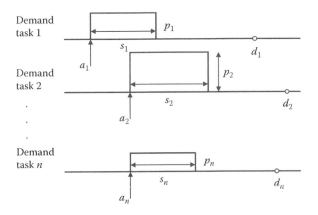

FIGURE 8.5   Power demand task parameters.

- An instantaneous power consumption requirement p_i, if activated
- Delay tolerance is activated and is captured by a deadline d_i at which point it needs to be activated

Hence, power demand task n is generated at time a_n, has a duration s_n and a power requirement p_n, and needs to be activated by d_n (Figure 8.5). The objective of the SG operator is to devise a power demand task scheduling policy that minimizes the operational costs.

8.5 WIRELESS MESH NETWORKING FOR THE SMART GRID

One of the most important challenges is to provide a reliable last mile communication that covers the connectivity from home gateways to the advanced metering infrastructure (AMI) headend. The design of such network depends not only on the application layer requirements but also on the nature of its PHY and MAC layers. There are several heterogeneous communication technologies that can be wireless and wired. WLAN technologies have been extensively deployed for home networking. The main issue is how to effectively apply this technology to handle the last mile communication [19]. An IEEE 802.11 WLAN, which operates in a single hop, can be used as a gateway, representing data aggregation point (DAP) between home gateways and the AMI headend. The use of multihop communications between distributed modes offers pathways around electromagnetic transmission obstacles that would otherwise prevent the formation of a long-range network [20]. This requires deploying mesh networking to meet the requirements of the SG. Also, it includes a high degree of reliability, self-configuration, and self-healing. Figure 8.6 represents a mesh network architecture for the last mile SG with communication between home appliances and their home gateways. Communication between meters and AMI headend through one of the DAPs is included too.

To deploy IEEE 802.11 [19] devices to represent home gateway as a mesh node, it would be necessary that these nodes operate as a router. To design a P2P multihop mesh topology, the most convenient case is when the nodes can operate as single-hop AP or in a distributed multihop mode. For the observed multigate network, the following types of nodes exist:

- Mesh relay station (MRS), which represents the relay node
- Mesh station with access point (MSAP), which represents a residential meter and operates as the gateway of the home network to the meter and from the meter toward its local DAP

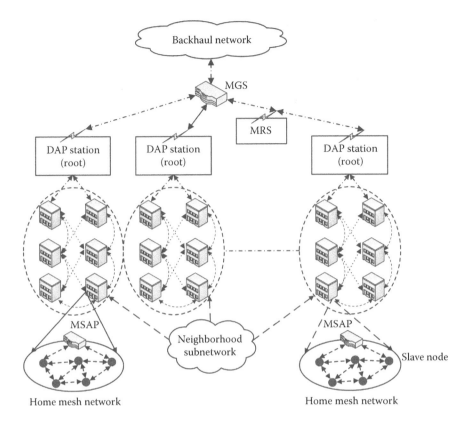

FIGURE 8.6 Mesh network architecture for the smart grid.

- Slave node, which represents in-home devices, that is, smart home appliances, and operates in an infrastructure mode
- DAP station, which represents the neighborhood gateway point (root of the tree in each network)
- Master gateway station (MGS), which represents the AMI head-end connected to the backhaul network.

A DAP, as the proactive routing entity, periodically broadcasts root announcements by increasing the sequence number each time such a message is generated. The root announcement allows proactive routing toward each MSAP in the residential area. When a meter receives such a message, it catches the MAC address of the transmitting node and then adds it to its list of parent nodes. Before rebroadcasting the DAP station announcement message, a meter should wait for a predefined time to check whether there are more of the same root announcements from other neighbor nodes. The DAP station will then unicast a path reply back to the meter. The tree formation process is continued by every new meter by rebroadcasting the root announcement to neighboring nodes. Because of the static nature of the network, the route announcement plays an important role in meeting the self-organization and self-healing requirements of the SG. In SG applications, where meters, DAPs, and relay nodes are mainly stationary, the on-demand part of routing mainly deals with the effect of interference that can cause path breakage. There exist many on-demand protocols. Dynamic source routing (DSR) and ad hoc on-demand distance vector (AODV) are the most popular and well tested [21,22]. In DSR, every packet must carry the IP addresses of all the nodes along the source to the destination route. On the other hand, in AODV, only the destination address is carried in each packet. For a large network where there is a possibility of a link with a high number of hops, DSR can lead to a significant increase in the packet overhead.

Example 8.4

Each subnetwork is managed independently by its local DAP between meters and AMI head-end. Because of the variable nature of the traffic, some DAPs may suffer from more congestion then others. Nodes from neighboring subnetworks cannot participate in traffic load reduction. To enable collective participation in the routing, it would be more convenient to combine

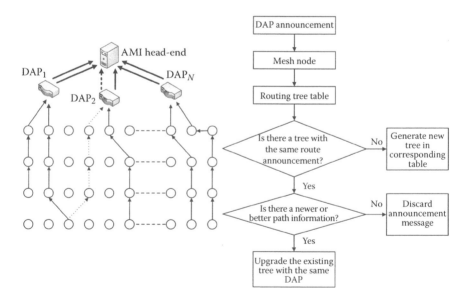

FIGURE 8.7 Routing tree table upgrading algorithm for multi-DAP wireless mesh network.

all subnetworks into a larger network with multiple DAPs [19]. In that case, all the meters (nodes) can access any of DAPs.

An expansion process of a single DAP network to a multi-DAP network together with routing table upgrading algorithm is shown in Figure 8.7. The MAC addresses of DAP_1, DAP_2, ..., DAP_N are used as the unique identification of the corresponding tree.

8.6 HETEROGENEOUS NETWORKS INTEGRATION TO COORDINATE THE SMART GRID FUNCTIONS

A heterogeneous network achieves an E2E integration of the corresponding technologies by using sensor network architecture [23]. Defining the interoperability with NGN as the SG backbone is of importance too [24]. Foremost, the main component of the SG is the WSN, which consists of a system of distributed nodes that interact among themselves and with the infrastructure to acquire, process, transfer, and provide information extracted from the physical world [25]. The processing of sensor information should allow the information of the SG behavior through intelligent actuators.

The smart meter is the bridge between user behavior and power consumption metering. The distribution management system (DMS) is required for analyzing, controlling, and providing enough useful information to the utility. The SG is also composed of legacy remote terminal units (RTUs) that can perform sensor network gateway functions acting as intermediate points in the medium voltage network. The sensor network gateway is the bridge between the sensor network and the back-end system. Therefore, it provides necessary interfaces to other sensor nodes as well as interfaces to existing ICT infrastructures.

AMI consists of smart meters, data management, communication network, and applications. Along with distributed energy resource (DER) and advanced distributed automation (ADA), AMI is one of the three main anchors of the SG. Usually, GIS as well as consumer information system (CIS) contribute with tools and important processes. The information recollected and processed by DMS must be reported to a supervisory control and data acquisition (SCADA) system.

Heterogeneous networks manage and collect information from established intelligent electronic devices for control and automation purposes in real time. Thus, the SG needs to communicate many different types of devices, with different needs for QoS over different physical media. Devices involved in these processes can be situated in different locations because of the decentralized architecture. As an example, electrical substation elements are connected to the substation's Ethernet network. As for sensors, they can be installed along electrical cables communicated through wireless sensor standards (e.g., IEEE 802.11s). Communications from the control center to energy meters and between substations can be carried out over a high variety of technologies such as PLC, UMTS, or WiMAX. Standardized, open information models and communication services for all data exchanges are needed in the SG. Thus, the SG will be supported by a highly heterogeneous data network.

PLC technology uses the power grid for transmitting data. However, the characteristics of the PLC medium make it especially difficult to ensure a given QoS. This technology has to overcome some problems such as unpredictable frequency and time dependence of impedance, attenuation and transmission characteristics, impulse and background noise and their wide variability, limited bandwidth, and harmonic interference. For example, the variability of the channel is especially troublesome for QoS because it can suddenly bring the bandwidth down.

Besides PLC, several access technologies can be integrated into the resulting SG architecture. Some options for integration are IEEE standards for mesh WPANs, WLANs, WMANs, and WRANs. Traditional wireless systems use point-to-point or point-to-multipoint technologies. Mesh networks are an alternative to these topologies. There are some reasons for it. For example, it is easy to associate new nodes in the network, thanks to the fact of self-configuration and self-organization capabilities. Second, a mesh network is a robust system as there will almost always be an alternative path to the destination. Taking into account the scenario in which the SG is going to be deployed, different technologies will be needed to cover all the areas.

Some WPAN standards are presented as wireless communication candidate technologies that work within mesh networks. IEEE 802.15.4 defines the PHY and MAC layers in low-rate PANs [26]. In 2008, the Smart Utility Networks Task Group 4g was created within the 802.15 group. The role of TG4g is to define new PHY layers to provide a global standard that facilitates very large-scale process control applications, such as the utility SG. IEEE 802.15.5 is the WPAN mesh standard approved in March 2009. This working group was established to define mesh architecture in IEEE 802.15.4-based WPAN. There exist different proposals regarding routing in low-rate WPAN networks. Nevertheless, these algorithms are not fully optimal.

A draft from the IEEE 802.11 working group for mesh networks defines how wireless devices can be connected to create ad hoc networks. The implementation should be over the IEEE 802.11n physical layer. A combination of IEEE 802.11n and IEEE 802.11s could also be a feasible solution for ubiquitous sensor networks (USNs).

IEEE 802.16 is a standard technology for wireless wideband access. Among its advantages, the ease of installation is by far the most important aspect. This technology supports either point-to-multipoint or mesh topologies. In mesh topologies, it is not necessary that all the nodes are connected to the central node. In this way, active nodes periodically announce mesh network configuration messages, which contain information about the base station identifier and channel in use.

IEEE 802.22 uses the existing gaps in the TV frequency spectrum between 54 and 862 MHz [27]. The development of this standard is based on the use of cognitive radio techniques to give broadband access in sparsely populated areas that cannot be economically served by wireline means or other wireless solutions at higher frequencies.

Example 8.5

The problem often arises is when a SG application wants to reach the sensor network through middleware to collect data [25]. The solution for middleware interaction is presented in Figure 8.8.

As shown in Figure 8.8, we have three layers from the top to the bottom: applications, middleware, and sensor networks. During the middleware interaction, active applications are functioning independently, but controlled by an integrated control center. If

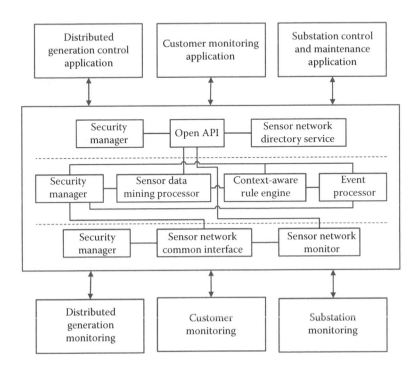

FIGURE 8.8 Ubiquitous sensor network middleware interaction.

each application has direct access to the sensor network, each application developer should know the details of each network and their interfaces. In the presented resolution, when using the USN middleware, each application developer only needs to know how to use the open application programming interface (API). All the applications communicate with the middleware, which has to exchange information appropriately with sensor network.

8.7   SMART MICROGRID NETWORK

A microgrid is a localized grouping of electrical sources and loads. The microgrid can operate connected to, and synchronous with, the main electrical power grid. However, in some exceeding situations (low power quality, electrical outage, etc.), the microgrid can be isolated from the main electrical power grid and function autonomously. Microgrid takes advantage of advances in electrical generating systems, for example, solar panels, wind turbines, and cogeneration, creating and distributing electrical power outside the traditional grid system. Because of independent control strategy and versatile power sources, this concept can at least decrease the local reliability of the power system, reduce the power loss, and enhance power utilization and efficiency. A microgrid is a relatively small-scale, self-contained, medium/low-voltage electric power system that houses various DERs (distributed generators) in a physical close location.

Microgrids have two operation modes: grid connected and islanded. In the grid-connected mode, the microgrid can be considered as a controllable load, or it can supply power and act like a generator from the grid's point of view. In the islanded mode, the microgrid is independent from the utility grid, and energy generation, storage, load, and power quality control are implemented in a standalone system [28]. The microgrid concept was favored for its ability to provide fault isolation and ease of distributed generator handling. Within the context of the SG, the definition of the microgrid is evolving into the smart microgrid (SMG). ICT is becoming integrated to load control tasks and being used for energy trading among communities [29]. Today, there is a growing interest in forming SMG for campuses, military bases, remote communities, and rural areas in developing countries. Also, several test beds have been implemented in the United States, Canada, Japan, and European Union.

Security is an important issue for the microgrid. To realize a high-secure system, security should be integrated into every system component [30,31]. Information security for the SMG should include confidentiality, authenticity, integrity, freshness, and privacy of data together with public key infrastructure, trusted computing, attack detection, and intelligent monitoring. For power flow, autonomous recovery is the main consideration for security. The SMG should have the capability of performing real-time monitoring and quick response. Smart meters together with many sensors should monitor the operation parameters or operator status of the microgrid [14]. On the basis of the monitored data, the SMG can adjust itself to maintain the optimal operation mode or to recover itself to the correct working condition.

Prediction capability is also needed in the SMG to ensure its security. The SMG should continuously search for hidden troubles or latent dangers in the system. Control in SMG refers to the activities, procedures, methods, and tools that relate to the operation, maintenance, administration, and provisioning of the microgrid. From the perspective of the microgrid system, central or cooperative control is necessary, whereas for the interest of each smart house, distributed or noncooperative control is preferred [32]. Also, control strategy for the SMG should be layered or hybrid.

8.8 SMART GRID DR

One of the approaches being used to reduce the peak demand and to improve the system reliability is DR, in which the end users modify their electricity consumption patterns in response to price variations. DR has been used in commercial and industrial sectors for some time to increase the health and stability of the grid. With the emerging SG, DR now has the potential to be expanded into residential electricity markets on a large scale. The SG adds intelligence and bidirectional communication capabilities to today's power grid, enabling utilities to provide RTP information to their customers and to collect the real-time usage from their customers [33].

The SG consists of sensing, communication, control, and actuation systems, which enable pervasive monitoring and control of the power grid. These features enable utilities to accurately predict, monitor, and control the electricity flows throughout the grid.

Various wireless schemes (GPRS, IEEE 802.11, etc.) [33] have been used to enable meters to automatically transmit data back to the utility. Automatic meter reading is unidirectional, without capability for the utility to send data to the meter, thereby limiting its usefulness. This deficiency led to the development of AMI, which provides bidirectional communication between the utility and the meters installed in residences.

HANs comprise smart appliances, which can communicate with one another, or a home energy controller (HEC) to enable residents to automatically monitor and control home energy usage. The widespread availability of low-cost wireless technologies has accelerated the deployment of HANs by facilitating the addition of communication capabilities to household

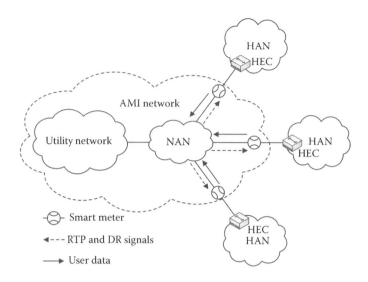

FIGURE 8.9   Home area networks and AMI interconnection.

appliances. In fact, smart appliances are home appliances that combine embedded computing, sensing, and communication capabilities to enable intelligent decision making. Sensing capabilities enable these devices to measure their usage, communication capabilities enable the reporting of their energy consumption to the HEC, and their actuation abilities enable them to respond to commands from the HEC.

The block scheme of system-interconnecting HANs and AMI is shown in Figure 8.9. RTP and DR signals are sent to the smart meter over the electric utilities NAN, with the neighborhood transformer serving as the aggregation point for all the smart meters in that region.

Smart devices report their current operating status to the HEC, whereas the operating modes of dumb appliances can be inferred using power measurements transmitted by attached smart plugs. These data enable the HEC to determine when loads are in standby mode and switch them off. The location of the scheduling intelligence in the HEC enables complex scheduling to be applied individually in each load. Authentication is a critical security service in SG networks, and it mainly involves the smart meters in different component networks of the SG.

Example 8.6

To answer the question how a SG communication system can be separated from the power transmission and distribution system, the SG hierarchical networking [34] can be applied, as shown in Figure 8.10.

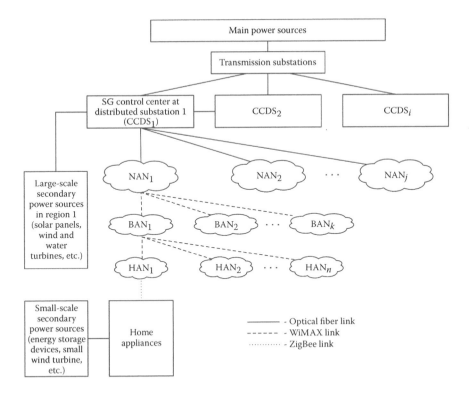

FIGURE 8.10   Smart grid communication system separation from the power transmission and distribution.

It can be seen that it is an information-sharing network comprising several hierarchical components. The transmission substation and the control centers of the distribution substations are connected with one another in a mesh-networking manner, which can be realized over optical fiber technology. The remaining components are divided into several networks based on real metropolitan topology.

Example 8.7

An important issue is how to implement nodes for SG test beds [35]. The functional architecture can be proposed as shown in Figure 8.11. The hardware abstraction layer is an interface for upper level function that screens hardware-specific details. Also, it provides data interfaces to both receiving data path and transmitting data path, as well as an access interface to other resources on the hardware platform. The spectrum and the channel manager control all the related resources, including links, frequencies,

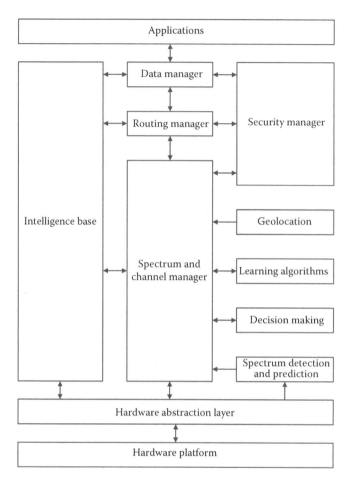

FIGURE 8.11   Node functional architecture for smart grid test beds.

and modulation methods. The spectrum detection and prediction module provides the information regarding the availability of some frequency bands.

The decision-making module uses decision algorithms to select which channel and when it will be used. Learning algorithms are implemented as an independent module. The spectrum and channel manager can use geolocation information (latitude and longitude) from the corresponding module to load prior information about the current location from the knowledge/policy/database. The routing manager uses routing algorithms to select the best route for sending and relaying data packages. The data manager organizes all the data from upper-level applications and the data to be relayed. The security manager provides encryption and decryption to the data manager, routing manager, and spectrum and channel manager. The intelligence base stores prior knowledge, policies, data, and experiences.

8.9   CONCLUDING REMARKS

In the different segments of a power grid, different communication technologies are applied to meet the unique specific requirements. In a power transmission segment, wired communications over power lines or optical cables are adopted to ensure the robustness of the backbone. However, in a power distribution segment that provides power directly to the end users, both wired and wireless communications should be considered.

To achieve a cost-effective and flexible control and monitoring of end devices, efficient dispatching of power, and dynamic integration of DERs with power grid, wireless communications and networking functionalities must be embedded into various electric equipment. The capability of wireless networking among various electric equipment is one of the key technologies that drive the evaluation of a conventional power distribution network into a SG.

The development of the SG means a drastic change in power use and administration. Users will become active participants in energy management and will be able to control their consumption. On the other hand, utilities will be able to control demand peaks and manage the grid efficiently from generation to distribution.

Different types of wireless networks are available, but which one is suitable for a SG depends on the system architecture and varieties of communication modules and links. The multihop wireless networking is definitely necessary, as electric equipment out of range of one another needs to exchange information. First, to simplify network organization and maintenance, the entire environment needs to be self-organized. Moreover, communication modules may pertain to heterogeneous nature in terms of coverage, computing power, and power efficiency. WMNs are considered an important networking for the SG. However, when WMNs are applied in the SG wireless infrastructure, a few challenging issues still remain (e.g., security level). Without effective measures to prevent security attacks, the privacy of users and the confidentiality of grid information cannot be guaranteed.

Wired communication technologies are not fully suitable for SG communications. Optical networks are reliable, but deploying fibers to connect all end devices is too expensive. PLCs are constrained by some shortcomings: (a) they are not flexible for peer-to-peer communications among electric devices, (b) throughput may not be sufficient for frequent data exchange, and (c) high-speed signals cannot pass through transformers.

In a SG, various electric devices need to have P2P communications, so a mesh-networking capability is a viable option. However, a WLAN can only support one-hop point-to-multipoint communications. Thus, WLAN is not enough for a smart distribution grid. The better solution is to build WMNs based on WLAN technologies. The next issue is that the architecture of SG communications needs to support different types of wireless applications. In this scenario, technologies based on IEEE 802.11 with high-gain antenna may be necessary. As a result, SG communication networks must be capable of integrating heterogeneous wireless networks. As for PLCs, they will be utilized as much as possible to enhance reliability and security. WMNs can easily integrate heterogeneous networks to fulfill different functions such as sensing, monitoring, data collection, control, and pricing. Although many of the benefits of wireless communications, such as untethered access to information, support for mobility, and reduced infrastructure, would be available to the grid, there are still several unanswered questions regarding network performance, suitability, and security.

9 Evolution of Embedded Internet

The growth of the Internet is the reason behind a new pervasive paradigm in computing and communications. This novel paradigm, named the Internet of things (IoT), is continuous with the concept of smart environments as well as with the deployment of numerous applications in many fields of future life. It utilizes low-cost information that facilitates interactions between the objects themselves in any place and at any time. From a wireless communication perspective, the IoT paradigm is strongly related to the effective utilization of wireless sensor networks (WSNs) and radio-frequency identification (RFID) systems. IoT will provide a wide range of smart applications and services like remote health care monitoring, intelligent transportation, smart distribution, home automation, systematic recycling, and others. Generally, IoT represents intelligent end-to-end systems that enable smart solutions and, as such, it covers a diverse range of technologies including sensing, communications, networking, computing, information processing, and intelligent control technologies. IoT is composed of a large number of nodes. This poses serious scalability requirements on any solution proposed for the IoT. Proposed solutions must be open and interactive with systems based on different technical solutions. This is a complex task given that most IoT nodes will have scarce capacity in terms of both energy and processing capabilities. Mobile management is another research issue that deserves particular investigation. Things will move and the system should be able to locate them at any moment. The major obstacles are related to the number of mobile nodes, which have an effect on scalability. This is another research issue to be addressed regarding traffic characterization that will traverse the IoT. Such traffic will have different properties from present-day Internet traffic. Most of it will be generated and directed to machines that communicate in a different way from humans. Differences in traffic characteristics, along with the energy constraints and the specific features of the IoT communication environment, will be the stimulus for research activities, modeling, and protocols design at both the network and transport layers.

9.1   INTRODUCTION

The term *Internet of things* was coined more than a decade ago by the Auto-ID Labs [1] (the leading global network of academic research laboratories in the field of networked RFID), where in parallel the concept of "ambient intelligence" and "ubiquitous computing" was developed. Since then, there have been considerable developments in both academia and industry, in the United States as well as in Europe and Asia. The developments have primarily been dedicated to applying RFID technology to the logistics value chain. The first trials to establish IoT-like applications for end users have been set up in "future stores" in Germany, Switzerland, and Japan.

The IoT can be considered as a convergence among a number of heterogeneous disciplines. This multidisciplinary domain covers a large number of topics from technical issues (routing protocols, semantic queries), to a mix of technical and societal issues (security, privacy, usability) including social and business themes. Pleasant user experiences are planned in the workplace and public areas as well as in the home environment by embedding computational intelligence into the nearby environment and simplifying human interactions with everyday service.

Overcoming the technical challenges and socioeconomic barriers of wide-scale IoT deployment requires a practical evaluation of corresponding solutions using interdisciplinary, multitechnology, large-scale, and realistic test beds. The test beds aim to design and deploy experimental environments that will allow [2]:

- The technical evaluation of IoT solutions under realistic conditions
- The assessment of the social acceptance of new IoT solutions
- The quantification of service usability and performance with end users in the loop

RFID is one of the key technologies because it not only permits a digital code to be associated with an object in a wireless modality but it also allows its physical status to be captured. RFID tags may be equipped with a large variety of sensors according to different modalities of integration exploiting a broad range of possible functionalities and costs. Active RFID tags use independent power supplies and a microcontroller, and dedicated electronics ensure long operating ranges. Also, it is possible to support high data rates and the greatest versatility in sensor interconnection. The main drawbacks of this solution are its high cost, limited lifetime, and large weight as well as size [3].

Sensor widespread deployment represents significant financial investment and technical achievement. The data they deliver is capable of supporting an almost unlimited set of high-value proposition applications. However, the main problem hampering success is that these sensors are locked into unimodal closed systems. Unlocking valuable sensor data from closed systems is a great task. Access to sensors should be opened such that their data and services can be integrated with data and services available in other information systems [4].

When it comes to wireless sensor technology, a variety of WSN approaches such as ZigBee, and other proprietary solutions have been proposed. If a trillion things are connected through a single open standard interface such as IP, they become transparent as general hosts and servers supporting seamless connectivity, unique addressability, and rich applicability. Because of that, IP-based WSNs have recently gained worldwide attention. The research in this field focuses on how to build a fundamental architecture that enables the IP to be used in a WSN space [5].

Nowadays, there is a clear need to develop a reference architectural model that will allow interoperability among different systems. With respect to the technological roadblocks, there is a need for action in three areas [6]:

a. An architectural reference model for the interoperability of IoT systems, outlining principles and guidelines for the design of its protocols, interfaces, and algorithms
b. Mechanisms for the efficient integration of the architecture into the service layer of a future Internet networking infrastructure
c. A novel resolution infrastructure, allowing scalable lookup and discovery of IoT resources, entities of the real world, and their associations

An emerging category of edge devices that will result in the evolution of the IoT are consumer-centric mobile sensing and computer devices connected to the Internet. They are equipped with various sensing facilities and wireless capabilities that allow producing data and uploading the data to the Internet [7]. Different from the IoT objects that lack computing capabilities, mobile devices have a variety of sensing, computing, and communication facilities. They can either serve as a bridge to other everyday objects, or generate information about the environment themselves. Based on the type of monitored phenomena, these applications can be classified into two categories: personal and community sensing. In personal sensing applications, the phenomena pertains to an individual (e.g., the monitoring of movement patterns of an individual, for personal record-keeping, or health care purposes). Community sensing is also known as participatory or opportunistic sensing. Participatory sensing requires the active involvement of individuals to contribute sensor data related to a large-scale phenomenon. Opportunistic sensing is more autonomous and user involvement is minimal. It means continuous sampling without the explicit action of the user [8]. Taking into account that community sensing spans a wide spectrum of user involvement, with participatory sensing and opportunistic sensing at two ends, the term mobile crowdsensing (MCS), which refers to a broad range of community sensing paradigms, was coined.

Machine-to-machine (M2M) is a promising technology for the development of IoT communications platforms with high potential to enable a wide range of applications in different domains. Providing suitable answers to this issue streaming from IoT platform design requires middleware level solutions to enable seamless interoperability between M2M-based applications and existing Internet-based services. Because of the growing demand for M2M-based services, various standardization bodies, such as 3GPP, the Open Mobile Alliance (OMA), IEEE, and ETSI have promoted various standardization activities on M2M [9]. In brief, M2M enables highly scalable direct communications among wireless heterogeneous terminals called M2M devices, and between M2M devices and control application servers (M2M servers) [10]. The ultimate goal of these standardization activities is to leverage widespread integration of M2M devices without any existing service.

Semantic Web of things [11] is a service infrastructure that makes the development and use of semantic applications involving Internet connected sensors almost as easy as searching and reading a web page. The IoT can be considered as a third ring of the present-day Internet, which already consists of a stable core (routers, servers) and a quickly growing fringe (personal computers and smartphones) [12]. This model is shown in Figure 9.1. As new embedded applications such as smart metering, building and industrial automation, personal sensing, and transportation become IP enabled, they add an entirely new dimension to the Internet with existing possibilities and challenges.

Networking alone does not make the Internet of practical use today. Applications depend on the web architecture using hypertext transfer protocol (HTTP) as well as extensible markup language (XML) to create web services. The interaction model of M2M applications differ greatly from how web services are used today. For example, multicast is commonly required in automation, whereas flow control is needed across the entire M2M network.

Finally, nanotechnology is enabling the development of devices in a scale ranging from one to a few hundred nanometers. At this scale, a nanomachine is defined as the most basic function unit, integrated by nanocomponents, and is able to perform simple tasks like sensing or actuation [13]. Coordination and information sharing among several nanomachines expands the potential applications of individual devices both in terms of

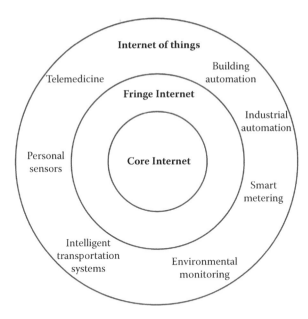

FIGURE 9.1   Ring model for the IoT.

complexity and range of operation [14,15]. The resulting nanonetwork will be able to cover larger areas, to reach new locations in a noninvasive way, and to perform additional in-network processing. The interconnection of nanoscale devices with classic networks and the Internet defines a new networking paradigm, which is referred to as the Internet of nanothings (IoNT) [13]. Two main alternatives for communication in the nanoscale can be envisioned: molecular communication and nanoelectromagnetic communication. Molecular communication is defined as the transmission and reception of electromagnetic (EM) waves from components based on novel nanomaterials. The unique properties observed in these materials will decide the specific bandwidth for electromagnetic emission, the time lag of the emission, and the magnitude of the emitted power for a given input energy.

9.2 MOBILE CROWDSENSING

The term "sensing" is considered ranging from the acquisition of an elementary status of an object (presence or absence with a given, region-localization) to multidimensional description to parameters of a thing with respect to the research environment and other things. Here, the term "thing" is capitalized to highlight the fact that the used tagging of an object is augmented with the physical interaction between the objects and the RFID tag itself to produce more information content [16]. The sensing modalities may fall into the stationary and nonstationary classes. Stationary sensing occurs when the measurement is performed in controlled conditions such as when the mutual position between the reader and the thing remains unchanged during the entire phenomenon to monitor, or when the same position can be replicated exactly in successive readings. On the other hand, we have the case of nonstationary sensing in which interrogation is performed at different times or the object is moving. The eventual change of the reader tag position as well as the change in the environment can be a further unknown of the sensing problem, making data retrieval more difficult. Generally speaking, additional independent data or functionalities are required to manage the sensing [3]. In what follows, we survey existing applications, identify characteristics of MCS, and discuss resource limitations, security and data integration, as well as architecture of MCS applications.

An emerging category of devices at the edge of the Internet are consumer-centric mobile sensing and computing devices such as smartphones, multimedia players, and in-vehicle sensors. These devices will fuel the evaluation of the IoT. In MCS, individuals with sensing and computing devices collectively share data and extract information to measure and map phenomena of common interest [7].

MCS applications provide a basis for illustrating various research challenges. They can be classified based on the type of phenomenon being measured or mapped as environmental, infrastructural, or social. In environmental MCS applications, the phenomena are those of the natural environment (measuring various air pollution levels in a city, water levels in rivers, etc.). Applications enable the mapping of various large-scale environmental phenomena by involving the common person. Infrastructural applications involve the measurement of large-scale phenomena related to the public infrastructure (real-time traffic congestion and road conditions, etc.). The third category is social applications, in which individuals share sensed information among themselves. As an example, individuals share their exercise data and compare their exercise levels with the community to improve daily routines. The functioning of typical MCS applications is presented in Figure 9.2.

Raw sensor data are collected on devices and processed by local analytical algorithms to produce adequate data for applications. These data may then be modified to preserve privacy and sent to the back end for aggregation and mining.

9.2.1 MCS APPLICATIONS ARCHITECTURE

MCS application has two specific components: one on the device for sensor data collection and propagation, and the second in the back

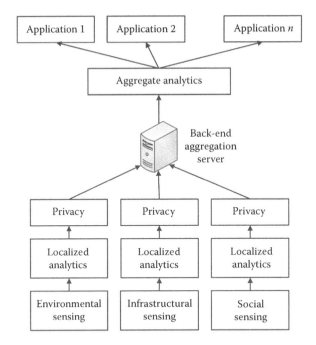

FIGURE 9.2   Function of typical MCS applications.

end or cloud for the analysis of the sensor data to drive the application [7]. This architecture is presented in Figure 9.3. Each application is built from the ground-up and independent from each other. There is no common component even though each application faces a number of common challenges in data collection, resource allocation, and

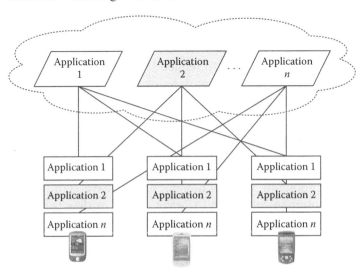

FIGURE 9.3   Architecture of existing MCS applications.

energy conservation. The presented architecture puts some limits to the development and deployment of MCS applications. First of all, the developer has to address challenges in energy, privacy, and data quality in an ad hoc manner. Second, he or she may need to develop different variants of local analytics if that person wants to run the application on heterogeneous devices. This approach is inefficient because there is a high likelihood of duplicating sensing and processing across multiple applications. For example, traffic sensing, air, and noise pollution all require location information, but these applications would each do its own sampling without reusing the same data samples. Also, there is no collaboration or coordination across devices. Namely, devices may not all be needed, especially when the device population is dense. Finally, the current architecture is not scalable because only a small number of applications can be installed on each device due to operating system limitations or user ability to keep track of a large number of applications. The data gathered from societal-scale sensing may overload network and back end server capacities, thus making the current architecture nonscalable.

Unifying architecture could address the current limitations of how MCS applications are developed and deployed. It will satisfy the common needs for different applications. It should allow application developers to specify their data needs in a high-level language. It should identify common data needs across applications to avoid duplicate sensing and processing activities on devices. Next, it should automatically identify the set of devices that can properly provide the desired data, and produce instructions to configure the sensing activities. For a given dynamic change, it should adapt the set of chosen devices and sensing instructions to ensure the desired data quality. Also, a layer that can shield the differences in physical sensor access application APIs and provide the same API upward is necessary. This means that it is possible to reduce some local analytics across different device platforms, assuming these platforms all support a common programming language.

9.2.2 Characteristics of MCS

MCS has a number of unique characteristics that bring both new opportunities and problems. Today's mobile devices have significantly more computing, communication, and storage resources than traditional sensors, and they are usually equipped with multimodality sensing capabilities. These will enable many applications that require resources and sensing modalities. Using these devices, we could potentially build large-scale sensing applications efficiently from the point of cost and time.

TABLE 9.1
Main Characteristics of MCS

MCS Issues	Description	Examples
Cost	There are already millions of user's devices that can collect data, so there is no need to install specialized sensors	Sensors on mobile phones can provide more information about traffic conditions than some specialized sensors [17]
Localized data processing	Depending on the nature of the raw data and the needs of applications, the physical readings from sensors may not be suitable for direct application. Often, raw data processing on the device is needed, producing intermediate results, which are sent to the aggregation server for further processing and application	In a pothole detection application, a local analytic computes spikes from three-axis acceleration sensor data to determine potential potholes [18]
Resource limitations	With different quality and resource consumption trade-offs, different types of data can be used for the same purpose. One of the challenges is leveraging these differences to improve the quality while minimizing resource consumption	Instead of GPS, location data can be provided using Wi-Fi and mobile systems [19], but with decreasing levels of accuracy. This approach represents trade-offs in data quality and accuracy for energy
Security	MCS applications potentially collect sensitive data for individuals. Because of that, privacy is one of the most sensitive subjects for IoT security [20]. The data availability expression has created entities that profile and truck users without their consent	GPS sensor readings can be utilized to inter private information about the individual, such as the routes they take during their daily communities and their home and work locations [21]
Data analyzing	MCS applications rely on analyzing the data from a collection of mobile devices, identifying spatiotemporal patterns. The challenge in identifying patterns from large amounts of data is usually application-specific and involves certain data mining algorithms	Data mining algorithms can provide reports to help prioritize and schedule the repair resources for MCS application in public work maintenance. Such algorithms take as input continuous data streams and identify patterns without the need to first store the data [7]

In traditional sensor networks, the population and the data they can generate are mostly known a priori. Because of that, controlling the data quality is much easier. In MCS, the population of mobile devices, the type of sensor data each can produce, and the quality in terms of accuracy, latency, and confidence can change all the time due to device mobility, variations in their energy levels and communication channels, and device user's preferences [7]. The main characteristics of MCS including description and examples are systematically analyzed in Table 9.1.

9.3　PERSPECTIVE PROTOCOL STACK FOR WIRELESS EMBEDDED INTERNET

The widespread sensor deployment represents significant investment and technical achievements. A crucial problem hampering success is that sensors are typically locked into unimodal closed systems. To solve this problem, sensor connections to the Internet and publishing outputs in well understood machine-processible formats on the web are indispensible.

Internet connectivity requires not only network level integration but also application level integration to enable structured access to sensor data. To enable sensor automatic reasoning, these sensors, their outputs, and their embedding into the real world must be described in a machine-readable format that is compatible with the data formats used to describe existing world knowledge in the Web. Not only must the syntax and semantics of such a description be defined but efficient mechanisms to annotate newly deployed sensors with appropriate descriptions are also required [4]. Finally, the users wish to search for real-world entities by their current state. Such search requests refer not only to the output of sensors but also to further machine-readable information that is available elsewhere in the web. The search engine needs to integrate these different data sources in a seamless manner.

Integrating resource-constrained sensors into the Internet is difficult because traditionally deployed protocols such as HTTP, TCP, or even IP are too complex and resource-demanding. To achieve integration, simple alternatives are required that can easily be converted from/to Internet protocols. Enormous progress has been made at the IETF over the past few years in specifying new protocols that connect smart objects to IP networks. The IETF has formed three WGs that define an adaptation layer (IPv6 in low-power WPAN—6LoWPAN) [22], routing over low-power and lossy networks (ROLL) [23], and a resource-oriented application protocol (Constrained Restful Environments—CoRE) [24]. The stack for wireless embedded Internet based on protocols proposed by corresponding IETF WGs is presented in Figure 9.4.

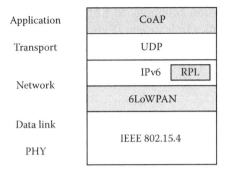

FIGURE 9.4 Perspective protocol stack for wireless embedded Internet.

9.3.1 ADAPTATION LAYER

The adoption of IP by wireless embedded devices is challenging due to several reasons [25]:

- Battery-powered wireless devices require low duty cycles, whereas IP is based on always connected devices
- Multicast is not supported natively in IEEE 802.15.4, but it is essential in many IPv6 operations
- Sometimes, it is difficult to route traffic in multihop wireless mesh networks to achieve the required coverage and cost efficiency
- Low-power wireless networks have low bandwidth (20–250 Kb/s) and frame size (IEEE802.15.4 packets are rather small, 127 bytes maximum at the physical layer, minus MAC/security and adaptation layer overhead). On the other hand, the minimum datagram size that all hosts must be prepared to accept for IPv6 is 1280 bytes. IPv6 requires that every link in the Internet has a maximum transmission unit (MTU) of 1280 bytes or greater. On any link that cannot convey a 1280-byte packet in one piece, link-specific fragmentation and reassembly must be provided at a layer below IPv6
- Standard protocols do not perform well in LoWPAN. For example, TCP performs very poorly in wireless networks due to its inability to distinguish between packet losses due to congestion and channel error.

Because IPv6 represents the backbone of NGN [26], the 6LoWPAN WG was chartered to standardize necessary adaptations of this network protocol for systems that use the IEEE 802.15.4 PHY layer, and has defined how to carry IP datagrams over IEEE 802.15.4 links and perform necessary configuration functions to form and maintain an IPv6 subnet. 6LoWPAN represents a lightweight IPv6 adaptation layer allowing sensors to exchange IP packets [27,28]. Core protocols for 6LoWPAN architecture have already been specified and some commercial products have been launched that implement this protocol suite.

9.3.2 ROUTING OVER LOW-POWER AND LOSSY NETWORKS

The IETF has significant experience in IP routing and it has specified a number of routing protocols over the past two decades (RIP, OSPF, etc.). On the other hand, routing in networks made of smart objects has unique characteristics. These characteristics led to the formation of a new WG called ROLL, whose objective is to specify a routing protocol for low-power and lossy networks (LLNs), known as RPL [29]. LLNs are formed by smart objects with limited processing power, memory, and energy. Unlike the MANET routing protocols, which perform well for ad hoc networks, RPL is optimized for upstream and downstream routing (to/from a root node), a paradigm appropriate for networks connected to the Internet. This routing protocol is essential for the deployment of the IoT because it enables traffic forwarding between low-power devices and the Internet. It has been designed assuming that the LLN scan comprises up to thousands of nodes interconnected by unstable links. Furthermore, RPL has been designed to operate over a variety of link layers such as IEEE 802.15.4, and it is a typical distance vector IPv6 routing protocol.

A directed acyclic graph (DAG) [29] is a directed graph having the property that all edges are oriented in such a way that no cycles exist. All edges are contained in paths oriented toward and terminating at one or more root nodes (traditionally called sinks in WSNs). RPL routes are optimized for traffic to or from one or more roots (sinks). As a result, RPL uses the DAG topology and is partitioned into one or more destination-oriented DAGs (DODAGs), one DODAG per sink. RPL specifies how to build the DODAG using an objective function. The objective function computes the optimal path according to certain routing metrics and constraints. In this way, DODAGs with different characteristics can be formed. For example, different DODAGs are constructed with the objective to (1) find the best path in terms of link throughput while avoiding battery-operated nodes, or (2) find the optimal path in terms of latency while avoiding nonencrypted links. There can be several objective functions operating at the same node depending on the different path requirements of a given traffic. In this way, it is possible to have multiple DODAGs active at the same time to carry traffic with different requirements.

Example 9.1

RPL specifies local and global repair mechanisms for recomputing routes when an inconsistency is detected or based on administrative decisions [30]. Local repair means detaching a node's sub-DODAG by increasing its rank value. Once a root initiates a global repair event, all the nodes in the DODAG recompute their rank values and reconfigure their parent sets. An example topology of a RPL network together with nonstoring and storing modes is presented in Figure 9.5.

Solid arrows represent each node's preferred parent (determined from the node's neighbors and their rank values) whereas dotted arrows point to the other nodes in the parent set. To support routing to various destinations within the DODAG, which is the root, RPL uses the destination advertisement object (DAO) message. RPL supports scenarios in which in-network nodes do not have enough memory to store routes to all possible destinations. In this case, the DAO messages, which contain information on the desired parent set of a destination node, are propagated up the DODAG until they reach the root. The root gathers DAOs from all nodes in the DODAG, and uses them to construct "down" routes to various destinations. Data to these advertised destinations is forwarded along a DODAG until it reaches the root, which then attaches a source routing header and sends it back down the DAG. Alternatively, nodes in the DODAG may store next-hops to downstream destinations. However, a key design simplification was not supporting "mixed-mode operation" in which storing and nonstoring nodes coexist because this

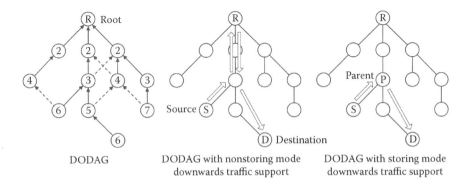

DODAG

DODAG with nonstoring mode
downwards traffic support

DODAG with storing mode
downwards traffic support

FIGURE 9.5 Example of RPL nodes that form a DAG rooted at a destination node supporting multipoint-to-point traffic.

was still considered a research issue; thus, all nodes in a DODAG must either store or not store routes.

9.3.3 APPLICATION PROTOCOL

In 2010, the IETF started a new WG, called CoRE [24], with the aim of extending the Web architecture to even the most constrained networks and embedded devices. Today's Web protocols work well between servers and clients running on PCs and handheld devices. However, constrained LLNs often mean high packet loss (5%–10%), frequent topology changes, low throughput (10–20 Kb/s), and useful payload sizes that are often less than 100 bytes. Embedded devices typically depend on cheap embedded microcontrollers, with processors running at several megahertz and limited memory. In addition, the interaction patterns in M2M applications are different, often requiring multicast support, asynchronous transactions, and push rather than pull. The CoRE WG has been chartered to develop a new Web transfer protocol and appropriate security setups for these M2M applications over constrained networks and nodes [31].

The WG is currently completing work on the constrained application protocol (CoAP) [32]. It provides a highlight alternative to HTTP using a binary representation and a subset of HTTP's methods (GET, POST, etc.). In addition, CoAP provides some transport reliability using acknowledgments and retransmissions. For seamless integration, reverse proxies may convert 6LoWPAN to IPv6 and UDP/CoAP to TCP/HTTP so that sensor data can be accessed using these omnipresent protocols. The integration of sensors into the Internet using CoAP/HTTP already enables many applications in which developers query and process data provided by a known set of sensors. A machine-understandable description of sensors and the data they generate are required. Semantic Web technologies fulfill this requirement as they enable machines to understand, process, and inter-link data using structured descriptions of service [33]. Linked open data as the framework makes this integration both immediate and meaningful through the inclusion of semantic links into a resource's machine-readable description.

Example 9.2

The use of UDP as transport protocol and the reduction of the packet header size significantly decrease power consumption in IoT. To evaluate the CoAP performance improvement compared with the HTTP, a simple experiment can be performed.

TABLE 9.2

HTTP and CoAP Comparison in Terms of Power Consumption

Protocol	Bytes/Transaction	Power Consumption (mW)
HTTP	1451	1.333
CoAP	154	0.744

A series of web service requests, first, between a CoAP client/server system and then between an HTTP client/server system are generated by Colitti et al. [34]. The CoAP system is based on a previously described protocol stack. Table 9.2 illustrates the results of the comparison between CoAP and HTTP in terms of bytes transferred per transaction and power consumption. It should be noted that the results have been taken in steady state conditions.

An HTTP transaction has a number of bytes nearly 10 times larger than the CoAP transaction. This is a consequence of the significant header compression executed in CoAP. In fact, CoAP uses a short fixed-length compact binary header of 4 bytes and a typical request has a total header of about 10 to 20 bytes. After being encapsulated in the UDP, 6LoWPAN, and MAC layer headers, the CoAP packet can be transferred into a single MAC frame, which has a size of 127 bytes. It is straightforward that the higher number of bytes transferred in an HTTP transaction implies a more intensive activity of the transceiver and CPU and, consequently, higher power consumption. The battery lifetime is unrealistically short in both cases as a consequence of the high number of client requests generated during the experiment. It is worth underlining that the results presented in this example do not exhaustively compare the two protocols. The simple experiment presented is only intended to illustrate how the UDP binding and the header compression introduced in CoAP improve the power consumption of IoT.

9.4 WSNs AND IoT

There have been many research projects in the field of IP-based WSN approach to the IoT from a dedicated IP stack for low–processing-power microprocessors to real deployments. Since the development of a micro-IP (μIP), as an open source TCP/IP stack capable of being used with tiny microcontrollers for smart sensors, a few approaches were carried out on top of TinyOS [35] and ContikiOS [36].

There are a wide range of technologies that will be involved in building the IoT. The enhancement of the communications network infrastructure, through heterogeneous technologies, is essential as well as the adoption of IPv6 to provide a unique address to each thing connected to the network. The technologies that allow the location and identification of physical objects will also be basic in this context. WSNs are able to provide an autonomous and intelligent connection between the physical and virtual worlds. Focusing on this type of network, a particular important challenge is the creation of a secure E2E channel between remote entities [37]. Therefore, it is necessary to allow the elements of a WSN to connect with other entities through the Internet. An increase in research efforts has lead to maturity in this field, yet it seems that there are some gaps that need to be filled.

As for protocols, they should be carefully designed in terms of the trade-off between interlayer, independence, and optimization. Some protocols may be vertically located through layers for optimization. Other protocols may stay beyond the adoption layer for protocol independence. The sensor networks for an all-IP world (SNAIL) [5] approach to the IoT protocols were designed to comply with IETF RFC 4944 [27]. The SNAIL adaptation layer includes all the necessary frame formats and operations, and is fully compliant with standards related to header compression, addressing, fragmentation and reassembly, and so on. To complete the IPv6 adaptation, it employs important protocols that are not specified in the standard. To obtain an IPv6 address in the start-up process, a bootstrapping network protocol is proposed. In this protocol, autoconfiguration is completed by combining a network prefix. This prefix is obtained from a neighbor discovery (ND) message from a response node with an interface identifier from an association process between two IEEE 802.15.4 nodes. During the bootstrapping process, a joining node registers its information, which includes an IP address, 64-bit extended unique identifier (EUI-64), type of routing protocol, sensors, as well as service for further network management by the gateway. Also, the registration information is used to advertise the gateway's liveness and the network prefix.

To serve rich Web content and to reduce the overhead of sensor nodes with no sacrifice of interoperability, the distributed resource-based simple web service (DRESS-WS) was designed [5]. It uses HTTP over TCP and distributes traffic to presentation servers and sensor nodes. As for web content, there exists two types of web content: real data and presentation templates. The real data are dynamic, whereas the templates are static and shareable accounting for the high traffic intensity. Only real data are served by nodes that host a web server, whereas the templates are served by presentation servers.

Example 9.3

Consider the DRESS-WS architecture shown in Figure 9.6. It consists of four components: distributed domain name system (DDNS) server, presentation server, gateway, and sensor node. The DDNS server manages the IP address information, which corresponds to a domain name to handle mobility of nodes. The presentation server serves templates, including multimedia and application codes to process the sensed data. The codes enable clients to control the periodic reporting of the sensed data. Each presentation server is assigned to sensor nodes, and each sensor node can have many distributed servers, depending on the manufacture of the node or the type of application in the node. The gateway performs HTTP/TCP/IP header compression and decompression to cover a limited network bandwidth with interoperability. The sensor nodes operate as a web server and serve the sensed data through Web services.

In DRESS-WS, when a client enters the domain name of a specific node in a common web browser, the browser is connecting to the DDNS server to get the corresponding IP address. While it requests web pages by using the address, the gateway compresses the headers and forwards the compressed packet to the sensor node. For the first access, the node replies with an HTTP redirect toward its presentation server (otherwise, it responds with the requested data). If the client receives this instruction,

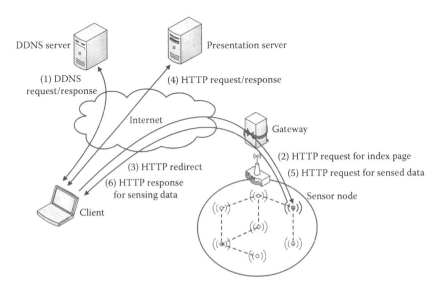

FIGURE 9.6   DRESS-WS architecture.

the browser automatically requests the template files. After the templates are downloaded, the application codes in the templates work to receive and process the sensed data. Because of that, this architecture enables a resource-constrained device to serve its information with a rich user interface and less overhead. Together with web enablement (previously analyzed), the main issues for WSN and IoT interaction are related to mobility managements, time synchronization, and security concepts.

9.4.1 MOBILITY MANAGEMENT IN EMBEDDED INTERNET

Mobility management is one of the most important research issues in 6LoWPAN based networks. Because typical mobility protocols are generally targeted for global mobility, they introduce significant network overhead in terms of increasing delay, packet loss, and signaling when mobile nodes change their point of attachment very frequently within small geographical areas. Having that fact in mind, methods for reducing handover delay are therefore essential for the IoT. The mobility protocol should be supported to a lightweight fashion because a thing's mobility behavior directly inherits the characteristics of portable devices.

SNAIL architecture uses a novel mobility management protocol called MARIO, which stands for mobility management protocol to support intra-PAN and inter-PAN handover with route optimization for 6LoWPAN [38]. The design of MARIO is based on MIPv6 and a fast and seamless handover scheme.

In the case of the inter-MARIO handover procedure illustrated in Figure 9.7, when a partner node detects MN's movement, it sends a pre-configuration message with the MN information. The information is stored by foreign agents (FAs) due to resource limitations. The partner node also gives the MN information, such as channel information, about the neighbor PAN. When orphaned, the MN can use the channel information to selectively scan a channel. When the MN associates with the new PAN, the FA performs a surrogate binding update simultaneously with the MN's IP operations, as in the case of the care-of address (CoA) generation. With this operation, the home agent (HA) creates a binding for the home-of address (HoA) of the MN to the FA. After the process of joining the new PAN is completed, the MN sends a binding update to the FA with a binding for the MN's HoA to the MN's CoA. This binding operation brings the MN to the end of the handover procedure. Because of the handover preconfiguration and surrogate binding update, MARIO can reduce the channel scan delay, the layer 2 association delay, and the binding message exchange delay. MARIO also provides an additional benefit to the solution of the route optimization problem.

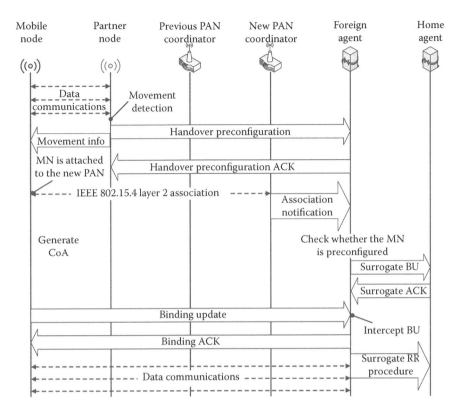

FIGURE 9.7   Inter-MARIO handover procedure.

This procedure does not provide an optimal route from the corresponding node to the MN because packets addressed to the MN may be delivered to the HA, and then it is forwarded to the new location of MN. To solve this problem, which is also known as triangle routing, the MIPv6 standard proposes a route optimization method, which is the return routability (RR) procedure [39].

9.4.2   GLOBAL TIME SYNCHRONIZATION

Network time protocol (NTP) [40] and simple NTP (SNTP) [41] are the most widely used time protocols on the Internet. They are application-level time protocols based on E2E communication. Also, they are unlikely to be used for resource-limited multihop-based networks because E2E communication for time correction causes substantial overhead. This overhead is due to the necessity for a number of periodic control messages to synchronize from each node.

Several time synchronization protocols and schemes are proposed for WSNs in the open literature [42]. These protocols only focus on clock

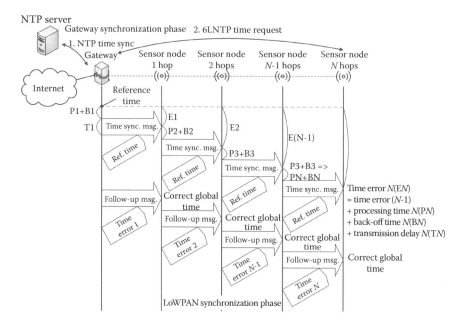

FIGURE 9.8   Multihop time synchronization protocol for IoT.

synchronization, not global time synchronization. This limitation is caused by protocols that are tightly coupled with energy conservation in the MAC layer. Thus, the synchronization process is often executed only between neighbors.

To ensure the consistency and precision of information, the IoT requires globally synchronized time. A multihop time synchronization protocol called 6LoWPAN NTP (6LNTP) was proposed by Hong et al. [5]. This protocol has two phases (Figure 9.8):

1. Gateway synchronization phase, which involves the use of the NTP, and
2. LoWPAN synchronization phase, which adjusts the time to the reference time from the gateway

In the first phase, the 6LoWPAN gateway corrects the time by exchanging request and response packets with an NTP server. Whereas in the second phase, the nodes synchronize their time with the gateway that synchronized with an NTP server.

9.4.3 SECURITY ISSUES IN EMBEDDED INTERNET

For WSN to become a part of the IoT, it is necessary to consider from the adaptation of existing network standards to the creation of interoperable protocols and the development of supporting mechanisms for composable services. One of the challenges is security, mainly because it is not possible to directly apply existing Internet-centric security mechanisms due to the nature of WSN. The relevant security challenges are related to the integration of WSN within the IoT. Although these challenges are tightly related to WSN, they can also be applicable to other relevant technologies of the IoT (embedded systems, MCS, etc.). Even if a WSN itself is protected with its own security mechanisms (e.g., using the link-layer security), the public nature of the Internet requires the existence of secure communication protocols for protecting the communications between two peers.

Sensor nodes can make use of the 6LoWPAN protocol to interact with IPv6-based networks. They are powerful enough to implement symmetric key cryptography standards such as AES-128 [43]. Due to the power constraints and limited computational capabilities of the nodes, there is currently no explicit support for the IPsec protocol suite in 6LoWPAN. As a consequence, it is necessary to study how other mechanisms can be used to create an E2E secure channel.

The creation of secure channels is just one of the steps in the creation of a securely integrated WSN [37]. To avoid unauthorized users from accessing the functions of the WSN, authentication and authorization mechanisms must be developed. Also, we need to create suitable and scalable identification mechanisms that can provide "unique identifiers" and "virtual identifiers" to all the different network elements. Finally, we have to take into account the survivability problem of IP-based WSN.

Other important challenges in this particular field are the integration of security mechanisms and data privacy [44]. The security of the IoT from a global perspective, regarding information, must be considered. Even if different wireless technologies are secure by their own, their integration will surely generate new security requirements that must be fulfilled. Also, it is necessary to analyze how the security mechanisms that protect one single technology will be able to coexist and interact with each other.

9.5 M2M COMMUNICATIONS AND EMBEDDED INTERNET

M2M communications is a new technology that provides the networks the ability to bring smart services to users. It is viewed as one of the next frontiers in wireless communications [45]. Different from the traditional human to human communications for which the current wireless networks are designed and optimized, the M2M concept is seen as a form of data communications between entities that do not necessarily need any form of human implication. It is different from current communication models in the sense that it involves new or different market scenarios, low cost and low effort, a potentially very large number of communicating terminals, and small and infrequent traffic transmission per terminal [46]. As for the industry, it has already been working on providing M2M communications and smart services offerings across a wide variety of market segments, including health care, manufacturing, utilities, distribution, and consumer products.

Because this concept brings very different requirements, and the number of communication devices may increase quickly, industry members have proposed enhanced wireless access networks for M2M communications. The topic of M2M communications attracted the attention of standardization bodies such as 3GPP LTE, whose objectives are looking into potential requirements to facilitate improvements in this field, and more efficient use of radio interface and network resources.

Advanced wireless multimedia networks are ready to deliver broadband data service at a significantly lower cost than in the past, thanks to diffusion standardization. These networks offer many of the features necessary to enable M2M services in the future embedded Internet [47]. Ubiquitous Wi-Fi and mobile systems, together with P2P communication, further extend the coverage of wireless networks whereas significantly reducing cost per bit transferred. For the wireless industry, there are also profound economic motivations for M2M agile implementation. The set of potential revenue-generating services includes M2M, cloud computing, and application stores.

M2M represents a future in which billions to trillions of everyday objects and the surrounding environment are connected and managed through a range of devices, networks, and cloud-based servers. The essential components to this IoT vision are the following [47]:

- A continuum of devices from low-cost/low-power to compute-rich/high-performance
- Ultrascalable connectivity, and
- Cloud-based mass device management and services

In the M2M market, large numbers of devices are expected to be embedded, requiring low prices and power consumption. On the other hand, the important challenge is to enable low-cost connectivity that addresses not only the massive network scale but also the vastly diverse requirements dictated by the device continuum. Finally, the vision of the future is for no one device acting alone, but many devices acting together. Thus, centralized decision making and management of many devices within the cloud will become an essential value of the IoT vision.

9.5.1 M2M System Architecture

There exists a need for a cost-effective, scalable M2M solution that will support a variety of applications and devices. The expected increase in M2M

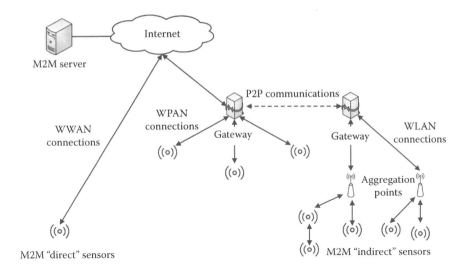

FIGURE 9.9   A high-level M2M system architecture.

devices poses a network capacity concern. A reasonable solution offers hierarchical network architectures. Multiple connectivity options are available to connect M2M devices to a server and to each other. When many devices are limited in range due to cost/size/power constraints, hierarchical deployments that provide reliable and efficient interworking between multiple communication protocols will be needed. A high-level M2M system architecture based on wireless communications is shown in Figure 9.9.

The M2M devices can be connected to the M2M server directly through a WWAN connection or an M2M gateway (aggregation point). Here, the gateway represents a smart M2M device that collects and processes data from simpler M2M devices and manages their operation. Connecting through a gateway is desirable when devices are sensitive to cost, power, or location. These devices can communicate using some lower-cost wireless interfaces (e.g., IEEE 802.11 or IEEE 802.15).

Many M2M applications will require connectivity between end devices. P2P connectivity can be supported in this architecture at various levels of hierarchy, depending on QoS requirements and the type of content. M2M systems need to be able to detect unusual events, such as changes in device location and device malfunction, and support appropriate levels of authentication for M2M devices and gateways. Enhanced monitoring and security may require changes to the network entry/re-entry procedure.

9.5.2   M2M STANDARDIZATION

Most standard bodies are taking a phased approach because the requirements and applications are still evolving. Fundamental M2M features are standardized and enabled quickly, with optimizations expected in later phases as the market grows.

In the first phase, only enhancements that require software changes (e.g., MAC modifications) are enabled. In the later phases, more extensive modifications to the PHY and MAC are expected. This will accommodate advanced requirements such as those for the M2M gateways. M2M is dependent on various technologies across multiple industries. Thus, the required scope of standardization is significantly greater than that of any traditional standards development. Unique challenges related to the M2M, which must be resolved by the wireless standards bodies, include the following [48]:

- A much larger number of devices need to be supported in a M2M network than an H2H. Optimizations are needed to avoid network congestion and system overload.
- Traffic patterns of M2M devices are quite different from those of H2H networks. M2M devices might frequently access the network, only to transmit small bursts of data.
- Many types of M2M devices, running various applications with different characteristics and requirements, all need to be supported.
- As many M2M devices are fixed devices, resource management and allocation for low-mobility devices need to be optimized.
- As M2M devices may be deployed without human supervision, advanced mechanisms for security and antivandalism need to be supported.
- It's crucial for network operators to be able to offer M2M services and devices at a low cost level for mass-market acceptance.
- Enhancements for minimizing battery power usage are important for low-power M2M devices.
- Other challenges exist, including subscription management and billing.

Collaboration among standards organizations across different industries is essential. The M2M community recognized this need, and joint efforts and collaborations among standards bodies are increasing. The current status of global M2M standards development is presented in Table 9.3.

Besides developing open interfaces and standard system architectures, M2M ecosystems also need to establish a set of common software and hardware platforms to substantially reduce development costs and improve time to market. Most of the existing proprietary vertical M2M solutions have difficulty scaling.

TABLE 9.3

Status of M2M Standards Development

Standards Development Organization (Project)	M2M Development
3GPP (Release 11)	Requirements and network optimization for features such as low power, congestion and overload control, identifiers, addressing, subscription control, and security
	Network improvements for M2M communication, network selection and steering, service requirements, and optimizations
ETSI (M2M network architecture)	Functional and behavioral requirements of each network element to provide an E2E view
GSM Alliance (GSM operation for M2M)	Define a set of GSM-based embedded modules that address operational issues, such as module design, radio interface, remote management, provisioning and authentication, and basic element costs
IEEE 802.16p	Optimize radio interface for low power, mass device transmission, small bursts, and device authentication. M2M gateway, cooperative M2M networks, and advanced M2M features
IEEE 802.11	Update radio interface to enable use of subgigahertz spectrum
IEEE 802.15.4	Radio interface optimization for smart grid networks
WiMAX Forum (network system architecture specification)	Define usages, deployment models, functional requirements based on IEEE 802.16 protocols, and performance guidelines for end-to-end M2M system
Wi-Fi Alliance (smart grid task group)	Promote the adoption of Wi-Fi within the smart grid through marketing initiatives, government and industry engagement, and technical/certification programs
Open Mobile Alliance (device manageability)	Define requirements for the gateway-managed object
Telecommunications Industry Association (M2M SW architecture TR50)	Develop and maintain access interface standards for monitoring and bidirectional communication of events and information between smart devices and other devices, applications, or networks

9.5.3 IP MULTIMEDIA SUBSYSTEM AND M2M INTEGRATION

In the next few years, IoT services and applications are likely to become an integral part of everyday life. Basic technologies that leverage seamless interaction between unconventional artifacts have already been developed and play a relevant role in different application domains (e.g., tracking and tracing, vehicular telematics, health care, remote maintenance, and control).

Early experimentation on the heterogeneous communication concept in IP multimedia subsystem (IMS) networks has shown that consolidation of strategies for harmonized data handling can greatly assist liberalization of rich content IP communication and guarantee QoS, integrity of mobility management, as well as uniformity of service charging [49].

IMS seems to be a natural solution to simplify the integration of M2M applications in a wider municipal service ecosystem. On the one hand, IMS represents an excellent integration framework that permits us to take full advantage of legacy solutions. On the other hand, IMS makes it possible to effectively interact with internal M2M mechanisms and coordinate M2M device communications to encompass relevant aspects of communication management. Also, the adoption of IMS simplifies the service management process, as involved technical staff requires only limited training to administer and maintain a new IMS-based system [50].

The need to seamlessly integrate the system with a broader service architecture suggested the adoption of IMS as the basic communication management support. IMS provides ubiquitous and production-level support for the development of novel services that are easy to integrate with existing wireless infrastructures, and further decrease costs. As a consequence, IMS is particularly well suited to realizing advanced management platforms that are able to integrate different infrastructures and service components according to specific application domain requirements.

Example 9.4

An example of IMS-enabled M2M-based management system is presented in Figure 9.10. This system consists of three main domains and is applicable in road traffic management [50].

Each retractable bollard hosts an M2M device that participates in the M2M device domain. The second domain is the network domain, which enables communication between the M2M device and the M2M server over the mobile operator network. This domain consists of the most important 3GPP evolved packet system nodes and IMS components such as

- Home subscriber server (HSS) is the database storing authentication data and profiles for clients, ranging from M2M devices to IMS-enabled clients.

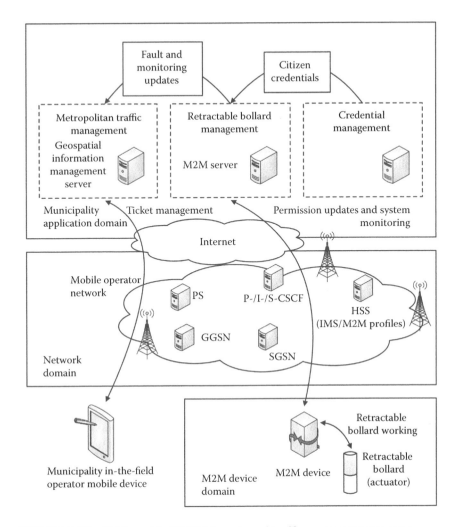

FIGURE 9.10   IMS-enabled M2M-based road traffic management system.

- Proxy-/interrogating-/serving-call session control functions (P-/I-/S-CSCF) core entities of IMS that realize several main functions, including localization, routing SIP messages, associating an IMS client with its S-CSCF (as indicated within the client profile), and modifying the routing of specific types of SIP messages to applications servers depending on filters/triggers specified by client profiles maintained by the HSS.
- Presence service (PS) that, following a publish/subscribe model, allows users and hardware/software components to publish data to interested entities previously subscribed to the IMS PS server.

- Serving GPRS support node (SGSN) acts as a local mobility anchor node.
- Gateway GPRS support node (GGSN) acts as an interface between the mobile operator network and different packet data networks.

Detailed architecture and functions of IMS components are analyzed by Bakmaz et al. [51]. In the third domain (municipality application domain), the M2M server represents the service integration core component. That is, it interacts with M2M devices over IMS, provides suitable support to authorize citizens to access restricted areas by interacting with a credential management server to obtain currently applicable citizen's credentials, and interacts with the traffic management system.

9.6 NANONETWORKS AND IoT

Nanotechnology promises new solutions for many applications in the biomedical, industrial, and homeland security fields as well as in consumer and industrial goods [13]. Nanomachines can be defined as the most basic functional units, integrated by nanocomponents and able to perform simple tasks such as sensing or actuation. Coordination and information sharing among several nanomachines will expand the potential applications of individual devices both in terms of complexity and range of operation. Traditional communication technologies are not suitable for nanonetworks mainly due to the size and power consumption of transceivers, receivers, and other components. The use of molecules, instead of electromagnetic or acoustic waves, to encode and transmit the information represents a new communication paradigm that demands novel solutions such as molecular transceivers, channel models, or protocols for nanonetworks [14,15].

Nanonetworks do not represent a simple extension of traditional communication networks at the nanoscale. They provide a complete new communication concept in which most of the processes are inspired by biological systems. The main differences between traditional communication networks and nanonetworks enabled by molecular communication are evaluated by Hiyama et al. [52] and summarized Table 9.4.

The interconnection of nanoscale devices with existing communication networks and, ultimately, the Internet defines a new networking paradigm that is further referred to as the IoNT. The interconnection of nanomachines with existing communication networks and eventually the Internet requires the development of new network architectures.

TABLE 9.4

Comparison of Traditional Communication Networks and Nanonetworks Enabled by Molecular Communication

	Communication	
	Traditional	**Molecular**
Signal type	EM waves	Chemical
Propagation speed	Speed of light	Extremely low
Medium conditions	Wired: Almost immune Wireless: Affects communication	Affects communication
Noise	Electromagnetic fields and signals	Particles and molecules in medium
Encoded information	Multimedia	Phenomena, chemical states, or processes
Energy consumption	High	Low

9.6.1 COMPONENTS OF IoNT ARCHITECTURE

Regardless of the final application, the following components such as nanonodes, nanorouters, nano-microinterface devices, and gateways can be identified (Figure 9.11).

Nanonodes are the simplest and smallest nanomachines. They can perform simple computation, have limited memory, and can only transmit over very short distances, mainly because of their reduced energy and limited communication capabilities. Good examples are biological nanosensor

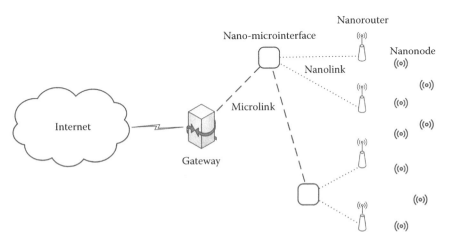

FIGURE 9.11 Network architecture for IoNT.

nodes inside the human body and nanomachines with communication capabilities integrated in all types of things such as books, keys, and so on.

Nanorouters as a type of nanodevice have comparatively larger computational resources than nanonodes and are suitable for aggregating information coming from underlying nanomachines. Also, nanorouters can control the behavior of nanonodes by exchanging very simple control commands such as on/off, standby, and so on. However, this increase in capabilities involves an increase in their size, and this makes their deployment more invasive.

Nano-microinterface devices are able to aggregate the information coming from nanorouters, to convey it to the microscale, and vice versa. They can be contemplated as hybrid devices able both to communicate in the nanoscale using the aforementioned nanocommunication techniques and to use classic communication paradigms in conventional communication networks.

In this architecture, a gateway enables the remote control of the entire system over the Internet. For example, in an intrabody network scenario [13], a smartphone can forward the information it receives from a nano-microinterface to the health care provider. On the other hand, in the interconnected office, a modem-router can provide this functionality.

9.6.2 NANONETWORKS COMMUNICATION

The IoNT communication concept begins at the networking of several nanomachines. Nanonetworks are not downscaled networks. There are several properties stemming from the nanoscale that require us to totally rethink well-established networking concepts. The main challenges from the communication perspective are presented in a bottom-up fashion, by starting from the physical nanoscale issues affecting a single nanomachine up to the nanonetworking protocols [13,53]. Figure 9.12 shows the design flow for the development of nanonetworks.

Communication at the nanoscale is strongly determined by the operating frequency band of future nanotransceivers and nanoantennas. Graphene-based nanoantennas have been proposed as prospective solutions for nanoscale communications [54]. The wave propagation velocity in graphene can be up to one hundred times below the speed of light. As a result, the resonant frequency of nanoantennas can be up to two orders of magnitude below that of nanoantennas built with noncarbon materials. The use of EM waves in the megahertz range can initially be more appealing than emission in the terahertz band, provided that by transmitting at lower frequencies, nanomachines can communicate over longer distances. However, the energy efficiency of the process to mechanically generate EM waves in a nanodevice is predictably very low [55].

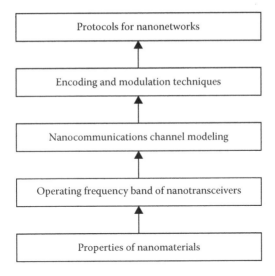

FIGURE 9.12   The design flow for nanonetworks communication.

9.7   CONCLUDING REMARKS

An emerging category of devices at the edge of the Internet are consumer-centric mobile sensing and computing devices. In a category of applications termed MCS, individuals with sensing and computing devices collectively share data and extract information to measure and map phenomena of common interest.

Existing semantic sensor web technologies enable the integration of sensors into the Web. It was difficult to foresee the wealth of current web applications back when the Web was first created, yet now we have seen how widely adopted the Web has become. Likewise, it is difficult to predict how people will come to use the semantic Web of things. Using sensor data is clearly beneficial, because then integration with knowledge from arbitrary services is possible. For example, sensor data can be linked to geographic data, user-generated data, scientific data, and so on. A strong indicator of whether this line of development will be successful in the long run is also provided by the exponential growing amount of linked data.

An important and big step toward the IoT would be to facilitate suitable IP-based WSN technologies to support the network of things. An increase in research efforts has led to maturity in this field, yet there seems to be gaps to be filled because of the focus on how to adopt the IP to the space of things. Considering the effect of billions of new internetworked devices, the emerging IETF protocol stack for IoT should be the cornerstone for research in this field.

M2M communications in the context of the mobile Internet has been a subject of intense discussions over the past 2 years. Some people see it as the next technology revolution after the computer and the Internet. As for M2M, it presents both challenges and opportunities to the industry. Although there are significant business and economic motivations for operators and equipment manufacturers to invest in future generations of M2M services, the highly fragmented markets risk the forecasted growth of M2M markets. Two things are needed for the embedded Internet vision to materialize: the development of new technologies that scale with the growth of M2M markets, and a broad standardization effort in system interfaces, network architecture, and implementation platforms.

Although the IoT concept has been around for several years now, there are still many crucial issues that have not been solved, including heterogeneity, scalability, security, and others. The complexity of these technical issues, especially in view of the resource-constrained nature of many components and of the use of wireless communications, calls for a unified architectural view that is able to address them in a coherent fashion.

Nanotechnology is enabling the development of advanced devices that are able to generate, process, and transmit multimedia content at the nanoscale. The wireless interconnection of pervasively deployed nanodevices with all sorts of devices and, ultimately, the Internet will enable a new networking paradigm, known as the IoNTs. This new concept will have a great effect on almost every field, starting from telemedicine to military purposes. To enable communication among nanomachines, it is necessary to readjust traditional communications and to define new alternatives stemming from the nature of the nanoscale. Although nanohardware is still being developed, the definition of new encoding and modulation for nanomachines, and the development of nanonetworking structures and protocols are major scientific research issues.

10 Flexible Future of the Internet

The evolution of the Internet has played a central and crucial role as the main enabler of digital era. Thus, the Internet has been successfully deployed for several decades due to its high flexibility in operating using different physical media and supporting different higher layer protocols and applications. Although future networks will require a multimedia transport solution that is more aware of a delivery network's requirements, the future Internet (FI) is expected to be more agile, scalable, secure, and reliable. Rapidly emerging applications with different requirements and implications for FI design pose a significant set of problems and challenges. In migrating toward the FI era, key characteristics are robust network infrastructure as well as numerous applications and services. Of course, the infrastructure must be highly pervasive, consisting of humans, smart objects, machines and the surrounding space, and embedded devices, which will result in a highly decentralized environment of resources interconnected by networks. These characteristics make it clear that management of the FI will have additional complexity to support multiple demanding and changing situations for the desired provisioning quality of service/experience (QoS/QoE). The FI will need to be intelligent and adaptive, continuously optimizing the use of its resources and recovering from faults and attacks without any effect on the demanding services and applications. Due to the rapid developments in mobile communications, and the increased quality and quantity of user-created contents, the role of the Internet has changed to a content-oriented data-sharing network. This new paradigm requires a major transformation of the current Internet architecture, which has lead to the birth of the information-centric networking concept.

10.1 INTRODUCTION

The existing Internet, principally based on the best effort service, has several structural inefficiencies and relies on significant overprovisioning of bandwidth to reduce congestion and achieve weak statistical QoS guarantees. The new ubiquitous multimedia communication services are radically changing the nature of the Internet: from a host-to-host communication service to a content-centric network, in which users access the network to find relevant content and to possibly modify it [1]. The user-generated content paradigm is further pushing toward this evolution, whereas the social platforms have an increasing role in the way users access, share, and modify the content. The radical departure from the objectives that have driven the Internet's original design is now pushing toward a redesign of the Internet's architecture and protocols to take account of completely new design requirements.

The FI is being addressed as a novel concept for future interoperable networks providing direct and ubiquitous access to information [2]. Generally, FI is an information exchange system that interfaces, interconnects, integrates, and expands today's Internet, as well as networks of any type and scale to provide efficiently, transparently, timely, and securely any type of service, from the best effort information retrieval to highly demanding performance services.

The key differences between the current and future Internet are as follows: time bandwidth, sharing of control, content distribution optimization, content identity management, network models with guaranteed QoE, and scalability. Time bandwidth will provide gigabits or even higher per second connectivity, thus enabling a wider range of services with higher quality. Sharing of control means that whereas the first generation Internet facilitates the sharing of information, the second generation Internet facilitates the sharing of control to manage the provisioning of services. Content distribution optimization understands new techniques to optimize content distribution from the users', network providers', and content providers' points of view. Content identity management ensures content veracity, whereas the rights of use are needed together with content protection. Accurate metrics and optimization mechanisms are the most appealing challenges related to the QoE monitoring and optimization in FI. For network models to guarantee scalability, it should be noted that new network paradigms have to be devised to cope with substantial scalability needs in different dimensions of the network.

The FI will experience two major shifts from the conventional Internet [3]. The first is a shift from wired to wireless communications. Secondly, the role of a network will be more rapidly shifted from communication between end users to content delivery, especially large media files. Based on these observations, it is expected that there would be a rapid growth in the consumption of multimedia content by mobile devices. At the same time, network resources are hard to be pre-allocated to accommodate the growth due to the unpredictable behavior of proliferated mobile devices. To accommodate such an environment adaptively and to provide any user the means to access contents efficiently, a new network architecture is highly anticipated.

This chapter starts with the main principles of FI architecture. Then, the delivery infrastructure for next generation services is presented. Next, information-centric networking is invoked and analyzed through the most representative approaches. The concept of scalable video delivery is briefly presented as key for efficient streaming in FI. Also, media search and retrieval in the FI is outlined. FI self-management scenarios conclude the chapter.

10.2   PRINCIPLES FOR FI ARCHITECTURE

Recently, significant demands to transform the Internet from a simple host-to-host packet delivery infrastructure into a more diverse paradigm built around the data, content, and users instead of the machines, are anticipated [4]. All these challenges have led to research on FI architecture, which is not a single improvement on a specific topic or goal. A clean slate solution on a specific topic may assume the other parts of the architecture to be fixed and unchanged. Thus, assembling different clean slate solutions targeting different aspects does not necessarily lead to a new Internet architecture. Instead, it has to be an overall redesign of the entire architecture, taking all the issues (security, mobility, performance reliability, etc.) into consideration. It also needs to be evolvable and flexible to accommodate future changes. Some of the other widely accepted key research topics specifically aimed at the design of the FI can be identified as follows.

10.2.1   INFORMATION-CENTRIC NETWORKING

The primary use of today's Internet has changed from host-to-host communication to content distribution. Thus, it is desirable to change the architecture from IP-based to the data/content-oriented distribution. This category of new paradigms introduces challenges in data and content security and privacy, scalability of naming and aggregation, compatibility and interoperability with IP, and efficiency of the new paradigm. Inspired by the fact that the Internet is increasingly used for information dissemination, rather than for pair-wise communication between end hosts, this paradigm aims to reflect current and future needs better than the existing TCP/IP architecture [5]. Instead of accessing and manipulating information only bypassing of servers hosting them, putting named information objects themselves at the center of networking is appealing, from the viewpoint of information flow and storage. This information-centric usage of the Internet raises various architectural challenges. Many of them are not handled effectively by the current network architecture. This makes information-centric networking an important research field in FI architecture. Storage for caching information is part of the basic network infrastructure, whereas network service is defined in terms of named information objects, independently of where and how they are stored or distributed. Using this approach, an efficient and application-independent large-scale information distribution is enabled [6].

10.2.2 NETWORK VIRTUALIZATION AND ADAPTIVE RESOURCE MANAGEMENT

Before the Internet gained its current popularity, single network providers usually owned the communication infrastructures. However, this situation is slowly transforming into a new business model in which a distinction between network providers and service providers is becoming apparent. This concept is also known as network virtualization [7], in which service providers lease the resources they need from network providers and are allowed to have a certain control over the use of these resources. The increased flexibility means that service providers may configure their virtual networks according to the services they are offering, whereas the network provider needs to safeguard the fair usage of the resources. However, the dynamics of services may change over short timescales, leading to the need for dynamic resource subscription policies from the network provider. Another crucial requirement of the FI is the adaptive use of resources through efficient routing. The majority of traditional routing solutions are based on optimization methods, in which previous knowledge of traffic demand exists, and the demand does not change frequently. However, as the number of services increases and evolves at a fast pace, more reactive and intelligent routing mechanisms are required.

10.2.3 FLEXIBLE AND EVOLVABLE OPEN INFRASTRUCTURE

A major factor behind the requirement of redesigning the Internet is the fact that the original Internet was designed mainly for accommodating data traffic with stable traffic patterns. However, it is neither feasible nor practical to perform a complete redesign of the Internet each time new requirements or drastic technological or social changes arise that do not fit the current architecture. Therefore, the design of the FI should be based on sustainable infrastructure that is able to support evolvability. This should enable new protocols to be introduced with minimal conflict to existing ones. At the same time, the design of architecture and protocols should be made in a modular way, in which protocol components can have cross-layer interactions. The evolvability of the FI should also allow for a certain degree of openness, wherein protocols with the same functionalities can be deployed by various entities to suit their own needs [8].

10.2.4 MOBILITY MANAGEMENT

Mobility is the norm of the architecture that potentially nurtures FI with new scenarios and applications. Convergence demands are increasing among heterogeneous networks that have different technical standards and business models. The mobility, as the norm of the architecture potentially nurtures FI architecture with new scenarios and applications. Also, there are challenges such as how to trade off mobility with scalability, security, and privacy protection of mobile users, mobile end point resource usage optimization, and others. Virtual mobility domains architecture [9] seems to be a good starting point in FI seamless design because it supports both interautonomous system (macro) and intra-autonomous system (micro) mobility by leveraging tiered addressing (a network cloud concept) and a unique packet-forwarding scheme. This architecture is distinct from traditional mobility approaches by not using IP addressing and classic routing protocols, and deploying user-centric overlapping mobility domains.

10.2.5 CLOUD COMPUTING-CENTRIC ARCHITECTURES

The cloud computing perspective has attracted considerable research effort and industry projects toward these goals. A major technical challenge is how to guarantee the trustworthiness of users while maintaining persistent service availability. Migrating storage and computation into the cloud and creating computing utility is a trend that demands new Internet services and applications. Data centers are the key components of such new architectures. The design of secure, trustworthy, extensible, and robust architecture to interconnect data, control, and management planes of data centers is an important issue.

10.2.6 SECURITY

Although security was added into the traditional Internet as an additional overlay instead of an inherent part of the architecture, it has now become an important design goal for the FI's architecture. The challenge is related to both the technical context and the economic and public policy context. From the technical aspect, it has to provide multiple granularities such as encryption, authentication, authorization, and others (see Chapter 7), for any potential use case. It needs to be open and extensible to future security-related solutions. As for the nontechnical aspect, it should ensure a trustworthy interface among the participants such as users, network providers, and content providers.

10.2.7 ENERGY EFFICIENCY

There is an increasing recognition of the importance of energy conservation on the Internet because of the realization that the exponential growth of energy consumption that follows the exponential increase in the traffic intensity is not sustainable [1]. A common approach toward saving energy today is switching devices off or putting them into sleep state. However, with the large number of nodes anticipated in the FI, this process should be performed in a collaborative manner while ensuring that end users' requirements are met.

10.2.8 EXPERIMENTAL TEST BEDS

The current Internet is controlled and managed by multiple stakeholders who may not be willing to expose their networks to the risk of experimentation. So the other goal of FI architecture research is to explore open virtual large-scale test beds without affecting existing services. Currently, test bed research includes multiple projects with different virtualization technologies, and the cooperation and coordination among them. These projects explore challenges related to large-scale hardware, software, distributed system test and maintenance, security and robustness, coordination, openness, and extensibility [4].

10.3 PHYSICAL LAYER–AWARE NETWORK ARCHITECTURE

Limitations and shortcomings of the current Internet's architecture are driving research trends toward a novel, secure, and flexible architecture. FI's architecture needs to provide the coexistence and cooperation of multiple networks on common platforms through the virtualization of network resources. Possible solutions embrace a full range of technologies, from fiber backbones to wireless access networks. The virtualization of physical networking resources will enhance the possibility of handling different profiles while providing the impression of mutual isolation [2]. This abstraction strategy implies the use of well-elaborated mechanisms to deal with channel impairments and requirements in access and core networks. In this context, a physical layer–aware perspective of the different issues involved in the entire virtualization process becomes a key aspect to properly assess how the required level of virtualization can be achieved, while being as transparent as possible to the user, independently of the type of service and technology being used.

In the 4WARD project [10], multiple ways of constituting a generic path, with transport functions adapted to the capabilities of the underlying network, were investigated and evaluated, taking physical and technological constraints into account. These generic paths will be the abstraction of all possible communication relationships between end points, irrespective of their physical realization across wireless and wireline technologies, thus hiding specific channel characteristics from the overall transport system.

10.3.1 PHYSICAL LAYER AWARENESS IN WIRELESS ACCESS NETWORKS

Although the 4WARD project examines all types of media for modern core and access networks, significant attention is provided to physical layer awareness in wireless access networks. Concerning the wireless propagation channel, the unreliability is usually caused by the radiofrequency propagation itself. The signal experiences different kinds of attenuation and losses due to several phenomena involved in the physical mechanism of wave propagation in multipath environments. Channel impairments that can influence the overall network behavior can be classified as

- User behavior, in which the operating environment and the mobile terminal speed are the strongest impairments
- Radio channel, which itself imposes several impairments due to transmitting power levels, time variability of propagation conditions, signal loss, fading, interference, etc.
- Possible impairments caused by the system are due to available bandwidth, channel access technique, handover schemes, and uplink/downlink channel asymmetry

To overcome these shortcomings, taking into account that a wireless link is subject to bursty errors and varying channel capacities, it is also desirable that higher layers (e.g., radio resource management and scheduling levels) have the following features [2]:

- Efficient link use to avoid slots to be assigned to currently bad links
- Delay bound, for supporting delay-sensitive applications
- Fair resource distribution according to different QoS requirements
- Low complexity algorithms, allowing real-time control of wireless link parameters
- Service adaptation to link quality, making it possible to dynamically update data rate and QoS requirements according to the link quality
- Isolation among different sessions, to maintain QoS requirements for a given session
- Low energy consumption for improved MT battery life and efficient radio networks
- Delay/bandwidth decoupling because delays are usually coupled to the available bandwidth
- Scalability, allowing for an efficient operation by fairly sharing the resources among different sessions

10.3.2 Physical Layer–Aware Architecture Framework

Current network architectures are commonly based on layered models. On the other hand, efficient and effective communications systems require cross-layer information. For example, integrating IP-based services with wireless networks has been a great challenge. Specifically, in a strictly layered system, it is difficult to incorporate several enhancements, such as IP security and multicasting, or to adapt to new communication systems with specific requirements, such as energy-constrained sensor networks and content centric ones (see Chapter 9). Regarding developments toward networks virtualization, it should become much easier to bring out new architectures with features and properties that are individually tailored to the respective requirements.

An architecture framework was proposed by Cardoso et al. [2] to represent perspective design approach for FI. This set of concepts and procedures can be used to model existing networks as well as future solutions, applying a design process that provides a systematic approach to guide the network architects. The architecture framework provides two views on network architectures:

- The macroscopic view focuses mainly on structuring the network at a higher level of abstraction, and introduces the concept of stratum as a flexible way to layer the services of the network. Here, extensions, amendments, as well as generalizations to the traditional layered system are incorporated.
- The microscopic view concentrates more on the functionalities needed within the network, their selection, and composition to so-called *netlets* (containers that provide a certain service) that are embedded in the physical nodes of the network. Also, it complements this design, concentrating on the functionalities needed within the participating network nodes, and on how this functionality can be organized within the architecture of such nodes.

The stratum is a structural element of the network architecture (Figure 10.1), being used for designing, realizing, and deploying distributed

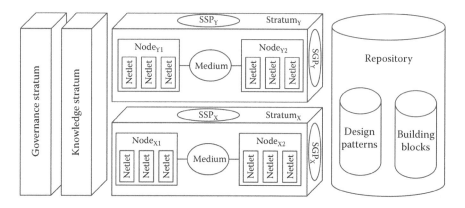

FIGURE 10.1 4WARD architecture framework.

functions in a communication system. It is modeled as a set of logical nodes that are connected through a medium [11]. The functionality of a stratum is encapsulated by and offered as a service to other strata through a stratum service point (SSP). A stratum implementing the same or a similar distributed function, but in a different domain (thus assuming independent implementations), can interoperate through a stratum gateway point (SGP).

Horizontal strata (i.e., $stratum_X$ and $stratum_Y$) provide different levels of transmission capabilities and communication among the different nodes that form them, whereas on the other hand, vertical strata have the responsibility of managing and monitoring the different aspects present within the network, and taking decisions on which action to take if that is required. Netlets consist of functionalities required to provide the end-to-end services. They contain protocols and, therefore, provide the medium for the strata they belong to. Inside the same netlet, there could be functionalities that are related to different strata. With the principle of hiding protocol details, yet providing a number of properties via its interfaces, netlets can be easily exchanged without the need to change application or network interfaces.

The two vertically oriented strata provide governance and knowledge for an entire network (i.e., a set of horizontal strata). The knowledge stratum provides and maintains a topology database as well as context and resource allocation status as reported by a horizontal stratum. The governance stratum uses this information, together with input provided via policies, to continuously determine an optimal configuration of horizontal strata to meet the performance criteria for a network. This stratum also establishes and maintains relations and agreements with other networks.

The repository contains the set of building blocks and design patterns for the composition of functionalities (i.e., to construct the strata and the netlets) for specific network architectures, including best practices and constraints to ensure interoperability among network architectures.

Signaling capabilities must be present in any stratum to coordinate their own internal operations, and to interoperate with other strata as well. In addition to mobility, the accumulation of control information in different system elements needs to be considered. Both are related to dynamics because the state of the network might have changed during the transmission of control information. Thus, signaling is needed to find out the current level of dynamics and react in real-time to minimize these overhead.

10.3.3 NETWORK VIRTUALIZATION

The application of network virtualization enables the coexistence of multiple network architectures on a shared infrastructure to meet the diverse requirements of the FI. In a virtual network (VNet) approach, a three-role model was chosen for splitting up provider roles into infrastructure and service providers [2]. The infrastructure provider (InP) is responsible for maintaining physical networking resources, such as routers, links, wireless infrastructure, and others, and enabling the virtualization of these resources. Also, the InP provides a resource control interface for the virtualized resources, through which InPs can make virtual resources and partial virtual topologies (slices) available to virtual network providers (VNPs), which are the customers of the InP.

The VNP constructs VNets using virtual resources and partial topologies, provided by one or more InPs or other VNPs. This role (in essence) adds the layer of indirection that virtualization provides. The resource control interface is used, provided by the InP who owns the resource, to request and configure these virtual resources. A newly constructed VNet can be made available to a virtual network operator (VNO) or to another VNP, who can recursively use it to construct an even larger VNet. Network virtualization process is shown in Figure 10.2.

The VNO controls and manages the VNet to provide services. Once the VNet has been constructed by a VNP, the VNO is given console access to the virtual resources setting up the VNet, allowing it to configure and manage them just like a traditional network operator manages physical network resources.

In the virtualization process, a relevant issue that has to be considered is the occupancy and management of resources in VNets, and their possible

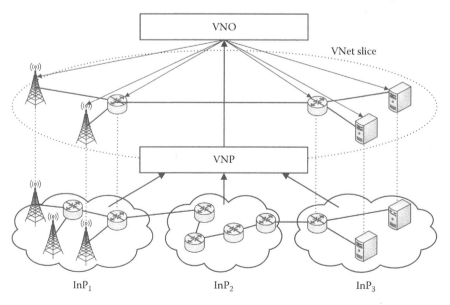

FIGURE 10.2 Network virtualization process.

(re)utilization by other services/flows belonging to another VNet. This can be used to optimize their efficiency. After the creation of a VNet, physical resources that have been allocated to it may not be in use, which can lead to an inefficient use of physical resources, hence, an entity in the VNet architecture must manage the resource occupancy.

The mobility dimension in the virtualization process comprises any physical entity that can affect virtual network performance due to its mobility. Physical resources comprise any physical entity of the network with the ability to be in motion (from the network resources to the end users' devices). VNets are directly related to physical resources underneath, and any change in the availability of resources (mobility generates many of them) affects the performance of the virtual networks that use them. Detection and reaction to any physical changes in the network topology are key issues in mobility management. A VNet, with the information regarding the changes provided by the management layer, will rearrange the corresponding virtual networks to the new physical topology.

10.4 INFORMATION-CENTRIC NETWORKING

Traditional communication between a pair of networked devices has evolved into a scenario in which new services generate unprecedented amounts of content (video streaming, cloud computing, etc.), and at the same time allow multimode mobile devices to access this information via different connectivity opportunities. The existing host-centric architecture has been upgraded to encompass content-oriented mechanisms such as content delivery networks (CDNs), peer-to-peer (P2P) overlays, and HTTP proxies deployed over the existing infrastructure. Recently, novel networking principle was proposed for efficient content delivery. In this approach, the Internet needs a fundamental paradigm shift from a traditional host-to-host conversation model to a content-centric communication model. Using this model, it is no longer necessary to connect to a server to obtain content. Instead, a user can directly send a request for content to the network with the content name without considering the original content location.

The focus on content distribution has led to the concept of information-centric networking (ICN), allowing content to be addressed by name and not by location or end point addresses. This principle has motivated various projects' approaches, such as data-oriented network architecture (DONA) [12], network of information (NetInf) [13], content-centric networking (CCN), also known as named data networking (NDN) [14], and so on. Although all these approaches differ in many specific details, they share many common assumptions, objectives, and architectural properties. The general goal is to develop a network architecture that is better suited for efficiently accessing and distributing content, and that better copes with disconnections, disruptions, and flash crowd effects in the communication service.

By naming information at the network layer, ICN favors the deployment of in-network caching and multicast mechanisms, thus facilitating the efficient and timely content delivery to the users. However, there is more to ICN than information distribution, with related research initiatives employing information awareness as the means for addressing a series of additional limitations in the current Internet architecture, for example, mobility management and the security to fulfill the entire spectrum of FI requirements and objectives. In this context, ICN has emerged as a promising candidate for the FI architecture.

10.4.1 Concept and Principles of Information-Centric Networking

Distinct from IP networking, the ICN concept has three main characteristics [15]. At first, an ICN node performs routing by content names and not by host locators. In this case, identifying hosts is replaced by identifying contents and the location of a content file is independent of its name. Because IP address has both the identifier and locator roles, IP networking has problems like mobility. By splitting these roles, ICN has location independence in content naming and routing, while being free from mobility and multihoming problems. Second, in the publish–subscribe communication model, the content generation and consumption can be decoupled in time and space to distribute contents efficiently and more scalably. Third, the authenticity of contents can easily be verified by leveraging public key cryptography. As for content authentication solution in ICN, either a self-certifying content name [12] or a signature in a packet [14] can be used.

10.4.1.1 Information Naming

In ICN, instead of specifying a source–destination host pair for communication, a piece of information itself is named, addressed, and matched independently of its location; therefore, it may be located anywhere in the network. An indirect implication and benefit of moving from the host-naming model to the information-naming model is that information retrieval becomes receiver-driven. In contrast to the current Internet in which senders have absolute control over the data exchanged, in ICN, no data can be received unless it is explicitly requested by the receiver. In ICN, after a request is sent, the network is responsible for locating the best source that can provide the desired information. Generally, two naming schemes have largely been proposed: one with a hierarchical structure and one with a flat namespace.

Hierarchical structure [14] is introduced to name a content file. Even though it is not mandatory, a content file is often named by an identifier, which is compatible with the current URL-based applications/services. Its hierarchical nature can help mitigate the routing scalability issue because routing entries for contents might be aggregated. Components in a hierarchical identifier have semantics, which prohibit persistent naming. Persistence refers to a property that once a content name is given, users would like to access the content file with the name as long as possible. If the ownership of a content file is changed, its name becomes misleading with the above naming. In some cases, the names are human-readable, which makes it possible for users to manually assign names and, to some extent, assess the relation between a name and what the user wants.

As for flat naming [12], a content identifier is defined as a cryptographic hash of a public key. Due to its flatness (i.e., a name is a random-looking series of bits with no semantics), persistence and uniqueness are achieved. However, flat-naming aggravates the routing scalability problem due to the absence of the possibility of aggregation. Because flat names are not human-readable, an additional resolution between human-readable names and content names may be needed in the application layer.

10.4.1.2 Name-Based Routing and Name Resolution

Efficient information dissemination should make use of any available data source to reduce network load and latency and increase information availability. ICN architectures have to solve the problem of how to retrieve data based on a location-independent identifier. There are two major solutions to this problem: name-based routing and name resolution. Name-based routing involves constructing a path for transferring the information from that provider to the requesting host, whereas name resolution involves matching an information name to a provider or source that can supply that information. The key issue is whether these two functions are integrated (coupled) or are independent (decoupled). In the coupled approach, the information request is routed to an information provider, which subsequently sends the information to the requesting host by following the reverse path over which the request was forwarded. In the decoupled approach, the name resolution function does not determine or restrict the path that the data will use from the provider to the subscriber. DONA and NDN follow the name-based routing approach, whereas NetInf follows the name resolution approach.

Whether it is a name-based routing or a name resolution approach, a list of common desirable properties can be identified as [16]

- Any ICN routing mechanism should provide low-latency network level primitive operations for content (original, replica, or cached) registration, metadata update, and deletion. Currently, none of the presented research projects explicitly consider metadata update or content deletion.

- The routing mechanism should be able to route a content request to the closest (based on some network metric) copy. This feature ensures the reduction of interdomain traffic.
- Message propagation for name resolution and retrieval should not leave the network domain that contains both the source and the content.
- The routing mechanism should provide guarantees on discovery of any existing content, regardless of the content's popularity and replication level.
- As the number of contents for ICN is in the order of trillions, any routing/name resolution scheme needs to scale to at least this many contents and possibly beyond to accommodate future growth. The trade-off between routing stretch (ratio between routing path length and minimum length path) and routing table size needs to be analyzed, while keeping in mind the huge number of names and physical limitations imposed by memory technologies.
- Ideally, the content retrieval process should be a one-step process, either by combining name resolution and routing in a single step or by completely eliminating the name resolution part.

Considering name-based routing, there are unstructured and structured approaches [15]. First, one assumes no structure to maintain routing tables. Thus, the routing advertisement for contents is mainly performed based on flooding. NDN (CCN) suggests inheriting IP routing, and thus has IP compatibility to a certain degree. However, NDN, being an unstructured-flooding–based routing protocol, neither guarantees content discovery nor ensures scalable routing table size and manageable update message overhead.

In structured routing, two structures have been proposed: a tree and a distributed hash table. DONA is the most representative tree-based routing scheme. Routers in DONA form a hierarchical tree, and each router maintains the routing information of all the contents published in its descendant routers. Thus, whenever a content file is newly published, replicated, or removed, the announcement will be propagated up along the tree until it encounters a router with the corresponding routing entry. This approach imposes an increasing routing burden as the level of a router becomes higher. The root router should have the routing information of all the contents in the network. Because DONA employs nonaggregatable content names, this scalability problem is severe. The flatness of a distributed hash table imposes an equal and scalable routing burden among routers. If the number of contents is C, each router should have $\log_2(C)$ routing entries. However, the distributed hash table is constructed by random and uniform placement of routers, and thus typically exhibits a few times longer paths than a tree that can exploit the information of network topology. Also, the flatness of a distributed hash table often requires forwarding traffic in a direction that violates the provider–customer relation among ISPs. An effective routing mechanism for ICN may require combining the advantages of structured and unstructured routing mechanisms, while still operating at the network layer without requiring any overlays [16].

10.4.1.3   Multisource Dissemination

New Internet services require one-to-many (1:N) and many-to-many (M:N) connectivities. As for 1:N connectivity, it represents content dissemination from a single source to multiple recipients (e.g., online streaming and IPTV). Compared with IP multicasting, ICN accommodates 1:N connectivity naturally by the publish–subscribe paradigm in terms of content naming and group management. However, its link efficiency is not different from IP multicasting. On the other hand, M:N connectivity takes place among multiple sources and multiple recipients. There are two kinds of M:N connectivity applications: M instances of 1:N connectivity (e.g., video-conference), and M sources disseminate different parts of a content file to N recipients. Substantiating M:N connectivity requires application-specific/service-specific overlays or relay mechanisms in the current Internet. However, ICN can disseminate contents more efficiently at the network level by spatial decoupling of the publish–subscribe paradigm and content awareness at network nodes.

Disseminating a content file from multiple sources is tightly coupled with name-based routing [15]. To exploit multiple sources in disseminating the same content, each ICN node may have to keep track of individual sources of the same content (e.g., DONA and NDN). In this case, an ICN node can seek to retrieve different parts of the requested content in parallel from multiple sources to expedite dissemination. Depending on RTTs and traffic dynamics of the path to each source, the ICN node should dynamically decide/adjust which part of the content file is to be received from each source. Another relevant issue is what routing information should be stored and advertised by each ICN node for multiple sources of the same content.

10.4.1.4   In-Network Caching

To alleviate the pressure that rapid traffic growth imposes on network bandwidth, a common approach of ICN is to provide transparent, ubiquitous in-network caching to speed up content distribution and improve network resource utilization [17]. Although caching is not a wholly new technique, the lack of a unique identification of identical objects makes it incompletely utilized in the current Internet architecture. The problem of protocol closeness and naming inconsistency can be gracefully addressed within the ICN infrastructure. As mentioned before, ICN names content in a unified, consistent, and network-aware way. This feature makes caching a general, open, and transparent service, independent of applications.

In the context of in-network caching, the major problems to address are related to the selection of the storage locations and of the specific content items to store in the different caches. In general, two opposite approaches can be distinguished to address such problems: coordinated and uncoordinated. In the coordinated case, routers exchange information to achieve a better estimation of the popularity of contents and to avoid the storage of too many content copies. In the uncoordinated case, each cache operates autonomously using transparent en route caching policy. Here, each router in the end-to-end (E2E) path between the interested user and the content source decides whether to cache transiting pieces of content and which cache replacing policy without interacting with other routers to perform. Although in most scenarios, in which the popularity of objects does not change frequently over a specific period, the least frequently used approach would achieve the highest performance, whereas most proposed solutions adopt the least recently used replacing policy [18].

Obviously, the effectiveness of in-network caching is higher when coordinated solutions are utilized. However, the overhead required to manage the coordination between caches may become extremely high. As a result, time uncoordinated solutions are now adopted more frequently. As a perspective solution, some trade-offs between the coordinated and uncoordinated approaches are proposed [19,20].

Example 10.1

As an illustrative example of an in-network caching concept, consider the network shown in Figure 10.3. Node A sends an interest

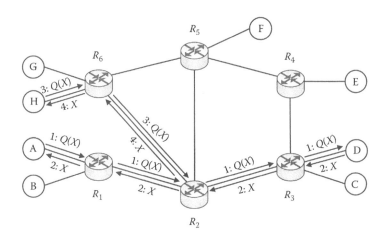

FIGURE 10.3   An example of in-network caching.

message requesting a piece of content X. Such content has been published by node D and therefore, the request message $Q(X)$ traverses routers R_1, R_2, and R_3. Node D receives the interest message issued by node A and sends it the desired piece of content X. Content data X will traverse nodes R_3, R_2 and R_1 back, which will store local copies of that piece of content in their caches.

Suppose that node H generates an interest message for the same piece of content X. The corresponding request message $Q(X)$ will traverse routers R_6 and R_2. The router R_2 can realize that there is a copy of X in its cache and therefore, it does not forward the interest message to the next hop toward D. Instead R_2 sends the content X to node H only via router R_6.

10.4.2 DATA-ORIENTED NETWORK ARCHITECTURE

In DONA, named data objects (NDOs), as the main abstraction of ICN (e.g., web pages, videos, and documents), are published into the network by the sources. Nodes that are authorized to serve data are registered to the hierarchical infrastructure consisting of resolution handlers (RHs). Requests (FIND packets) are routed by name toward the appropriate RH, as illustrated in Figure 10.4 (steps 1–4). Data is sent back in response, either through the reverse RH path (steps 5–8), enabling caching, or over the direct route (step 9).

Content providers can perform a wildcard registration of their principal in the RH, so that queries can be directed to them without needing to

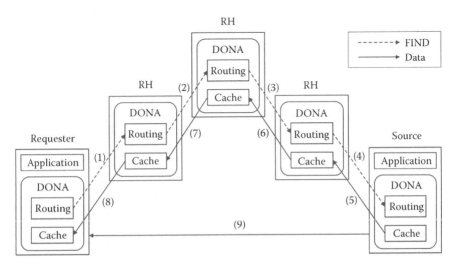

FIGURE 10.4   DONA approach.

register specific objects. It is also possible to register NDO names before the corresponding content is created and made available. Register commands have expiry times. When the expiry time is reached, the registration needs to be renewed. The RH resolution infrastructure routes request by name in a hierarchical fashion and tries to find a copy of the content closest to the client. DONA's anycast name resolution process allows clean support for network-imposed middleboxes (e.g., firewalls and proxies).

DONA names NDOs with a flat namespace in the form $P:L$, where P is the globally unique principal field, which contains the cryptographic hash of the publisher's public key, and L is the unique object label. Because P identifies the publisher (and not the owner), republishing the same content by a different publisher (e.g., by an in-network cache) generally results in a different name for the same content. Although this can be circumvented through specific means in DONA (e.g., via wildcard queries or principal delegation), it might complicate benefiting from all available content copies [5].

10.4.3 NETWORK OF INFORMATION

The NetInf approach [21] targets global-scale communication and supports many different types of networks and deployments, including traditional access and core network configurations, data centers, as well as challenged and infrastructure-less networks. NetInf's approach to connecting different technology and administrative domains into a single information-centric network is based on a hybrid name-based routing and name resolution scheme.

NetInf offers two models for retrieving NDOs, via name resolution and via name-based routing, thereby allowing adaptation to different network environments. Depending on the model used in the local network, sources publish NDOs by registering a name/locator binding with a name resolution service (NRS), or announcing routing information in a routing protocol. A NetInf node, holding a copy of an NDO, can optionally register its copy with an NRS, thereby adding a new name/locator binding. If an NRS is available, a receiver can first resolve an NDO's name into a set of available locators and can subsequently retrieve a copy of the data from the "best" available source(s), as illustrated in Figure 10.5 (steps 1–4). Alternatively, the receiver can directly send out a GET request with the NDO's name, which will be forwarded toward an available NDO's copy using name-based routing (steps 5–8). As soon as a copy is reached, the data will be returned to the receiver. The two models are merged in a hybrid resolution/routing approach in which a global resolution system provides mappings in the form of routing hints that enable aggregation of routing information.

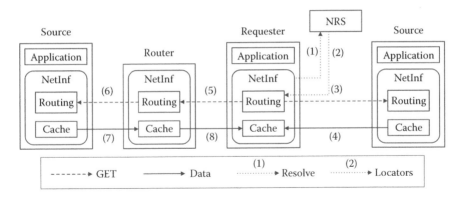

FIGURE 10.5   NetInf approach.

Generally, NetInf employs a flat namespace with some structures similar to the DONA namespace. To accommodate different ICN deployment requirements, NetInf distinguishes between a common naming format (understandable to all the nodes) and name semantics and name-object–binding validation mechanisms. The common NetInf naming format is based on containing hash digests in the name [22], and different hashing schemes are supported. The hash digest of the owner's public key can also be contained in the name to support dynamic data. NetInf names can be transformed to different representations, including a URI representation and a binary representation.

10.4.4   NAMED DATA NETWORKING

NDN (advanced CCN) has received wide attention due to its simple and efficient content delivery mechanism. Here, NDOs are published at nodes, and routing protocols are employed to distribute information about content location. Routing in NDN can leverage aggregation through a hierarchical naming scheme. Because of hierarchical naming, name aggregation and longest prefix matching are available in routing. Requests (interest packets) for a NDO are forwarded to a publisher location, as illustrated in Figure 10.6 (steps 1–3). Interest packets are routed toward the publisher of the name prefix using longest-prefix matching in the forwarding information base (FIB) of each node. The FIB can be built using routing protocols similar to those used in today's Internet. An NDN router maintains a pending interest table (PIT) for outstanding forwarded requests, which enables request aggregation. The PIT maintains the state for all interests and maps them to a network interface from which corresponding requests were received. Data is then routed back on the reverse request path using this state (steps 4–6). As NDN natively supports on-path caching, NDOs

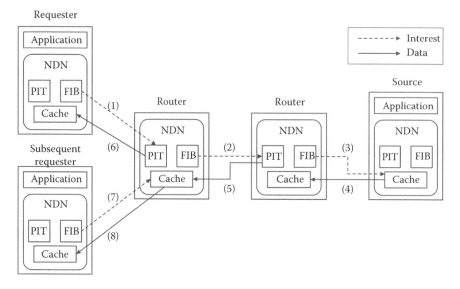

FIGURE 10.6 NDN approach.

that a NDN router receives (in responses to requests) can be stored so that subsequent received requests for the same object can be answered from that cache (steps 7–8). From a NDN node's perspective, there is balance of requests and responses, that is, every single sent request is answered by one response (or no response).

The NDN namespace is hierarchical to achieve better routing scalability through name–prefix aggregation. The names are rooted in a prefix unique to each publisher. The publisher prefix makes it possible for clients to construct valid names for data that does not yet exist, and publishers can respond with dynamically generated data. NDN names are used for both naming information and routing purposes.

10.5 STREAMING OF SCALABLE VIDEO FOR FI

Scalable video delivery over P2P networks seems to be key for efficient streaming in emerging and FI applications [23]. To cope with varying bandwidth capacities inherent to P2P systems, the underlying video coding/ transmission technology needs to support bit rate adaptation according to available bandwidth. Moreover, displaying devices at the user side may range from small handsets to large high-definition displays. Therefore, video streams need to be transmitted at a suitable spatiotemporal resolution supported by the user's display device. The above-mentioned issues cannot be solved efficiently using conventional video coding technologies.

Scalable video coding (SVC) techniques [24] address these problems because they allow encoding a sequence once and decoding it in many different versions. Scalable coded bitstreams can efficiently adapt to the application requirements. The adaptation is performed in the compressed domain by directly removing parts of the bitstream. The SVC encoded bitstream can be truncated to lower resolution, frame rate, or quality. In P2P environments, such real-time low-complexity adaptation results in a graceful degradation of perceived quality, avoiding the interruption of the streaming service in case of congestion or bandwidth narrowing.

Spatial, temporal, and quality modes are the most common modes of scalability. In the spatial mode, the difference between the base layer and enhancement layers is in the decoded frame size. Different frame rates (temporal resolution) appear in the temporal scalability mode. For example, the frame rate is 15 and 30 fps in the base layer and the enhancement layer, respectively. In the quality mode, the video layers have the same spatiotemporal resolution, but each enhancement quality layer appears with higher fidelity. In this way, the complexity of adaptation is very low, in contrast with the adaptation complexity of nonscalable bitstreams. The SVC scheme gives flexibility and adaptability to video transmission over resource-constrained networks. By adjusting one or more of the scalability parameters, it selects a layer containing an appropriate spatiotemporal resolution and quality according to current network conditions. An example of video distribution through links supporting different transmission speeds and display devices is shown in Figure 10.7. At each point where

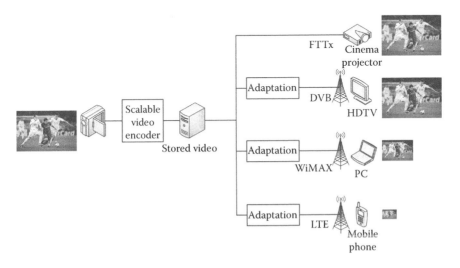

FIGURE 10.7   Scalable video streaming through links with different transmission speeds.

video quality/resolution needs to be adjusted, an adaptation is invoked. Because the adaptation complexity is very low, the video can be efficiently streamed in such an environment. SVC also provides a natural solution with a truncatable bitstream for error-prone transmissions inherent to FI. In addition, channel coding methods can be adaptively used to attach different degrees of protection to different bit-layers according to their relevance in terms of decoded video quality [25].

Although some of the earlier video standards (e.g., H.262/MPEG-2 and MPEG-4 Part 2) included limited support for scalability, the use of scalability in these solutions came at a significant increase in terms of the decoder complexity and coding efficiency. The video coding standard, H.264/MPEG-4 AVC [26], provides a fully scalable extension, SVC, which achieves significant compression gain and complexity reduction when scalability is required compared with the previous video coding standards. The scalable bitstream is organized into a base layer and one or several enhancement layers. SVC provides temporal, spatial, and quality scalabilities with a small increase of bit rate relative to the single-layer H.264/MPEG-4 AVC. It should be noted that high-efficiency video coding (HEVC), the latest standard that is also exploring SVC as an extension, will be finalized by Sullivan et al. 2014 [27].

10.5.1 Wavelet-Based SVC

Although a hybrid-based technology was chosen for standardization within MPEG, a great amount of research also continued on wavelet-based SVC (W-SVC) [28]. W-SVC systems have shown very good performance in different types of application scenarios, especially when fine-grained scalability is required.

A typical structure of wavelet-based scalable video encoder is shown in Figure 10.8. First, the input video is subjected to a spatiotemporal

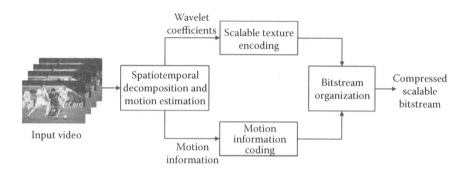

FIGURE 10.8   Typical structure of wavelet-based scalable video encoder.

decomposition, which is based on wavelet transform. The purpose of the decomposition is to decorrelate the input video content and provide the basis for spatial and temporal scalabilities. This results in two distinctive types of data: wavelet coefficients representing the texture information remaining after the wavelet transform, and motion information (obtained from motion estimation), which describes spatial displacements between blocks in neighboring frames. Generally, the wavelet transform performs very well in the task of video content decorrelation, but some amount of redundancies still remains between the wavelet coefficients after the decomposition. Moreover, a strong correlation also exists between motion vectors. For these reasons, further compression of the texture and motion vectors is performed.

Texture coding is performed in conjunction with so-called embedded quantization (bitplane coding) to provide the basis for quality scalability. Finally, the resulting data are mapped into the scalable stream in bitstream organization module, which creates a layered representation of the compressed data. This layered representation provides the basis for low-complexity adaptation of the compressed bitstream.

To achieve efficient layered extraction, scalable video bitstream consists of packets of data called atoms [25]. An atom represents the smallest entity that can be added or removed from the bitstream. After such an organization, the extractor simply discards atoms from the bitstream that are not required to obtain the video of the desired spatial resolution, temporal resolution, or quality. Each atom can be represented by its coordinates in three-dimensional temporal–spatial–quality space, denoted as (T, S, Q). The maximum coordinates are denoted as T_M, S_M, Q_M, where these values represent the number of refinement layers in temporal, spatial and quality directions, respectively. Except for refinement layers, a basic layer exists in each direction, which is denoted as zeroth layer and cannot be removed from the bitstream. If desired temporal resolution, spatial resolution, and quality are denoted as $m \in \{0, 1, \ldots, T_M\}$, $n \in \{0, 1, \ldots, S_M\}$, and $i \in \{0, 1, \ldots, Q_M\}$, respectively, then the atoms with the coordinates $T > m$, $S > n$, and $Q > i$ are discarded from the bitstream during the extraction process. For simplicity, an equal number of atoms is assumed, corresponding to each spatiotemporal resolution. Generally, this is not the case as the number of atoms corresponding to different spatiotemporal resolutions may be different. Moreover, the atoms corresponding to different qualities can be truncated at any byte to achieve fine granular scalability. However, the principle of extraction remains the same.

10.5.2 Event-Based Scalable Coding

Event-based scalable encoding can be used in FI to perform rate optimization for transmission and storage according to the event significance in a video sequence [25]. For this purpose, the temporal segments of the video sequence are classified into two types:

- Temporal segments representing an essentially static scene (in which only random environmental motion is present, e.g., swaying trees or flags moving in the wind)
- Temporal segments containing nonrandomized motion activity (e.g., moving of objects in a forbidden area)

To enable the above classification, a background subtraction and tracking module [29] is used for video content analysis (VCA). It uses a mixture of Gaussians to separate the foreground from the background. Each pixel of a sequence is matched with each weighted Gaussian of the mixture. If the pixel value is not within 2.5 standard deviations of any Gaussians representing the background, then the pixel is declared as the foreground. Because the mixture of Gaussians is adaptive and more than one Gaussian is allowed to represent the background, this module is able to deal robustly with light changes, bimodal background such as swaying trees, and the introduction or removal of objects from a scene. The output of the module defines the parameters of compressed video, which is encoded with the W-SVC framework.

At each time instance, the W-SVC encoder communicates with the VCA module, as shown in Figure 10.9. When the input video is essentially static, the output of the background subtraction does not contain any foreground pixels. This is a signal to the W-SVC encoder to adapt the captured video at low spatiotemporal resolution and quality. This allows, for instance, storing or transmitting (or both) of portions of the video containing long, static scenes using low-quality frame rate and spatial resolution. On the other hand, when some activity in the captured video is detected, the VCA module notifies the W-SVC encoder to automatically switch its

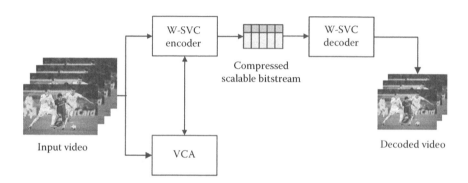

FIGURE 10.9   Event-based SVC framework.

output to a desired much higher spatiotemporal resolution and quality video. Therefore, decoding and use of the video at different spatiotemporal resolutions and qualities corresponding to different events is achieved from a single bitstream, without multicasting or complex transcoding.

Moreover, additional optional adaptation to lower bit rate is also possible without re-encoding the video. This is very useful in cases in which video has to be delivered to a device with a low display capability. Using this approach, the bit rate of video portions that are of low interest is kept low whereas the bit rate of important parts is kept high. Because in many realistic applications, it can be expected that large portions of the captured video have no events of interest, the proposed model leads to significant reduction of resources without jeopardizing the quality of any off-line event detection module that may be present at the decoder.

10.6   MEDIA SEARCH AND RETRIEVAL IN FI

The system concerning personalized media search and retrieval in the FI performs the operations cooperating with one another, as well as with external systems and user's inputs [29]. This system offers several advantages as it is able to enrich the content and metadata in every process by improving the search, retrieval, and content distribution operations simultaneously; improve the personalization of the results to the users' profiles; and take advantage of ICN to work with content as a native information piece. It aims to improve the logical process dealing with semantic audiovisual content search, the development of new FI technologies, and the integration of a broad spectrum of high-quality services for innovative search and retrieval applications.

The major novelty of the system lies in the creation of an architecture consisting of different modules around a media component that joins together the information needed to perform the proposed operations (e.g., semantic search, automatic selection, and composition of media), dealing with both the user context and preferences or smart content adaptation for existing network architectures. Within the scope of this media component is the data storage from all the processes involved in the semantic search (as low-level and high-level descriptors from media inputs), a vector repository from user queries, and a variety of user-related data as well.

As an example of system functionality, an algorithm of personalized search and retrieval is shown in Figure 10.10. First, the user sends a query to the system by inputting multimedia objects (e.g., text, image, video, or audio). These input objects are semantically enriched by the ontology, which takes into account the three dimensions: multilanguage, multidomain, and multimedia. Depending on the nature of the user's query, the annotation module applies the most suitable metadata extraction procedure. Then, the

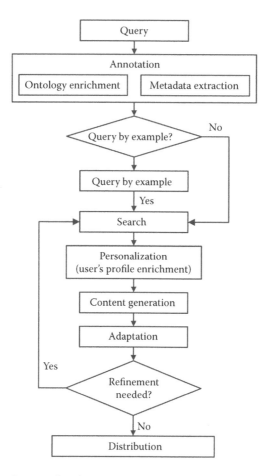

FIGURE 10.10 Personalized search and retrieval algorithm.

search module uses the generated metadata and, according to their nature, starts a specific search process over the multidimensional index stored in the multimedia component. Depending on the user's search target, the process looks in the suitable multimedia component layer to retrieve the required multimedia elements. The results of the process are the objects' identifiers found by the search methods, and a weight factor that indicates their potential importance. The results of the search module are sent to the personalization module, which combines the information of multimedia objects with the user information. The recommendation engines adapt the weight accuracy index provided by the search module to the user preferences. These engines sort the results depending on the user preferences and search context (time, location, etc.). Moreover, the personalization module enriches and stores the metadata in the multimedia component in a structured way.

In the next step, the content generation module enriches both the results and the multimedia objects. This module adds a new piece of content to the previous results, which is automatically generated using the users' preferences and considering the desired characteristics. This process is performed to coherently generate pieces of audiovisual content. This newly generated resource represents a multimedia summary of the obtained results that emphasize the users' interests. The adaptation module distributes the results to the user, allowing the consumption of the audiovisual content regardless of the current location, network, or device. The overall process ends with the distribution stage. It is possible to perform iterative queries to refine the results or to help the user extend the search.

Providing media content with searchable and accessible (metadata generation and structuring) capabilities is one of the major challenges of the FI. The presented system is intended to work over an ICN environment in a possible design of the FI. It becomes a powerful solution due to the architecture and functionalities of the described modules, which provide improved functionalities for not only the information-centric but also the user-centric approach followed in the personalization and content generation modules [29]. In addition, the multimedia component of the system provides an enriched description of objects thanks to the layered metadata structure, which is continuously enriched in every process. This layered metadata structure adds the necessary interaction and is perfectly adaptable to the ICN concept.

Example 10.2

To implement personalized search and retrieval systems in ICN environment, modules can be allocated in the network cloud. The multimedia component is linked to specific nodes called content nodes, which allow the network to access and route both media essence and metadata information. An example of the system configuration over ICN architecture is presented in Figure 10.11.

The system is protocol agnostic (messages, naming, etc.), and adaptable to some of the ICN architectures already presented [14]. A user starts a search query composed of one or more multimedia objects (step 1). These multimedia objects are annotated by the annotation module service (step 2), which gives a metadata vector with the features of a multimedia object. This vector forms a new search query, which is flooded to the ICN router and will reach every content node (step 3). When a content node receives a metadata search query, it has to perform search algorithms over the multimedia component to find the objects that

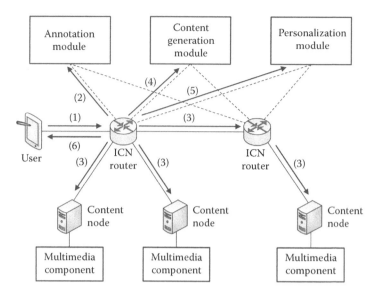

FIGURE 10.11 Configuration of search and retrieval system in information-centric architecture.

satisfy the user query. As a reply, the content node returns the metadata of the list of multimedia objects. The ICN router sends this reply to the content generation module service (step 4). First, the content generation module creates a discursive or narrative structure (depending on the communicative purpose) using only the descriptive and functional metadata of the objects. Second, once the narrative structure has been created and the necessary multimedia objects to create it have been decided upon, the generation module sends a query to get only the necessary objects for the final edition. In parallel to this step, the ICN router generates a request to the personalization module to obtain a list of recommendations over the search results (step 5). This module does not need to use the original multimedia objects. Moreover, it is enough to take advantage of the descriptive and functional metadata stored in the multimedia component and the user information stored in the user profiles. The personalization module implements a hybrid recommender (content-based and social), which is possible because of the presence of the multimedia component and the ICN architecture. The ICN contributes knowledge of the multimedia objects users' consumption because these objects can be unambiguously identified. This knowledge allows the development of automatic and transparent recommendation algorithms based on social techniques such as collaborative filtering. The final recommended results and the automatic summary are then sent back to the user (step 6).

10.7 FI SELF-MANAGEMENT SCENARIOS

One of the key research challenges in FI design is the embedding of management intelligence into the network. This new discipline is also known as *autonomic networking* [30]. The objective of autonomic networking is to enable the autonomous formation and parameterization of nodes and networks by letting protocols sense and adapt to the networking environment at run time. The intelligent substrate concept was proposed by Charalambides et al. [31], with the aim of forming the natural self-managed network environment through parallel and continuous resource management functions, with each substrate supporting a specific network management task by optimizing a specific resource. The term *in-network substrate* emphasizes the fact that whereas substrates are essential for optimized network operations, they are hidden within the network and, as such, invisible to network users and applications. In the FI self-management concept, there are some typical scenarios, such as adaptive resource management, energy-aware network management, and cache management [31].

10.7.1 ADAPTIVE RESOURCE MANAGEMENT

To cope with unexpected traffic variations and highly changeable network conditions, approaches that can dynamically adapt routing configuration and traffic distribution are required. Existing on-line approaches have mainly focused on solutions by which deciding entities act independently from one another.

The deployment of this substrate aims at achieving optimum network performance in terms of resource utilization by dynamically adapting the traffic distribution according to real-time network conditions. Reconfigurations occur at network source (ingress) nodes, which change the splitting ratios of traffic flows across multiple paths between source–destination (S–D) pairs. This functionality is provided by the resource management substrate, embedded in ingress nodes, which execute a reconfiguration algorithm with the objective of transferring traffic from the most utilized links toward less loaded parts of the network. Performing a reconfiguration involves adjusting the traffic splitting ratios of some flows for which traffic is routed across the link with the maximum utilization in the network. As a result, more traffic is being assigned to alternative, less loaded, paths for a S–D pair.

The formation of the resource management substrate is based on the identification of ingress nodes in the physical network. Each node of the substrate is associated with a set of neighbors' nodes that are connected with direct communication only possible between neighboring nodes. Different models can be used for the organization of the substrate, the choice of which can be driven by parameters related to the physical network, such as its topology and the number of source nodes, but also by the constraints of the coordination mechanism and the associated communication protocol. Factors that influence the choice of model are, for example, the number and frequency of exchanged messages.

Example 10.3

An example of a network and its associated full-mesh in-network substrate is given in Figure 10.12.

A direct logical link (dashed lines) exists between the four ingress nodes $(I_1–I_4)$ of the physical network implementing the substrate. This model offers flexibility in the choice of neighbors with which to communicate because all source nodes belong to the set of neighbors.

10.7.2 ENERGY-AWARE NETWORK MANAGEMENT

An important goal toward the design of FI is to achieve the best ratio of performance to energy consumption and at the same time assure manageability. Recently, various proposals have been made toward the realization of energy-aware network infrastructures. For example, network devices

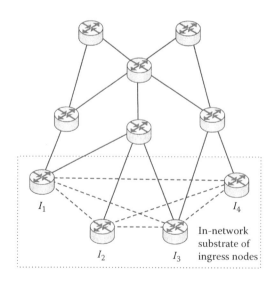

FIGURE 10.12 Network and its associated full-mesh in-network substrate of ingress nodes.

such as routers or L3 switches can adaptively reduce their transmission rates, or even enter sleep mode on low traffic load conditions to conserve energy during idle periods [32,33]. Some constraints need to be taken into account by self-managed network elements that are able to make "green" decisions for minimizing energy consumption during operation. First, the operating network topology should remain connected after some devices go to sleep mode. Second, the reduced network capability should not incur deteriorating service and network performances (e.g., traffic congestion).

Example 10.4

Consider the router level coordination shown in Figure 10.13. Both core routers C and D have detected very low incoming traffic intensity from their own upstream routers A and B, respectively. In the case when both routers take the opportunity to enter sleep mode without any knowledge of each other's decisions, the topology will become disconnected and user traffic from ingress nodes (l_1–l_3) will not be able to reach egress router E. To avoid such a situation, the two routers need to coordinate with each other for conflict-free decision making. For example, to allow only one of them to go to sleep mode, or both routers to simultaneously reduce their transmission rates while still remaining awake.

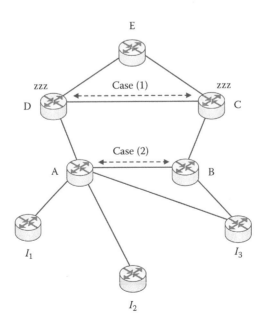

FIGURE 10.13   Router level coordination in energy-aware management scenario.

In the second case, assume that routers C and D are currently in sleep mode. Their upstream neighbors A and B have both detected traffic bursts from ingress routers and hence both may decide to trigger their own downstream neighbors to wake up. If the traffic upsurge is not sufficiently high, it might be the case that the wakeup of only one of the two downstream sleeping nodes will be able to accommodate the burst, while leaving the other in sleep mode. Routers A and B may also need to coordinate with each other to make optimal decisions for maximizing energy savings.

10.7.3 Cache Management

As previously stated, in the emerging ICN proposals, content is replicated almost ubiquitously throughout the network with subsequent optimal content delivery to the requesting users. Efficient placement and replication of content to caches installed in network nodes is the key to delivering on this promise. Management of such environments entails managing the placement and assignment of content in caches available in the network with objectives such as minimizing the content access latency, and maximizing the traffic intensity served by caches, thus minimizing bandwidth cost and network congestion [31]. Approaches applied to current ICN architectures follow static off-line approaches with algorithms that decide the optimal location of caches and the assignment of content objects and their replicas to those caches based on predictions of content requests by users. On the other hand, the deployment of intelligent substrate architecture will enable the assignment of content objects to caches to take place in real-time based on changing user demand patterns. Distributed managers will decide the objects every cache stores by forming a substrate that can be organized either in a hierarchical or in a P2P organizational structure. Communication of information related to request rates, popularity/locality of content objects, and current cache configurations will take place between the distributed cache managers through the intelligent substrate functionality.

As presented in Figure 10.14, every cache manager should decide in a coordinated manner with other managers whether to store an object that will probably lead to replacing another item already stored, depending on the cache size. The decision of this swapping of stored items can be based on maximizing an overall network-wide utility function, such as the gain in network traffic. This means that every node should calculate the gain the replacement of an object will achieve. Here, it is assumed that every cache manager has a holistic network-wide view of all the cache configurations and relevant request patterns. When a manager changes the configuration

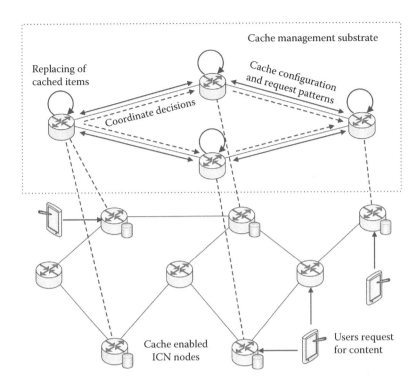

Cache management substrate

Replacing of cached items

Cache configuration and request patterns

Coordinate decisions

Cache enabled ICN nodes

Users request for content

FIGURE 10.14 Cache management substrate in information-centric networks.

of its cache, this information should be exchanged periodically or in an event-based manner.

It should be noted that uncoordinated decisions can lead to suboptimal and inconsistent configurations. On the other hand, coordinated decision making of a distributed cache management solution can be achieved through the substrate mechanisms by ensuring that managers change the configuration of their cache in an iterative manner until convergence to an equilibrium state is achieved.

10.8 CONCLUDING REMARKS

Nowadays, the Internet is primarily used for transporting content/media, where a high volume of both user-generated and professional digital content is delivered to users who are usually only interested in the content itself, rather than the location of the content sources. In fact, the Internet is rapidly becoming a superhighway for massive content dissemination. In the FI, broadband wireless networks will have a key role. The number of users accessing the Internet through mobile devices is continuously growing at a fast rate. It is easy to estimate that mobile users will be highly predominant in the FI.

Limitations and shortcomings of current Internet and network architecture have driven research trends at a global scale to properly define novel flexible Internet architectures. Concepts such as physical layer awareness, virtualization, information-centric networking, search and retrieval of content, as well as self-management scenarios are topics of current research.

Today's Internet architecture is not optimal for multimedia services, and current solutions for content delivery over IP are not particularly efficient. Therefore, future architectures such as ICN will provide a better environment for multimedia delivery, but will also require a transport solution that is more aware of a delivery networks' requirements. By expending the ICN to support multimedia similarity content search, the ability to retrieve multimedia content already available in the network without knowing where this content is stored or the name of which content in question, is provided.

ICN considers pieces of information as main entities of a networking architecture, rather than only indirectly identifying and manipulating them via a node hosting that information. In this way, information becomes independent from the devices they are stored in, enabling efficient and application-independent information caching in the network. Major challenges include global scalability, cache management, congestion control, and deployment issues. By providing access to the NDOs as a first-class networking service, ICN is beneficial to all applications and network interactions that can be modeled after this paradigm, including content distribution as well as M2M applications with comparatively small objects. ICN puts accessing NDOs, name-based routing and name resolution, in-network storage, and data object security at the center of networking, while removing the need for application-specific overlays.

In P2P networks, video is streamed to the user in a fully distributed fashion. Network resources are distributed among users instead of being handled by a single entity. However, due to the diversity of users' displaying devices and the available bandwidth levels in the Internet, the underlying coding and transmission technology needs to be highly flexible. Such flexibility can easily be achieved by SVC, in which bitstreams can be adapted in the compressed domain according to available bandwidth or users' preferences. Considering the flexibility given by scalable bitstreams, it is obvious that P2P streaming systems supporting SVC technology will play an important role in the FI.

The personalized media search and retrieval system provides a useful solution for the FI, applicable to both evolutionary and purely content-oriented networks, able to handle, process, deliver, personalize, find, and retrieve digital content. The advantages of this system go further with the provision of a complete framework to deal with media objects in a smarter way by proposing a multimedia component-centered modular architecture that is able to enrich the content dynamically. Another meaningful feature relevant to the FI is the user-centric approach. This system offers a comprehensive solution that involves the user from the beginning and is able to personalize media assets according to their needs, tastes, and preferences.

Current practices for the configuration of networks rely mainly on off-line predictive approaches, with management systems being external to the network. These are incapable of maintaining optimal configurations in the face of changing or unforeseen traffic demands and network conditions, and, due to their rigidity, they cannot easily support the requirements of emerging applications and future network operations. Self-management has been proposed as a potential solution to these challenges bringing intelligence into the network, and thus enabling customized management tasks in a flexible and scalable manner.

References

CHAPTER 1 NEXT GENERATION WIRELESS TECHNOLOGIES

1. S. Y. Hui and K. H. Yeung. Challenges in the Migration to 4G Mobile Systems. *IEEE Communications Magazine* 41, no. 12 (December 2003): 54–9.
2. E. Gustafsson and A. Jonsson. Always Best Connected. *IEEE Wireless Communications* 10, no. 1 (February 2003): 49–55.
3. I. F. Akyildiz, S. Mohanty and J. Xie. A Ubiquitous Mobile Communication Architecture for Next-Generation Heterogeneous Wireless Systems. *IEEE Communications Magazine* 43, no. 6 (June 2005): S29–36.
4. Z. Bojkovic, Z. Milicevic and D. Milovanovic. Next Generation Cellular Networks. *Proc. 15th WSEAS ICC 2011*. Corfu Island, Greece (July 2011): 233–9.
5. K. R. Rao, Z. Bojkovic and D. Milovanovic. *Wireless Multimedia Communications: Convergence, DSP, QoS and Security*. Boca Raton, FL: CRC Press, Taylor & Francis Group (2009).
6. K. R. Rao, Z. Bojkovic and D. Milovanovic. Service Platform Technology for Next Generation Wireless Multimedia: An Overview. *Proc. TELSIKS 2011*. Nis, Serbia (October 2011): 71–5.
7. Y. M. Fang. Looking into the Future of the Wireless World... [Message from the editor-in-chief]. *IEEE Wireless Communications* 19, no. 1 (February 2012): 2–4.
8. G. Tsoulos, ed. *MIMO System Technology for Wireless Communications*. Electrical Engineering & Applied Signal Processing Series. Boca Raton, FL: CRC Press, Taylor & Francis Group (2006).
9. W. Lee et al. Multi-BS MIMO Cooperation: Challenges and Practical Solutions in 4G Systems. *IEEE Wireless Communications* 19, no. 1 (February 2012): 89–96.
10. A. E. Khandani et al. Cooperative Routing in Static Wireless Networks. *IEEE Transactions on Communications* 55, no. 11 (November 2007): 2185–92.
11. A. Nosratinia, T. E. Hunter and A. Hedayat. Cooperative Communication in Wireless Networks. *IEEE Communications Magazine* 42, no. 10 (October 2004): 74–80.
12. Y. Tian, K. Xu and N. Ansari. TCP in Wireless Environments: Problems and Solutions. *IEEE Communications Magazine* 43, no. 3 (March 2005): S27–32.
13. H. Nishiyama, N. Ansari and N. Kato. Wireless Loss-Tolerant Congestion Control Protocol Based on Dynamic AIMD Theory. *IEEE Wireless Communications* 17, no. 2 (April 2010): 7–14.
14. D. Datla et al. Wireless Distributed Computing: A Survey of Research Challenges. *IEEE Communications Magazine* 50, no. 1 (January 2012): 144–52.

15. P. Cheolhee and T. S. Rappaport. Short-Range Wireless Communications for Next-Generation Networks: UWB, 60 GHz Millimeter-Wave WPAN, and ZigBee. *IEEE Wireless Communications* 14, no. 4 (August 2007): 70–8.

16. J. Karaoguz. High-Rate Wireless Personal Area Networks. *IEEE Communications Magazine* 39, no. 12 (December 2001): 96–102.

17. T. M. Siep et al. Paving the Way for Personal Area Network Standards: An Overview of the IEEE P802.15 Working Group for Wireless Personal Area Networks. *IEEE Personal Communications* 7, no. 1 (February 2000): 37–43.

18. IEEE Std. 802.15.1-2005. Wireless Medium Access Control (MAC) and Physical Layer (PHY) Specifications for Wireless Personal Area Networks (WPANs) (June 2005). Revision of IEEE 802.15.1-2002.

19. IEEE Std. 802.15.3-2003. Wireless Medium Access Control (MAC) and Physical Layer (PHY) Specifications for High-Rate Wireless Personal Area Networks (WPANs) (September 2003).

20. IEEE Std. 802.15.4-2006. Wireless Medium Access Control (MAC) and Physical Layer (PHY) Specifications for Low-Rate Wireless Personal Area Networks (WPANs) (September 2006). Revision of IEEE Std. 802.15.4-2003.

21. K. A. Ali and H. T. Mouftah. Wireless Personal Area Networks Architecture and Protocols for Multimedia Applications. *Ad Hoc Networks* 9, no. 4 (June 2011): 675–86.

22. L. Yang and G. B. Giannakis. Ultra-Wideband Communications: An Idea Whose Time Has Come. *IEEE Signal Processing Magazine* 21, no. 6 (November 2004): 26–54.

23. FCC. First Report and Order: Revision of Part 15 of the Commission's Rules Regarding Ultra-Wideband Transmission Systems. ET Docket 98–153 (April 2002).

24. H. Nikookar and R. Prasad. *Introduction to Ultra Wideband for Wireless Communications*. Series: Signals and Communication Technology. Dordrecht: Springer Science+Business Media B.V. (2009).

25. P. Smulders. 60 GHz Radio: Prospects and Future Directions. *Proc. 10th IEEE SCVT 2003*. Eindhoven, Netherlands (November 2003): 1–8.

26. IEEE Std. 802.15.3c™-2009. Wireless Medium Access Control (MAC) and Physical Layer (PHY) Specifications for High Rate Wireless Personal Area Networks (WPANs) Amendment 2: Millimeter- Wave-Based Alternative Physical Layer Extension (October 2009).

27. A. Sadri. Summary of the Usage Models for 802.15.3c. IEEE 802.15-06-0369-09-003c (November 2006).

28. T. Baykas et al. IEEE 802.15.3c: The First IEEE Wireless Standard for Data Rates over 1 Gb/s. *IEEE Communications Magazine* 49, no. 7 (July 2011): 114–21.

29. ZigBee Alliance. ZigBee Specification. Doc. 053474r06, v. 1.0 (December 2004).

30. P. Baronti et al. Wireless Sensor Networks: A Survey on the State of the Art and the 802.15.4 and ZigBee Standards. *Computer Communications* 30, no. 7 (May 2007): 1655–95.

31. Y. Xiao and H. Li. Evaluation of Distributed Admission Control for the IEEE 802.11e EDCA. *IEEE Communications Magazine* 42, no. 9 (September 2004): S20–4.

32. IEEE Std. 802.11a-1999. Wireless LAN Medium Access Control (MAC) and Physical Layer (PHY) Specifications High-Speed Physical Layer in the 5 GHz Band (September 1999). Supplement to IEEE Std. 802.11-1999.

33. IEEE Std. 802.11b-1999. Wireless LAN Medium Access Control (MAC) and Physical Layer (PHY) Specifications: Higher-Speed Physical Layer Extension in the 2.4 GHz Band (September 1999). Supplement to IEEE Std. 802.11-1999.

34. IEEE Std. 802.11g-2003. Wireless LAN Medium Access Control (MAC) and Physical Layer (PHY) Specifications Amendment 4: Further Higher Data Rate Extension in the 2.4 GHz Band (June 2003).

35. IEEE Std. 802.11n-2009. Wireless LAN Medium Access Control (MAC) and Physical Layer (PHY) Specifications Amendment 5: Enhancements for Higher Throughput (October 2009).

36. Y. Xiao. IEEE 802.11n: Enhancements for Higher Throughput in Wireless LANs. *IEEE Wireless Communications* 12, no. 6 (December 2005): 82–91.

37. E. Perahia. IEEE 802.11n Development: History, Process, and Technology. *IEEE Communications Magazine* 46, no. 7 (July 2008): 48–55.

38. E. Perahia and R. Stacey. *Next Generation Wireless LANs: Throughput, Robustness, and Reliability in 802.11n*. Cambridge University Press, New York (2008).

39. IEEE Std. 802.11s-2011. Wireless LAN Medium Access Control (MAC) and Physical Layer (PHY) Specifications Amendment 10: Mesh Networking (September 2011).

40. R. C. Carrano et al. IEEE 802.11s Multihop MAC: A Tutorial. *IEEE Communications Surveys & Tutorials* 13, no. 1 (First Quarter 2011): 52–67.

41. Y.-D. Lin et al. Design Issues and Experimental Studies of Wireless LAN Mesh. *IEEE Wireless Communications* 17, no. 2 (April 2010): 32–40.

42. IEEE Draft Std. P80211ac/D6.0. Wireless LAN Medium Access Control and Physical Layer Specifications Amendment 4: Enhancements for Very High Throughput for Operation in Bands below 6 GHz (July 2013).

43. IEEE Std. 802.11ad-2012. Wireless LAN Medium Access Control (MAC) and Physical Layer (PHY) Specifications Amendment 3: Enhancements for Very High Throughput in the 60 GHz Band (December 2012).

44. C. Cordeiro. The Pursuit of Tens of Gigabits per Second Wireless Systems. *IEEE Wireless Communications* 20, no. 1 (February 2013): 3–5.

45. IEEE Std. 802.16. Air Interface for Fixed Broadband Wireless Access Systems (April 2002).

46. B. Li et al. A Survey on Mobile WiMAX [Wireless Broadband Access]. *IEEE Communications Magazine* 45, no. 12 (December 2007): 70–5.

47. Z. Bojkovic and D. Milovanovic. Coexistence Goals of VoIP and TCP Traffic in Mobile WiMAX Networks: Performance of Flat Architecture. *Proc. 9th WSEAS AIC '09*. Moscow, Russia (August 2009): 409–14.

48. IEEE Std. 802.16m-2011. Air Interface for Broadband Wireless Access Systems Amendment 3: Advanced Air Interface (May 2011). Amendment to IEEE Std. 802.16-2009.

49. A. Osseiran et al. The Road to IMT-Advanced Communication Systems: State-of-the-Art and Innovation Areas Addressed by the WINNER+ Project. *IEEE Communications Magazine* 47, no. 6 (June 2009): 38–47.

50. H. Cho et al. Physical Layer Structure of Next Generation Mobile WiMAX Technology. *Computer Networks* 55, no. 16 (November 2011): 3648–58.
51. I. Papapanagiotou et al. A Survey on Next Generation Mobile WiMAX Networks: Objectives, Features and Technical Challenges. *IEEE Communications Surveys & Tutorials* 11, no. 4 (Fourth Quarter 2009): 3–18.
52. R. Y. Kim, J. S. Kwak and K. Etemad. WiMAX Femtocell: Requirements, Challenges, and Solutions. *IEEE Communications Magazine* 47, no. 9 (September 2009): 84–91.
53. Y. Li et al. Overview of Femtocell Support in Advanced WiMAX Systems. *IEEE Communications Magazine* 49, no. 7 (July 2011): 122–30.
54. X. Li et al. The Future of Mobile Wireless Communication Networks. *Proc. ICCSN '09*. Macau, China (February 2009): 554–7.
55. H. Yanikomeroglu. Towards 5G Wireless Cellular Networks: Views on Emerging Concepts and Technologies. *Proc. 20th SIU 2012*. Fethiye, Muğla, Turkey (April 2012): 1–2.
56. I. F. Akyildiz, D. M. Gutierrez-Estevez and E. C. Reyes. The Evolution to 4G Cellular Systems: LTE-Advanced. *Physical Communication* 3, no. 4 (December 2010): 217–44.
57. 3rd Generation Partnership Project (3GPP). Available at http://www.3gpp.org (2013).
58. B. A. Bjerke. LTE-Advanced and the Evolution of LTE Deployments. *IEEE Wireless Communications* 18, no. 5 (October 2011): 4–5.
59. GSA. Evolution to LTE Report. Available at http://www.gsacom.com, accessed July 2012.
60. S. Parkvall, A. Furuskär and E. Dahlman. Evolution of LTE toward IMT-Advanced. *IEEE Communications Magazine* 49, no. 2 (February 2011): 84–91.
61. D. Astely et al. LTE: The Evolution of Mobile Broadband. *IEEE Communications Magazine* 47, no. 4 (April 2009): 44–51.
62. A. Ghosh et al. LTE-Advanced: Next-Generation Wireless Broadband Technology. *IEEE Wireless Communications* 17, no. 3 (June 2010): 10–22.
63. IEEE Std. 802.20-2008. Air Interface for Mobile Broadband Wireless Access Systems Supporting Vehicular Mobility—Physical and Media Access Control Layer Specification (August 2008).
64. W. Bolton, Y. Xiao and M. Guizani. IEEE 802.20: Mobile Broadband Wireless Access. *IEEE Wireless Communications* 14, no. 1 (February 2007): 84–95.
65. B. Bakmaz et al. Mobile Broadband Networking Based on IEEE 802.20 Standard. *Proc. 8th TELSIKS 2007*. Nis, Serbia (September 2007): 243–6.
66. A. Greenspan et al. IEEE 802.20: Mobile Broadband Wireless Access for the Twenty-First Century. *IEEE Communications Magazine* 46, no. 7 (July 2008): 56–63.
67. FCC. Notice of Proposed Rule Making in the Matter of Unlicensed Operation in the TV Broadcast Bands. ET Docket 04-186 (May 2004).
68. IEEE Std. 802.22-2011. Cognitive Wireless RAN Medium Access Control (MAC) and Physical Layer (PHY) Specifications: Policies and Procedures for Operation in the TV Bands (July 2011).
69. W. Hu et al. Dynamic Frequency Hopping Communities for Efficient IEEE 802.22 Operation. *IEEE Communications Magazine* 45, no. 5 (May 2007): 80–7.

70. C. Stevenson et al. IEEE 802.22: The First Cognitive Radio Wireless Regional Area Network Standard. *IEEE Communications Magazine* 47, no. 1 (January 2009): 130–8.

71. IEEE Std 802.1D-2004. Media Access Control (MAC) Bridges (June 2004). Revision of IEEE Std 802.1D-1998.

72. B. Xie, A. Kumar and D. P. Agrawal. Enabling Multiservice on 3G and Beyond: Challenges and Future Directions. *IEEE Wireless Communications* 15, no. 3 (June 2008): 66–72.

73. K. S. Munasinghe and A. Jamalipour. Interworked WiMAX-3G Cellular Data Networks: An Architecture for Mobility Management and Performance Evaluation. *IEEE Transactions on Wireless Communications* 8, no. 4 (April 2009): 1847–53.

74. R. Agrawal and A. Bedekar. Network Architectures for 4G: Cost Considerations. *IEEE Communications Magazine* 45, no. 12 (December 2007): 76–81.

75. R. Ferrus, O. Sallent and R. Agusti. Interworking in Heterogeneous Wireless Networks: Comprehensive Framework and Future Trends. *IEEE Wireless Communications* 17, no. 2 (April 2010): 22–31.

76. B. Aboba et al. Extensible Authentication Protocol (EAP). *IETF RFC 3748* (June 2004).

77. P. Calhoun et al. Diameter Base Protocol. *IETF RFC 3588* (September 2003).

78. K.-S. Kong et al. Mobility Management for All-IP Mobile Networks: Mobile IPv6 vs. Proxy Mobile IPv6. *IEEE Wireless Communications* 15, no. 2 (April 2008): 36–45.

79. L. Taylor, R. Titmuss and C. Lebre. The Challenges of Seamless Handover in Future Mobile Multimedia Networks. *IEEE Personal Communications* 6, no. 2 (April 1999): 32–7.

80. M. Z. Siam and M. Krunz. An Overview of MIMO-Oriented Channel Access in Wireless Networks. *IEEE Wireless Communications* 15, no. 1 (February 2008): 63–9.

81. A. J. Paulraj and T. Kailath. Increasing Capacity in Wireless Broadcast System Using Distributed Transmission/Directional Reception. U.S. Patent no. 5345599 (September 1994).

82. M. Krunz, A. Muqattash and S. J. Lee. Transmission Power Control in Wireless Ad Hoc Networks: Challenges, Solutions, and Open Issues. *IEEE Network* 18, no. 5 (September/October 2004): 8–14.

83. K. Sundaresan et al. Medium Access Control in Ad Hoc Networks with MIMO Links: Optimization Considerations and Algorithms. *IEEE Transactions on Mobile Computing* 3, no. 4 (October/December 2004): 350–65.

84. K. Sundaresan and R. Sivakumar. Routing in Ad Hoc Networks with MIMO Links. *Proc. IEEE ICNP 2005*. Boston (November 2005): 85–98.

85. M. Hu and J. Zhang. MIMO Ad Hoc Networks: Medium Access Control, Saturation Throughput, and Optimal Hop Distance. *Journal of Communications and Networks* 6, no. 4 (December 2004): 317–30.

86. D. Hoang and R. A. Iltis. An Efficient MAC Protocol for MIMO-OFDM Ad Hoc Networks. *Proc. 48th IEEE ACSSC*. Pacific Grove, CA (October 2006): 814–8.

87. M. Z. Siam and M. Krunz. Throughput-Oriented Power Control in MIMO-Based Ad Hoc Networks. *Proc. IEEE ICC*. Glasgow, Scotland (June 2007): 3686–91.

88. Q. Li et al. MIMO Techniques in WiMAX and LTE: A Feature Overview. *IEEE Communications Magazine* 48, no. 5 (May 2010): 86–92.

89. L. Liu et al. Downlink MIMO in LTE-advanced: SU-MIMO vs. MU-MIMO. *IEEE Communications Magazine* 50, no. 2 (February 2012): 140–7.

90. D. J. Love et al. An Overview of Limited Feedback in Wireless Communication Systems. *IEEE Journal on Selected Areas in Communications* 26, no. 8 (October 2008): 1341–65.

91. Y. Zhang, H.-H. Chen and M. Guizani, eds. *Cooperative Wireless Communications*. Wireless Networks and Mobile Communications Series. Boca Raton, FL: Auerbach Publications, Taylor & Francis Group (2009).

92. W. Zhuang and M. Ismail. Cooperation in Wireless Communication Networks. *IEEE Wireless Communications* 19, no. 2 (April 2012): 10–20.

93. L. Cai et al. User Cooperation in Wireless Networks. *IEEE Wireless Communications* 19, no. 2 (April 2012): 8–9.

94. J. N. Laneman, D. N. C. Tse and G. W. Wornell. Cooperative Diversity in Wireless Networks: Efficient Protocols and Outage Behavior. *IEEE Transactions on Information Theory* 50, no. 12 (December 2004): 3062–80.

95. T. Zhou et al. A Novel Adaptive Distributed Cooperative Relaying MAC Protocol for Vehicular Networks. *IEEE Journal on Selected Areas in Communications* 29, no. 1 (January 2011): 72–82.

96. M. Ismail and W. Zhuang. A Distributed Multi-Service Resource Allocation Algorithm in Heterogeneous Wireless Access Medium. *IEEE Journal on Selected Areas in Communications* 30, no. 2 (February 2012): 425–32.

97. H. Choi and D. Cho. On the Use of Ad Hoc Cooperation for Seamless Vertical Handoff and Its Performance Evaluation. *Mobile Networks and Applications* 15, no. 5 (October 2010): 750–66.

98. M. Ismail and W. Zhuang. Network Cooperation for Energy Saving in Green Radio Communications. *IEEE Wireless Communications* 18, no. 5 (October 2011): 76–81.

99. L. Badia et al. Cooperation Techniques for Wireless Systems from a Networking Perspective. *IEEE Wireless Communications* 17, no. 2 (April 2010): 89–96.

100. K. Lee and L. Hanzo. Resource-Efficient Wireless Relaying Protocols. *IEEE Wireless Communications* 17, no. 2 (April 2010): 66–72.

101. S. Chieochan and E. Hossain. Cooperative Relaying in Wi-Fi Networks with Network Coding. *IEEE Wireless Communications* 19, no. 2 (April 2012): 57–65.

102. M. Peng et al. Cooperative Network Coding in Relay-Based IMT-Advanced Systems. *IEEE Communications Magazine* 50, no. 4 (April 2012): 76–84.

103. S. Zhang and S.-C. Liew. Channel Coding and Decoding in a Relay System Operated with Physical-Layer Network Coding. *IEEE Journal on Selected Areas in Communications* 27, no. 5 (June 2009): 788–96.

104. A. Afanasyev et al. Host-to-Host Congestion Control for TCP. *IEEE Communications Surveys & Tutorials* 12, no. 3 (Third Quarter 2010): 304–42.

105. D. Katabi, M. Handley and C. Rohrs. Congestion Control for High Bandwidth-Delay Product Networks. *ACM SIGCOMM Computer Communication Review* 32, no. 4 (October 2002): 89–102.

106. M. Lestas et al. Adaptive Congestion Protocol: A Congestion Control Protocol with Learning Capability. *Computer Networks* 51, no. 13 (September 2007): 3773–98.

107. C. P. Fu and S. C. Liew. TCP Veno: TCP Enhancement for Transmission over Wireless Access Networks. *IEEE Journal on Selected Areas in Communications* 21, no. 2 (February 2003): 216–28.

108. E. H.-K. Wu and M.-Z. Chen. JTCP: Jitter-Based TCP for Heterogeneous Wireless Networks. *IEEE Journal on Selected Areas in Communications* 22, no. 4 (May 2004): 757–66.

109. C. Casetti et al. TCP Westwood: End-to-End Congestion Control for Wired/Wireless Networks. *Wireless Networks* 8, no. 5 (September 2002): 467–79.

110. K. Xu, Y. Tian and N. Ansari. TCP-Jersey for Wireless IP Communications. *IEEE Journal on Selected Areas in Communications* 22, no. 4 (May 2004): 747–56.

111. T. R. Newman. Designing and Deploying a Building-Wide Cognitive Radio Network Testbed. *IEEE Communications Magazine* 48, no. 9 (September 2010): 106–12.

112. D. Datla et al. Wireless Distributed Computing in Cognitive Radio Networks. *Ad Hoc Networks* 10, no. 5 (July 2012): 845–57.

CHAPTER 2 COGNITIVE RADIO NETWORKS

1. J. Mitola III and G. Q. Maguire, Jr. Cognitive Radio: Making Software Radios More Personal. *IEEE Personal Communications* 6, no. 4 (August 1999): 13–8.

2. P. Pawelczak et al. Cognitive Radio: Ten Years of Experimentation and Development. *IEEE Communications Magazine* 49, no. 3 (March 2011): 90–100.

3. F. Granelli et al. Standardization and Research in Cognitive and Dynamic Spectrum Access Networks: IEEE SCC41 Efforts and Other Activities. *IEEE Communications Magazine* 48, no. 1 (January 2010): 71–9.

4. Q. Zhao and B. M. Sadler. A Survey of Dynamic Spectrum Access. *IEEE Signal Processing Magazine* 24, no. 3 (May 2007): 79–89.

5. I. F. Akyildiz et al. A Survey on Spectrum Management in Cognitive Radio Networks. *IEEE Communication Magazine* 46, no. 4 (April 2008): 40–8.

6. I. F. Akyildiz et al. NeXt Generation/Dynamic Spectrum Access/Cognitive Radio Wireless Networks: A Survey. *Computer Networks* 50, no. 13 (September 2006): 2127–59.

7. M. Mueck et al. ETSI Reconfigurable Radio Systems: Status and Future Directions on Software Defined Radio and Cognitive Radio Standards. *IEEE Communications Magazine* 48, no. 9 (September 2010): 78–86.

8. E. K. Au et al. Advances in Standards and Testbeds for Cognitive Radio Networks: Part I [Guest Editorial]. *IEEE Communications Magazine* 48, no. 9 (September 2010): 76–7.

9. IEEE DySPAN Standards Committee. Available at http://grouper.ieee.org/groups/dyspan (2013).

10. Standard ECMA-392. *MAC and PHY for Operation in TV White Space*, 2nd edition. ECMA International, Geneva (June 2012).

11. ETSI TR 102 838 V1.1.1. Reconfigurable Radio Systems (RRS); Summary of Feasibility Studies and Potential Standardization Topics (October 2009).

12. C. Cormio and K. R. Chowdhury. A Survey on MAC Protocols for Cognitive Radio Networks. *Ad Hoc Networks* 7, no. 7 (September 2009): 1315–29.

13. K. R. Chowdhury and I. F. Akyildiz. Cognitive Wireless Mesh Networks with Dynamic Spectrum Access. *IEEE Journal on Selected Areas in Communications* 26, no. 1 (January 2008): 168–81.

14. I. F. Akyildiz, W.-Y. Lee and K. R. Chowdhury. CRAHNs: Cognitive Radio Ad Hoc Networks. *Ad Hoc Networks* 7, no. 5 (July 2009): 810–36.

15. Encyclopedia Britannica. Available at http://www.britannica.com (2013).

16. P. S. Hall, P. Gardner and A. Faraone. Antenna Requirements for Software Defined and Cognitive Radios. *Proceedings of the IEEE* 100, no. 7 (July 2012): 2262–70.

17. W. A. Baan and P. Delogne. Spectrum Congestion. In *Modern Radio Science 1999*, M. A. Stuchly, ed. Hoboken, NJ: Wiley-IEEE Press (1999): 309–27.

18. S. Haykin. Cognitive Dynamic Systems: Radar, Control, and Radio. *Proceedings of the IEEE* 100, no. 7 (July 2012): 2095–103.

19. J. Mitola III. Software Radios—Survey, Critical Evaluation and Future Directions. *Proc. IEEE NTC-92*. George Washington University, Virginia Campus, Washington, DC (May 1992): 13/15–13/23.

20. Wireless Innovation Forum. Available at http://www.wirelessinnovation.org (2013).

21. W. Tuttlebee ed. *Software Defined Radio: Enabling Technologies*. Hoboken, NJ: Wiley (2002).

22. ITU-R SM.2152. Definitions of Software Defined Radio (SDR) and Cognitive Radio System (CRS) (September 2009).

23. S. Filin et al. International Standardization of Cognitive Radio Systems. *IEEE Communications Magazine* 49, no. 3 (March 2011): 82–9.

24. ETSI TR 102 683 V1.1.1. Reconfigurable Radio Systems (RRS); Cognitive Pilot Channel (CPC) (September 2009).

25. P. Ballon and S. Delaere. Flexible Spectrum and Future Business Models for the Mobile Industry. *Telematics and Informatics* 26, no. 3 (August 2009): 249–58.

26. IEEE Std. 1900.4-2009. Architectural Building Blocks Enabling Network-Device Distributed Decision Making for Optimized Radio Resource Usage in Heterogeneous Wireless Access Networks (February 2009).

27. IEEE Std. 1900.6-2011. Spectrum Sensing Interfaces and Data Structures for Dynamic Spectrum Access and Other Advanced Radio Communication Systems (April 2011).

28. IEEE Std. 802-11y-2008. Wireless LAN Medium Access Control (MAC) and Physical Layer (PHY) Specifications; Amendment 3: 3650–3700 MHz Operation in USA (November 2008).

29. IEEE P802.11af. Wireless LAN Medium Access Control (MAC) and Physical Layer (PHY) Specifications; Amendment: TV White Spaces Operation (October 2013).

30. IEEE P802.19.1. Telecommunications and Information Exchange between Systems—Local and Metropolitan Area Networks—Specific Requirements—Part 19: TV White Space Coexistence Methods (October 2013).

31. ETSI TR 102 682 V1.1.1. Reconfigurable Radio Systems (RRS); Functional Architecture (FA) for Management and Control of Reconfigurable Radio Systems (July 2009).

32. D. Cabric, S. M. Mishra and R. W. Brodersen. Implementation Issues in Spectrum Sensing for Cognitive Radios. *Conference Record of the 38th Asilomar Conference on Signals, Systems and Computers*. Pacific Grove, CA (November 2004): 772–6.

33. K. Ishizu, H. Murakami and H. Harada. Feasibility Study on Spectrum Sharing Type Cognitive Radio System with Outband Pilot Channel. *Proc. 6th CROWNCOM 2011*. Osaka, Japan (June 2011): 286–90.

34. H. A. B. Salameh and M. Krunz. Channel Access Protocols for Multihop Opportunistic Networks: Challenges and Recent Developments. *IEEE Network* 23, no. 4 (July–August 2009): 14–9.

35. M. Fitch et al. Wireless Service Provision in TV White Space with Cognitive Radio Technology: A Telecom operator's Perspective and Experience. *IEEE Communications Magazine* 49, no. 3 (March 2011): 64–73.

36. FCC. Second Report and Order in the Matter of Unlicensed Operation in the TV Broadcast Bands (ET Docket no. 04-186), Additional Spectrum for Unlicensed Devices Below 900 MHz and in 3 GHz Band (EC Docket no. 02-380). Available at http://www.fcc.org, accessed November 2008.

37. Ofcom. Digital Dividend: Cognitive Access. Statement on Licence-exempting Cognitive Devices Using Interleaved Spectrum. Available at http://www.ofcom.org.uk, accessed July 2009.

38. B. Bakmaz et al. Selection of Appropriate Technologies for Universal Service. *Proc. ICEST 2008*. Nis, Serbia, (June 2008): 8–11.

39. M. Nekovee. A Survey of Cognitive Radio Access to TV White Spaces. *International Journal of Digital Multimedia Broadcasting*, no. 2010 (2010): Article ID 236568.

40. M. Nekovee, T. Irnich and J. Karlsson. Worldwide Trends in Regulation of Secondary Access to White Spaces Using Cognitive Radio. *IEEE Wireless Communications* 19, no. 4 (August 2012): 32–40.

41. H. R. Karimi. Geolocation Databases for White Space Devices in the UHF TV Bands: Specification of Maximum Permitted Emission Levels. *Proc. IEEE DySPAN 2011*. Aachen, Germany (May 2001): 443–53.

42. R. Yu et al. Secondary Users Cooperation in Cognitive Radio Networks: Balancing Sensing Accuracy and Efficiency. *IEEE Wireless Communications* 19, no. 2 (April 2012): 30–7.

43. B. Cao et al. Toward Efficient Radio Spectrum Utilization: User Cooperation in Cognitive Radio Networking. *IEEE Network* 26, no. 4 (July–August 2012): 46–52.

44. O. Simeone et al. Spectrum Leasing to Cooperating Secondary Ad Hoc Networks. *IEEE Journal on Selected Areas in Communications* 26, no. 1 (January 2008): 203–13.

45. K. B. Letaief and W. Zhang. Cooperative Communications for Cognitive Radio Networks. *Proceedings of the IEEE* 97, no. 5 (May 2009): 878–93.

46. M. Pan, P. Li and Y. Fang. Cooperative Communication Aware Link Scheduling for Cognitive Vehicular Networks. *IEEE Journal on Selected Areas in Communications* 30, no. 4 (May 2012): 760–8.

47. J. Wang, M. Ghosh and K. Challapali. Emerging Cognitive Radio Applications: A Survey. *IEEE Communications Magazine* 49, no. 3 (March 2011): 74–81.

48. A. Alsarhan and A. Agarwal. Optimizing Spectrum Trading in Cognitive Mesh Network Using Machine Learning. *Journal of Electrical and Computer Engineering,* no. 2012 (2012): Article ID 562615.

49. H. Bogucka et al. Secondary Spectrum Trading in TV White Spaces. *IEEE Communications Magazine* 50, no. 11 (November 2012): 121–9.

50. H. W. Kuhn and A. W. Tucker. Nonlinear Programming. *Proc. 2nd Berkeley Symposium on Mathematical Statistics and Probability.* University of California Press, Berkeley and Los Angeles (July–August 1950): 481–92.

51. M. Cesana, E. Ekici and Y. Bar-Ness. Networking over Multi-Hop Cognitive Networks [Guest Editorial]. *IEEE Network* 23, no. 4 (July–August 2009): 4–5.

52. H. Khalife, N. Malouch and S. Fdida. Multihop Cognitive Radio Networks: To Route or Not to Route. *IEEE Network* 23, no. 4 (July–August 2009): 20–5.

53. E. Alotaibi and B. Mukherjee. A Survey on Routing Algorithms for Wireless Ad-hoc and Mesh Networks. *Computer Networks* 56, no. 2 (February 2012): 940–65.

54. M. Cesana, F. Cuomo and E. Ekici. Routing in Cognitive Radio Networks: Challenges and Solutions. *Ad Hoc Networks* 9, no. 3 (May 2011): 228–48.

55. L. Hesham et al. Distributed Spectrum Sensing with Sequential Ordered Transmissions to a Cognitive Fusion Center. *IEEE Transactions on Signal Processing* 60, no. 5 (May 2012): 2524–38.

56. T. Luo et al. Multicarrier Modulation and Cooperative Communication in multihop Cognitive Radio Networks. *IEEE Wireless Communications* 18, no. 1 (February 2011): 38–45.

57. S. C. Jha et al. Medium Access Control in Distributed Cognitive Radio Networks. *IEEE Wireless Communications* 18, no. 4 (August 2011): 41–51.

58. X. Wang et al. Common Control Channel Model on MAC Protocols in Cognitive Radio Networks. *Proc. ICCSNT 2011.* Harbin, China (December 2011): 2230–4.

59. H. Su and X. Zhang. Cross-Layer based Opportunistic MAC Protocols for QoS Provisionings over Cognitive Radio Wireless Networks. *IEEE Journal on Selected Areas in Communications* 26, no. 1 (January 2008): 118–29.

60. J. So and N. Vaidya. Multi-Channel MAC for Ad Hoc Networks: Handling Multi-Channel Hidden Terminals Using A single Transceiver. *Proc. 5th ACM MobiHoc.* Tokyo, Japan (May 2004): 222–33.

61. Q. Zhao et al. Decentralized Cognitive MAC for Opportunistic Spectrum Access in Ad Hoc Networks: A POMDP Framework. *IEEE Journal on Selected Areas in Communications* 25, no. 3 (April 2007): 589–600.

62. Y. R. Kondareddy and P. Agrawal. Synchronized MAC Protocol For Multi-Hop Cognitive Radio Networks. *Proc. IEEE ICC '08.* Beijing, China (May 2008): 3198–202.

63. J. Jia, Q. Zhang and X. Shen. HC-MAC: A Hardware-Constrained Cognitive MAC for Efficient Spectrum Management. *IEEE Journal on Selected Areas in Communications* 26, no. 1 (January 2008): 106–17.

64. L. Le and E. Hossain. OSA-MAC: A MAC Protocol for Opportunistic Spectrum Access in Cognitive Radio Networks. *Proc. IEEE WCNC '08*. Las Vegas, NV (April 2008): 1426–30.

65. C. Cordeiro and K. Challapali. C-MAC: A Cognitive MAC Protocol for Multi-Channel Wireless Networks. *Proc. IEEE DySPAN '07*. Dublin, Ireland (April 2007): 147–57.

66. L. Ma, X. Han and C.-C. Shen. Dynamic Open Spectrum Sharing MAC Protocol for Wireless Ad Hoc Networks. *Proc. IEEE DySPAN '05*. Baltimore (November 2005): 203–13.

67. T. Chen et al. CogMesh: A Cluster-Based Cognitive Radio Network. *Proc. IEEE DySPAN '07*. Dublin, Ireland (April 2007): 168–78.

68. S. C. Jha, M. M. Rashid and V. K. Bhargava. OMC-MAC: An Opportunistic Multichannel MAC for Cognitive Radio Networks. *Proc. IEEE VTC '09-Fall*. Anchorage, AK (September 2009): 1–5.

CHAPTER 3 MOBILITY MANAGEMENT IN HETEROGENEOUS WIRELESS SYSTEMS

1. Z. Bojkovic, B. Bakmaz and M. Bakmaz. Multimedia Traffic in New Generation Networks: Requirements, Control and Modeling. *Proc. 13th WSEAS CSCC*. Rodos, Greece (July 2009): 124–30.

2. A. Damnjanovic et al. A Survey on 3GPP Heterogeneous Networks. *IEEE Wireless Communications Magazine* 18, no. 3 (June 2011): 10–21.

3. D. López-Pérez, İ. Güvenc and X. Chu. Mobility Management Challenges in 3GPP Heterogeneous Networks. *IEEE Communications Magazine* 50, no. 12 (December 2012): 70–8.

4. Ł. Budzisz et al. Towards Transport-Layer Mobility: Evolution of SCTP Multihoming. *Computer Communications* 31, no. 5 (March 2008): 980–98.

5. M. R. HeidariNezhad et al. Mobility Support across Hybrid IP-Based Wireless Environment: Review of Concepts, Solutions, and Related Issues. *Annales des Télécommunications* 64, no. 9–10 (October 2009): 677–91.

6. I. F. Akyildiz, J. Xie and S. Mohanty. A Survey of Mobility Management in Next-Generation All-IP-Based Wireless Systems. *IEEE Wireless Communications* 11, no. 4 (August 2004): 16–28.

7. A. Dhraief, I. Mabrouki and A. Belghith. A Service-Oriented Framework for Mobility and Multihoming Support. *Proc. 16th IEEE MELECON*. Yasmine Hammamet, Tunisia (March 2012): 489–93.

8. N. Banerjee, W. Wu and S. K. Das. Mobility Support in Wireless Internet. *IEEE Wireless Communications* 10, no. 5 (October 2003): 54–61.

9. F. M. Chiussi, D. A. Khotimsky and S. Krishnan. Mobility Management in Third-Generation All-IP Networks. *IEEE Communications Magazine* 40, no. 9 (September 2002): 124–35.

10. S. Mohanty and I. F. Akyildiz. Performance Analysis of Handoff Techniques Based on Mobile IP, TCP-Migrate, and SIP. *IEEE Transactions on Mobile Computing* 6, no. 7 (July 2007): 731–47.

11. C. Perkins. IP Mobility Support for IPv4. *IETF RFC 3344* (August 2002).

12. D. Johnson, C. Perkins and J. Arkko. Mobility Support in IPv6. *IETF RFC 3775* (June 2004).

13. I. Al-Surmi, M. Othman and B. M. Ali. Mobility Management for IP-Based Next Generation Mobile Networks: Review, Challenge and Perspective. *Journal of Network and Computer Applications* 35, no. 1 (January 2012): 295–315.

14. C. Makaya and S. Pierre. An Analytical Framework for Performance Evaluation of IPv6-Based Mobility Management Protocols. *IEEE Transactions on Wireless Communications* 7, no. 3 (March 2008): 972–83.

15. R. Koodli. Fast Handovers for Mobile IPv6. *IETF RFC 4068* (July 2005).

16. H. Soliman et al. Hierarchical Mobile IPv6 (HMIPv6) Mobility Management. *IETF RFC 4140* (August 2005).

17. S. Gundavelli et al. Proxy Mobile IPv6. *IETF RFC 5213* (August 2008).

18. I. Ali et al. Network-Based Mobility Management in the Evolved 3GPP Core Network. *IEEE Communications Magazine* 47, no. 2 (February 2009): 58–66.

19. H. Yokota et al. Fast Handovers for PMIPv6. *IETF RFC 5949* (September 2010).

20. J.-M. Chung et al. Enhancements to FPMIPv6 for Improved Seamless Vertical Handover between LTE and Heterogeneous Access Networks. *IEEE Wireless Communications* 20, no. 3 (June 2013): 112–9.

21. V. Devarapalli et al. Network Mobility (NEMO) Basic Support Protocol. *IETF RFC 3963* (January 2005).

22. S. Céspedes, X. Shen and C. Lazo. IP Mobility Management for Vehicular Communication Networks: Challenges and Solutions. *IEEE Communications Magazine* 49, no. 5 (May 2011): 187–94.

23. A. Z. M. Shahriar, M. Atiquzzaman and W. Ivancic. Route Optimization in Network Mobility: Solutions, Classification, Comparison, and Future Research Directions. *IEEE Communications Surveys & Tutorials* 12, no. 1 (First Quarter 2010): 24–38.

24. I. Soto et al. NEMO-Enabled Localized Mobility Support for Internet Access in Automotive Scenarios. *IEEE Communications Magazine* 47, no. 5 (May 2009): 152–9.

25. A. de la Oliva et al. The Costs and Benefits of Combining Different IP Mobility Standards. *Computer Standards & Interfaces* 35, no. 2 (February 2013): 205–17.

26. Z. Yan et al. Design and Implementation of a Hybrid MIPv6/PMIPv6-Based Mobility Management Architecture. *Mathematical and Computer Modelling* 53, no. 3–4 (February 2011): 421–42.

27. W. M. Eddy. At What Layer Does Mobility Belong? *IEEE Communications Magazine* 42, no. 10 (October 2004): 155–9.

28. R. Stewart et al. Stream Control Transmission Protocol. *IETF RFC 2960* (October 2000).

29. R. Stewart et al. Stream Control Transmission Protocol (SCTP) Dynamic Address Reconfiguration. *IETF RFC 5061* (September 2007).

30. M. Handley et al. SIP: Session Initiation Protocol. *IETF RFC 2543* (March 1999).
31. J. Zhang, H. C. B. Chan and V. C. M. Leung. A SIP-Based Seamless-Handoff (S-SIP) Scheme for Heterogeneous Mobile Networks. *Proc. IEEE WCNC 2007*. Hong Kong (March 2007): 3946–50.
32. S. Salsano et al. SIP-Based Mobility Management in Next Generation Networks. *IEEE Wireless Communications* 15, no. 2 (April 2008): 92–9.
33. O. A. El-Mohsen, H. A. M. Saleh and S. Elramly. SIP-Based Handoff Scheme in Next Generation Wireless Networks. *Proc. 6th NGMAST 2012*. Paris, France (September 2012): 131–6.
34. A. Dutta et al. Fast-Handoff Schemes for Application Layer Mobility Management. *Proc. 15th IEEE PIMRC*. Barcelona, Spain (September 2004): 1527–32.
35. N. Banerjee, A. Acharya and S. K. Das. Seamless SIP-Based Mobility for Multimedia Applications. *IEEE Network* 20, no. 2 (March/April 2006): 6–13.
36. Y. C. Yee et al. SIP-Based Proactive and Adaptive Mobility Management Framework for Heterogeneous Networks. *Journal of Network and Computer Applications* 31, no. 4 (November 2008): 771–92.
37. B. S. Ghahfarokhi and N. Movahhedinia. A Survey on Applications of IEEE 802.21 Media Independent Handover Framework in Next Generation Wireless Networks. *Computer Communications* 36, no. 10–11 (June 2013): 1101–19.
38. A. Achour et al. A SIP-SHIM6-Based Solution Providing Interdomain Service Continuity in IMS-Based Networks. *IEEE Communications Magazine* 50, no. 7 (July 2012): 109–19.
39. E. Nordmark and M. Bagnulo. Shim6: Level 3 Multihoming Shim Protocol for IPv6. *IETF RFC 5533* (June 2009).
40. A. García-Martínez, M. Bagnulo and I. van Beijnum. The Shim6 Architecture for IPv6 Multihoming. *IEEE Communications Magazine* 48, no. 9 (September 2010): 152–7.
41. J. Arkko and I. van Beijnum. Failure Detection and Locator Pair Exploration Protocol for IPv6 Multihoming. *IETF RFC 5534* (June 2009).
42. H. Chan. Requirements of Distributed Mobility Management. *IETF Internet-Draft* (July 2012).
43. J. C. Zúñiga et al. Distributed Mobility Management: A Standards Landscape. *IEEE Communications Magazine* 51, no. 3 (March 2013): 80–7.
44. Available at http://datatracker.ietf.org/wg/dmm (2013).
45. B. Sarikaya. Distributed Mobile IPv6. *IETF Internet-Draft* (February 2012).
46. D.-H. Shin et al. Distributed Mobility Management for Efficient Video Delivery over All-IP Mobile Networks: Competing Approaches. *IEEE Network* 27, no. 2 (March/April 2013): 28–33.
47. J. Korhonen, T. Savolainen and S. Gundavelli. Local Prefix Lifetime Management for Proxy Mobile IPv6. *IETF Internet-Draft* (July 2013).
48. C. J. Bernardos, A. de la Oliva and F. Giustet. A PMIPv6-based Solution for Distributed Mobility Management. *IETF Internet-Draft* (July 2013).
49. P. McCann. Authentication and Mobility Management in a Flat Architecture. *IETF Internet-Draft* (March 2012).

50. 3GPP TS 23.401 V12.1.0. General Packet Radio Service (GPRS) Enhancements for Evolved Universal Terrestrial Radio Access Network (E-UTRAN) Accessed June 2013.

51. 3GPP TS 23.859 V12.0.1. Local IP Access (LIPA) Mobility and Selected IP Traffic Offload (SIPTO) at the Local Network (April 2013).

CHAPTER 4 NETWORK SELECTION IN HETEROGENEOUS WIRELESS ENVIRONMENT

1. M. Kassar, B. Kervella and G. Pujolle. An Overview of Vertical Handover Decision Strategies in Heterogeneous Wireless Networks. *Computer Communications* 31, no. 10 (June 2008): 2607–20.

2. E. Gustafsson and A. Jonsson. Always Best Connected. *IEEE Wireless Communications* 10, no. 1 (February 2003): 49–55.

3. B. Bakmaz. Network Selection in Heterogeneous Wireless Environment. PhD Thesis. University of Belgrade, Serbia (2011).

4. ITU-R Rec. M.1645. Framework and Overall Objectives of the Future Development of IMT-2000 and System beyond IMT-2000 (June 2003).

5. 3GPP TS 24.312 V11.3.0. Access Network Discovery and Selection Function (ANDSF) Management Object (MO) (June 2012).

6. IEEE Std. 802.21-2008. Media Independent Handover Services (January 2009).

7. M. Stemm and R. H. Katz. Vertical Handoffs in Wireless Overlay Networks. *Mobile Networks and Applications* 3, no. 4 (1998): 335–50.

8. N. Nasser, A. Hasswa and H. Hassanein. Handoffs in Fourth Generation Heterogeneous Networks. *IEEE Communications Magazine* 44, no. 10 (October 2006): 96–103.

9. A. Sgora and D. D. Vergados. Handoff Prioritization and Decision Schemes in Wireless Cellular Networks: A Survey. *IEEE Communications Surveys and Tutorials* 11, no. 4 (Fourth Quarter 2009): 57–77.

10. D. Wong and T. J. Lim. Soft Handoffs in CDMA Mobile Systems. *IEEE Personal Communications* 4, no. 6 (December 1997): 6–17.

11. J. McNair and F. Zhu. Vertical Handoffs in Fourth-generation Multinetwork Environments. *IEEE Wireless Communications* 11, no. 3 (June 2004): 8–15.

12. P. Chan et al. Mobility Management Incorporating Fuzzy Logic for a Heterogeneous IP Environment. *IEEE Communications Magazine* 39, no. 12 (December 2001): 42–51.

13. L.-S. Lee and K. Wang. Design and Analysis of a Network-Assisted Fast Handover Scheme for IEEE 802.16e Networks. *IEEE Transactions on Vehicular Technology* 59, no. 2 (February 2010): 869–83.

14. B. Bakmaz, Z. Bojkovic and M. Bakmaz. Network Selection Algorithm for Heterogeneous Wireless Environment. *Proc. IEEE PIMRC 2007.* Athens, Greece (September 2007): 1007–10.

15. A. de la Oliva et al. An Overview of IEEE 802.21: Media Independent Handover Service. *IEEE Wireless Communications* 15, no. 4 (August 2008): 96–103.

16. D. Johnson, C. Perkins and J. Arkko. Mobility Support in IPv6. *IETF RFC 3775* (June 2004).

17. A. de la Oliva et al. IEEE 802.21: Media Independence beyond Handover. *Computer Standards & Interfaces* 33, no. 6 (November 2011): 556–64.

18. R. Rouil, N. Golmie and N. Montavont. Media Independent Handover Transport Using Cross-Layer Optimized Stream Control Transmission Protocol. *Computer Communications* 33, no. 9 (June 2010): 1075–85.

19. K. Taninchi et al. IEEE 802.21: Media Independent Handover: Features, Applicability, and Realization. *IEEE Communications Magazine* 47, no. 1 (January 2009): 112–20.

20. Z. Bojkovic, B. Bakmaz and M. Bakmaz. Potential of IEEE 802.21 as Backbone Standard in Heterogeneous Environment. *Proc. 15th WSEAS CSCC '11*. Corfu Island, Greece (July 2011): 220–7.

21. L. Eastwood et al. Mobility Using IEEE 802.21 in a Heterogeneous IEEE 802.16/802.11-Based, IMT-Advanced (4G) Network. *IEEE Wireless Communications* 15, no. 2 (April 2008): 26–34.

22. D. E. Charilas and A. D. Panagopoulos. Multiaccess Radio Network Environments. *IEEE Vehicular Technology Magazine* 5, no. 4 (December 2010): 40–9.

23. R. Trestian, O. Ormond and G. Muntean. Game Theory—Based Network Selection: Solutions and Challenges. *IEEE Communications Surveys & Tutorials* 14, no. 4 (Fourth Quarter 2012): 1212–31.

24. D. He et al. A Simple and Robust Vertical Handoff Algorithm for Heterogeneous Wireless Mobile Networks. *Wireless Personal Communications* 59, no. 2 (July 2011): 361–73.

25. ITU-T Rec. Y.1541. Network Performance Objectives for IP-Based Services (December 2011).

26. B. Bakmaz et al. Mobile IPTV over Heterogeneous Networks: QoS, QoE and Mobility Management. *Proc. 18th ERK2009*. Portoroz, Slovenia (September 2009): 101–4.

27. B. Bakmaz, M. Bakmaz and Z. Bojkovic. Elements of Security Aspects in Wireless Networks: Analysis and Integration. *International Journal of Applied Mathematics and Informatics* 1, no. 2 (2007): 70–5.

28. D. Niyato and E. Hossain. Competitive Pricing in Heterogeneous Wireless Access Networks: Issues and Approaches. *IEEE Network* 22, no. 6 (November/December 2008): 4–11.

29. Y. Ji et al. CPC-Assisted Network Selection Strategy. *Proc. 16th IST Mobile and Wireless Communications Summit*. Budapest, Hungary (July 2007): 1–5.

30. W. Shen and Q.-A. Zeng. Cost-Function-Based Network Selection Strategy in Integrated Wireless and Mobile Networks. *IEEE Transactions on Vehicular Technology* 57, no. 6 (November 2008): 3778–88.

31. E. H. Ong and J. Y. Khan. Dynamic Access Network Selection with QoS Parameters Estimation: A Step Closer to ABC. *Proc. 67th IEEE VTC Spring 2008*. Marina Bay, Singapore (May 2008): 2671–6.

32. F. Bari and V. C. M. Leung. Automated Network Selection in a Heterogeneous Wireless Network Environment. *IEEE Network* 21, no. 1 (January/February 2007): 34–40.

33. Q. Song and A. Jamalipour. Network Selection in an Integrated Wireless LAN and UMTS Environment Using Mathematical Modeling and Computing Techniques. *IEEE Wireless Communications* 12, no. 3 (June 2005): 42–8.

34. Y. Wei, Y. Hu and J. Song. Network Selection Strategy in Heterogeneous Multi-Access Environment. *The Journal of China Universities of Posts and Telecommunications* 14, Supplement 1 (October 2007): 16–20.

35. D. Charilas et al. Packet-Switched Network Selection with the Highest QoS in 4G Networks. *Computer Networks* 52, no. 1 (January 2008): 248–58.

36. P. Zhang et al. A Novel Network Selection Mechanism in an Integrated WLAN and UMTS Environment Using AHP and Modified GRA. *Proc. 2nd IEEE IC-NIDC 2010*. Beijing, China (September 2010): 104–9.

37. T. L. Saaty. *The Analytic Hierarchy Process: Planning, Priority Setting, Resource Allocation*. New York: McGraw Hill (1980).

38. B. Bakmaz, Z. Bojkovic and M. Bakmaz. Vertical Handover Techniques Evaluation. *Proc. 10th WSEAS EHAC '11*. Cambridge, UK (February 2011): 259–64.

39. B. Bakmaz and M. Bakmaz. Network Selection Heuristics Evaluation in Vertical Handover Procedure. *Proc. ICEST 2011*. Nis, Serbia (June 2011): 583–6.

40. X. Yan, Y. A. Şekercioĝlu and S. Narayanan. A Survey of Vertical Handover Decision Algorithms in Fourth Generation Heterogeneous Wireless Networks. *Computer Networks* 54, no. 11 (August 2010): 1848–63.

41. B. Bakmaz. Network Selection Equilibrium in Heterogeneous Wireless Environment. *Elektronika ir Elektrotechnika (Electronics and Electrical Engineering)* 19, no. 4 (April 2013): 91–6.

42. L. Wang and G. Kuo. Mathematical Modeling for Network Selection in Heterogeneous Wireless Networks—A Tutorial. *IEEE Communications Surveys & Tutorials* 15, no. 1 (First Quarter 2013): 271–92.

43. L. Pirmez et al. SUTIL—Network Selection based on Utility Function and Integer Linear Programming. *Computer Networks* 54, no. 13 (September 2010): 2117–36.

44. I. Malanchini and M. Cesana. Modelling Network Selection and Resource Allocation in Wireless Access Networks with Non-Cooperative Games. *Proc. 5th IEEE MASS*. Atlanta, GA (October 2008): 404–9.

45. C.-H. Yeh. A Problem-based Selection of Multi-attribute Decision-making Methods. *International Transactions in Operational Research* 9, no. 2 (March 2002): 169–81.

46. R. Tawil, G. Pujolle and O. Salazar. A Vertical Handoff Decision Scheme in Heterogeneous Wireless Systems. Proc. *IEEE 67th VTC Spring 2008*. Marina Bay, Singapore (May 2008): 2626–30.

47. Q.-T. Nguyen-Vuong, Y. Ghamri-Doudane and N. Agoulmine. On Utility Models for Access Network Selection in Wireless Heterogeneous Networks. Proc. *IEEE NOMS 2008*. Salvador, Bahia, Brazil (April 2008): 144–51.

48. C. L. Hwang and K. Yoon. *Multiple Attribute Decision Making: Methods and Applications*. New York: Springer-Verlag (1981).

49. L. Ren et al. Comparative Analysis of a Novel M-TOPSIS and TOPSIS. *Applied Mathematics Research Express* 2007, (January 2007): Article ID abm005.

50. B. Bakmaz, Z. Bojkovic and M. Bakmaz. Traffic Parameters Influences on Network Selection in Heterogeneous Wireless Environment. *Proc. 19th IWSSIP 2012*. Vienna, Austria (April 2012): 306–9.

51. L. A. Zadeh. Fuzzy Sets. *Information and Control* 8, no. 3 (June 1965): 338–53.
52. S. Kher, A. K. Somani and R. Gupta. Network Selection Using Fuzzy Logic. *Proc. IEEE BroadNets 2005*. Boston (October 2005): 876–85.
53. M. Kassar, B. Kervella and G. Pujolle. An Intelligent Handover Management System for Future Generation Wireless Networks. *EURASIP Journal on Wireless Communications and Networking*, no. 2008 (2008): Article ID 791691.
54. B. Kröse and P. van der Smagt. *An Introduction to Neural Networks*. University of Amsterdam, Amsterdam, The Netherlands (1996).
55. N. Nasser, S. Guizani and E. Al-Masri. Middleware Vertical Handoff Manager: A Neural Network-based Solution. *Proc. IEEE ICC'07*. Glasgow, Scotland (June 2007): 5671–6.
56. Z. Bojkovic, M. Bakmaz and B. Bakmaz. To the Memory of Agner K. Erlang: Originator of Teletraffic Theory. *Proceedings of the IEEE* 98, no. 1 (January 2010): 123–7.
57. B. Bakmaz and M. Bakmaz. Solving Some Overflow Traffic Models with Changed Serving Intensities. *International Journal of Electronics and Communications (AEÜ: Archiv fuer Elektronik und Übertragungstechnik)* 66, no. 1 (January 2012): 80–5.

CHAPTER 5 WIRELESS MESH NETWORKS

1. I. F. Akyildiz, X. Wang and W. Wang. Wireless Mesh Networks: A Survey. *Computer Networks* 47, no. 4 (March 2005): 445–87.
2. E. Hossain and K. K. Leung, eds. *Wireless Mesh Networks: Architectures and Protocols*. New York: Springer (2008).
3. I. F. Akyildiz and X. Wang. *Wireless Mesh Networks*. Hoboken, NJ: Wiley (2009).
4. Y. Cheng et al. Wireless Mesh Network Capacity Achievable over the CSMA/CA MAC. *IEEE Transactions on Vehicular Technology* 61, no. 7 (September 2012): 3151–65.
5. N. A. Abu Ali et al. IEEE 802.16 Mesh Schedulers: Issues and Design Challenges. *IEEE Network* 22, no. 1 (January–February 2008): 58–65.
6. M. E. M. Campista et al. Routing Metrics and Protocols for Wireless Mesh Networks. *IEEE Network* 22, no. 1 (January–February 2008): 6–12.
7. H. Hu, Y. Zhang and H.-H. Chen. An Effective QoS Differentiation Scheme for Wireless Mesh Networks. *IEEE Network* 22, no. 1 (January–February 2008): 66–73.
8. K. R. Rao, Z. S. Bojkovic and D. A. Milovanovic. *Wireless Multimedia Communications: Convergence, DSP, QoS and Security*. Boca Raton, FL: CRC Press (2009).
9. D. Benyamina, A. Hafid and M. Gendreau. Wireless Mesh Networks Design—A Survey. *IEEE Communications Surveys & Tutorials* 14, no. 2 (Second Quarter 2012): 299–310.
10. A. Esmailpour, N. Nasser and T. Taleb. Topological-Based Architectures for Wireless Mesh Networks. *IEEE Wireless Communications* 18, no. 1 (February 2011): 74–81.

11. M. Kodialam and T. Nandagopal. Characterizing the Capacity Region in Multi-Radio Multi-Channel Wireless Mesh Networks. *Proc. 11th MobiCom 2005*. Cologne, Germany (August–September 2005): 73–87.

12. M. Jaseemuddin et al. Integrated Routing System for Wireless Mesh Networks. *Proc. CCECE '06*. Ottawa, Canada (May 2006): 1003–7.

13. A. A. Pirzada and M. Portmann. High Performance AODV Routing Protocol for Hybrid Wireless Mesh Networks. *Proc. MobiQuitous 2007*. Philadelphia, PA (August 2007): 1–5.

14. A. Esmailpour et al. Ad-Hoc Path: An Alternative to Backbone for Wireless Mesh Networks. *Proc. IEEE ICC 2007*. Glasgow, UK (June 2007): 3752–7.

15. D. Benyamina, N. Hallam and A. Hafid. On Optimizing the Planning of Multi-hop Wireless Networks using a Multi Objective Evolutionary Approach. *International Journal of Communications* 2, no. 4 (2008): 213–21.

16. B. Alawieh et al. Improving Spatial Reuse in Multihop Wireless Networks— A Survey. *IEEE Communications Surveys & Tutorials* 11, no. 3 (3rd Quarter 2009): 71–91.

17. D.-W. Huang, P. Lin and C.-H. Gan. Design and Performance Study for a Mobility Management Mechanism (WMM) Using Location Cache for Wireless Mesh Networks. *IEEE Transactions on Mobile Computing* 7, no. 5 (May 2008): 546–56.

18. L. Qiu et al. Troubleshooting Wireless Mesh Networks. *ACM SIGCOMM Computer Communication Review* 36, no. 5 (October 2006): 17–28.

19. B. Aoun et al. Gateway Placement Optimization in Wireless Mesh Networks With QoS Constraints. *IEEE Journal on Selected Areas in Communications* 24, no. 11 (November 2006): 2127–36.

20. C.-Y. Hsu et al. Survivable and Delay-Guaranteed Backbone Wireless Mesh Network Design. *Journal of Parallel and Distributed Computing* 63, no. 3 (March 2008): 306–20.

21. M. Mitchell. *An Introduction to Genetic Algorithms*. Cambridge, MA: MIT Press (1998).

22. E. W. Dijkstra. A Note on Two Problems in Connexion with Graphs. *Numerische Mathematik* 1 (1959): 269–71.

23. F. Li, Y. Wang and X.-Y. Li. Gateway Placement for Throughput Optimization in Wireless Mesh Networks. *Proc. IEEE ICC '07*. Glasgow, Scotland (June 2007): 4955–60.

24. Y. Yan et al. CORE: A Coding-Aware Opportunistic Routing Mechanism for Wireless Mesh Networks. *IEEE Wireless Communications* 17, no. 3 (June 2010): 96–103.

25. D. S. J. de Couto. High-Throughput Routing for Multi-Hop Wireless Networks. PhD Thesis. Cambridge, MA: MIT Press (2004).

26. D. Passos et al. Mesh Network Performance Measurements. *Proc. 5th I2TS 2006*. Cuiabá, MT, Brazil (December 2006): 48–55.

27. C. E. Koksal and H. Balakrishnan. Quality-Aware Routing Metrics for Time-Varying Wireless Mesh Networks. *IEEE Journal on Selected Areas in Communications* 24, no. 11 (November 2006): 1984–94.

28. A. P. Subramanian, M. M. Buddhikot and C. Miller. Interference Aware Routing in Multi-Radio Wireless Mesh Networks. *Proc. 2nd IEEE Workshop on Wireless Mesh Networks*. Reston, VA (September 2006): 55–63.

29. P. H. Pathak and R. Dutta. A Survey of Network Design Problems and Joint Design Approaches in Wireless Mesh Networks. *IEEE Communications Surveys & Tutorials* 13, no. 3 (Third Quarter 2011): 396–428.

30. T. Clausen and P. Jacquet. Optimized Link State Routing Protocol (OLSR). *IETF RFC 3626* (October 2003).

31. M. Ikeda et al. A BAT in the Lab: Experimental Results of New Link State Routing Protocol. *Proc. 22nd AINA 2008*. Gino-wan, Japan (March 2008): 295–302.

32. K. Ramachandran et al. On the Design and Implementation of Infrastructure Mesh Networks. *IEEE 1st IEEE WiMesh 2005*. Santa Clara, CA (September 2005).

33. A. Subramanian, M. Buddhikot and S. Miller. Interference Aware Routing in Multi-Radio Wireless Mesh Networks. *2nd IEEE WiMesh 2006*. Reston, VA (September 2006): 55–63.

34. B. Shin, S. Y. Han and D. Lee. Dynamic Link Quality Aware Routing Protocol for Multi-Radio Wireless Mesh Networks. *Proc. 26th AINA 2012*. Fukuoka, Japan (March 2012): 44–50.

35. X. Hu, M. J. Lee and T. N. Saadawi. Progressive Route Calculation Protocol for Wireless Mesh Networks. *Proc. IEEE ICC '07*. Glasgow, Scotland (June 2007): 4973–8.

36. S. Biswas and R. Morris. ExOR: Opportunistic Multi-Hop Routing for Wireless Networks. *Proc. SIGCOMM '05*. Philadelphia, PA (August 2005): 133–44.

37. Y. Yuan et al. ROMER: Resilient Opportunistic Mesh Routing for Wireless Mesh Networks. *IEEE 1st IEEE WiMesh 2005*. Santa Clara, CA (September 2005).

38. E. Rozner et al. Simple Opportunistic Routing Protocol for Wireless Mesh Networks. *2nd IEEE WiMesh 2006*. Reston, VA (September 2006): 48–54.

39. S. Chachulski et al. Trading Structure for Randomness in Wireless Opportunistic Routing. *Proc. SIGCOMM '07*. Kyoto, Japan (August 2007): 169–80.

40. J. J. Gálvez, P. M. Ruiz and A. F. G. Skarmeta. Multipath Routing with Spatial Separation in Wireless Multi-Hop Networks Without Location Information. *Computer Networks* 55, no. 3 (February 2011): 583–99.

41. E. Alotaibi and B. Mukherjee. A Survey on Routing Algorithms for Wireless Ad-Hoc and Mesh Networks. *Computer Networks* 56, no. 2 (February 2012): 940–65.

42. N. S. Nandiraju, D. S. Nandiraju and D. P. Agrawal. Multipath Routing in Wireless Mesh Networks. *Proc. IEEE MASS '06*. Vancouver, Canada (October 2006): 741–6.

43. J. Tsai and T. Moors. A Review of Multipath Routing Protocols: From Wireless Ad Hoc to Mesh Networks. *ACoRN Early Career Researcher Workshop on Wireless Multihop Networking*. Sydney, Australia (July 2006).

44. S. Lee, B. Bhattacharjee and S. Banerjee. Efficient Geographic Routing in Multihop Wireless Networks. *Proc. 6th ACM MobiHoc '05*. Urbana-Champaign, IL (May 2005): 230–41.

45. P. N. Thai and H. Won-Joo. Hierarchical Routing in Wireless Mesh Network. *Proc. 9th International Conference on Advanced Communication Technology*. Phoenix Park, Korea (February 2007): 1275–80.

46. P. Tingrui et al. An Improved Hierarchical AODV Routing Protocol for Hybrid Wireless Mesh Network. *Proc. NSWCTC '09*. Wuhan, Hubei China (April 2009): 588–93.

47. P. Tingrui et al. A Cognitive Improved Hierarchical AODV Routing Protocol for Cognitive Wireless Mesh Network. *Information Technology Journal* 10, no. 2 (2011): 376–84.

48. T. Liu and W. Liao. On Routing in Multichannel Wireless Mesh Networks: Challenges and Solutions. *IEEE Network* 22, no. 1 (January–February 2008): 13–8.

49. R. Draves, J. Padhye and B. Zill. Routing in Multi-Radio, Multi-Hop Wireless Mesh Networks. *Proc. 10th ACM MobiCom 2004*. Philadelphia, PA (September 2004): 114–28.

50. S. Roy et al. High-Throughput Multicast Routing Metrics in Wireless Mesh Networks. *Ad Hoc Networks* 6, no. 6 (August 2008): 878–99.

51. J. Qadir. *Efficient Broadcasting for Multi-Radio Mesh Networks: Improving Broadcast Performance in Multi-Radio Multi-Channel Multi-Rate Wireless Mesh Networks*. Saarbrücken: VDM Verlag Dr. Müller (2009).

52. X. Zhang and H. Su. Network-Coding-Based Scheduling and Routing Schemes for Service-Oriented Wireless Mesh Networks. *IEEE Wireless Communications* 16, no. 4 (August 2009): 40–6.

53. W. Wang, X. Liu and D. Krishnaswamy. Robust Routing and Scheduling in Wireless Mesh Networks under Dynamic Traffic Conditions. *IEEE Transactions on Mobile Computing* 8, no. 12 (December 2009): 1705–17.

54. N. Chakchouk and B. Hamdaoui. Traffic and Interference Aware Scheduling for Multiradio Multichannel Wireless Mesh Networks. *IEEE Transactions on Vehicular Technology* 60, no. 2 (February 2011): 555–65.

55. Y. Bejerano, S.-J. Han and A. Kumar. Efficient Load-Balancing Routing for Wireless Mesh Networks. *Computer Networks* 51, no. 10 (July 2007): 2450–66.

56. J. B. Ernst. Scheduling Techniques in Wireless Mesh Networks. PhD Thesis. University of Guelph, Canada (2009).

57. J. Thomas. Cross-Layer Scheduling and Routing for Unstructured and Quasi-Structured Wireless Networks. *Proc. IEEE MILCOM 2005*. Atlantic City, NJ (October 2005): 1602–8.

58. N. B. Salem and J.-P. Hubaux. A Fair Scheduling for Wireless Mesh Networks. *IEEE 1st IEEE WiMesh 2005*. Santa Clara, CA (September 2005).

59. S. Singh, U. Madhow and E. M. Belding. Beyond Proportional Fairness: A Resource Biasing Framework for Shaping Throughput Profiles in Multihop Wireless Networks. *Proc. 27th IEEE INFOCOM 2008*. Phoenix, AZ (April 2008): 2396–404.

60. D. Chafekar et al. Approximation Algorithms for Computing Capacity of Wireless Networks with SINR Constraints. *Proc. 27th IEEE INFOCOM 2008*. Phoenix, AZ (April 2008): 1166–74.

61. Y. Li et al. Content-Aware Distortion-Fair Video Streaming in Congested Networks. *IEEE Transactions on Multimedia* 11, no. 6 (October 2009): 1182–93.

62. J. C. Fernandez et al. Bandwidth Aggregation-Aware Dynamic QoS Negotiation for Real-Time Video Streaming in Next-Generation Wireless Networks. *IEEE Transactions on Multimedia* 11, no. 6 (October 2009): 1082–93.

63. Z. He, J. Cai and C. W. Chen. Joint Source Channel Rate-Distortion Analysis for Adaptive Mode Selection and Rate Control in Wireless Video Coding. *IEEE Transactions on Circuits and Systems for Video Technology* 12, no. 6 (June 2002): 511–23.

64. Z. He and H. Xiong. Transmission Distortion Analysis for Real-Time Video Encoding and Streaming over Wireless Networks. *IEEE Transactions on Circuits and Systems for Video Technology* 16, no. 9 (September 2006): 1051–62.

65. J. Walrand and P. Varaiya. *High-Performance Communications Networks*, 2nd edition. San Francisco: Morgan Kaufmann Publishers (2000).

66. Y. Tu et al. An Analytical Study of Peer-to-Peer Media Streaming Systems. *ACM Transactions on Multimedia Computing Communications and Applications* 1, no. 4 (November 2005): 354–76.

67. J. Lik, S. Rao and H. Zhung. Opportunities and Challenges of Peer-to-Peer Internet Video Broadcast. *Proceedings of the IEEE* 96, no. 1 (January 2008): 11–24.

68. L. Kleinrock. *Queueing Systems: Volume 1: Theory*. New York: Wiley (1975).

69. H.-P. Shaing and M. van der Schaar. Multi-User Video Streaming over Multi-Hop Wireless Networks: A Distributed Cross-Layer Approach Based on Priority Queuing. *IEEE Journal on Selected Areas in Communications* 25, no. 4 (May 2007): 770–85.

70. E. Setton, X. Zhu and B. Girod. Minimizing Distortion for Multipath Video Streaming over Ad Hoc Networks. *Proc. 2nd IEEE ICIP-04*. Singapore (October 2004): 1751–4.

71. Y. Wang, A. R. Reibman and S. Lin. Multiple Description Coding for Video Delivery. *Proceedings of the IEEE* 93, no. 1 (January 2005): 57–70.

72. S. Mao et al. Multimedia-Centric Routing for Multiple Description Video in Wireless Mesh Networks. *IEEE Network* 22, no. 1 (January–February 2008): 19–24.

73. M. Alasti et al. Multiple Description Coding in Networks with Congestion Problem. *IEEE Transactions on Information Theory* 47, no. 3 (March 2001): 891–902.

74. L. Ozarow. On a Source Coding Problem with Two Channels and Three Receivers. *Bell System Technical Journal* 59, no. 10 (December 1980): 84–91.

CHAPTER 6 WIRELESS MULTIMEDIA SENSOR NETWORKS

1. S. Ehsan and B. Hamdaoui. A Survey on Energy-Efficient Routing Techniques with QoS Assurances for Wireless Multimedia Sensor Networks. *IEEE Communications Surveys & Tutorials* 14, no. 2 (Second Quarter 2012): 265–78.

2. E. Gürses and Ö. B. Akan. Multimedia Communication in Wireless Sensor Networks. *Annales des Télécommunications* 60, no. 7–8 (August 2005): 872–900.

3. N. M. Freris, H. Kowshik and P. R. Kumar. Fundamentals of Large Sensor Networks: Connectivity, Capacity, Clocks, and Computation. *Proceedings of the IEEE* 98, no. 11 (November 2010): 1828–46.

4. P. Naik and K. M. Sivalingam. A Survey of MAC Protocols for Sensor Networks. In *Wireless Sensor Networks*, C. S. Raghavendra, K. M. Sivalingam and T. Znati, eds. Norwell: Kluwer Academic Publishers (2004): 93–107.

5. J. O. Kephart and D. M. Chess. The Vision of Autonomic Computing. *Computer* 36, no. 1 (Januray 2003): 41–50.

6. J. Hao, B. Zhang and H. T. Mouftah. Routing Protocols for Duty Cycled Wireless Sensor Networks: A Survey. *IEEE Communications Magazine* 50, no. 12 (December 2012): 116–23.

7. O. B. Akan, M. T. Isik and B. Baykal. Wireless Passive Sensor Networks. *IEEE Communications Magazine* 47, no. 8 (August 2009): 92–9.

8. I. F. Akyildiz, T. Melodia and K. R. Chowdhury. A Survey on Wireless Multimedia Sensor Networks. *Computer Networks* 51, no. 4 (March 2007): 921–60.

9. G. A. Shah, W. Liang and O. B. Akan. Cross-Layer Framework for QoS Support in Wireless Multimedia Sensor Networks. *IEEE Transactions on Multimedia* 14, no. 5 (October 2012): 1442–55.

10. J. Yick, B. Mukherjee and D. Ghosal. Wireless Sensor Network Survey. *Computer Networks* 52, no. 12 (August 2008): 2292–330.

11. Z. Bojkovic and B. Bakmaz. A Survey on Wireless Sensor Networks Deployment. *WSEAS Transactions on Communications* 7, no. 12 (December 2008): 1172–81.

12. I. F. Akyildiz, T. Melodia and K. R. Chowdhury. Wireless Multimedia Sensor Networks: A Survey. *IEEE Wireless Communications* 14, no. 6 (December 2007): 32–9.

13. I. F. Akyildiz and M. C. Vuran. *Wireless Sensor Networks*. Hoboken, NJ: Wiley (2010).

14. I. T. Almalkawi et al. Wireless Multimedia Sensor Networks: Current Trends and Future Directions. *Sensors* 10, no. 7 (July 2010): 6662–717.

15. C. E. R. Lopes et al. A Multi-tier, Multimodal Wireless Sensor Network for Environmental Monitoring. *Proc. 4th International Conference UIC 2007*. Hong Kong, China (July 2007): 589–98.

16. P. Baronti et al. Wireless Sensor Networks: A Survey on the State of the Art and the 802.15.4 and ZigBee Standards. *Computer Communications* 30, no. 7 (May 2007): 1655–95.

17. A. Kerhet et al. A Low-Power Wireless Video Sensor Node for Distributed Object Detection. *Journal of Real-Time Image Processing* 2, no. 4 (December 2007): 331–42.

18. I. F. Akyildiz, T. Melodia and K. R. Chowdhury. Wireless Multimedia Sensor Networks: Applications and Testbeds. *Proceedings of the IEEE* 96, no. 10 (October 2008): 1–18.

19. T. Melodia and I. F. Akyildiz. Cross-Layer Quality of Service Support for UWB Wireless Multimedia Sensor Networks. *Proc. 27th IEEE INFOCOM 2008*. Phoenix, AZ (April 2008): 2038–46.

20. M. A. Yigitel, O. D. Incel and C. Ersoy. Design and Implementation of a QoS-aware MAC Protocol for Wireless Multimedia Sensor Networks. *Computer Communications* 34, no. 16 (October 2011): 1991–2001.

21. W. Ye, J. Heidemann and D. Estrin. An Energy-Efficient MAC Protocol for Wireless Sensor Networks. *Proc. 21st IEEE INFOCOM 2002*. New York (June 2002): 1567–76.

22. T. van Dam and K. Langendoen. An Adaptive Energy-Efficient MAC Protocol for Wireless Sensor Networks. *Proc. ACM SenSys '03*. Los Angeles (November 2003): 171–80.

23. C. Li et al. A Cluster based On-demand Multi-Channel MAC Protocol for Wireless Multimedia Sensor Networks. *Proc. IEEE ICC '08*. Beijing, China (May 2008): 2371–6.

24. M. Yaghmaee and D. Adjeroh. A Model for Differentiated Service Support in Wireless Multimedia Sensor Networks. *Proc. 17th ICCCN '08*. St. Thomas, Virgin Islands (August 2008): 1–6.

25. S. Floyd and V. Jacobson. Random Early Detection Gateways for Congestion Avoidance. *IEEE/ACM Transactions on Networking* 1, no. 4 (August 1993): 397–413.

26. A. Silberschatz, P. B. Galvin and G. Gagne. *Operating System Concepts*, 8th edition. Hoboken, NJ: Wiley (2008).

27. M. C. Vuran and I. F. Akyildiz. Cross-Layer Analysis of Error Control in Wireless Sensor Networks. *Proc. 3rd IEEE SECON '06*. Reston, VA (September 2006): 585–94.

28. M. Y. Naderi et al. Error Control for Multimedia Communications in Wireless Sensor Networks: A Comparative Performance Analysis. *Ad Hoc Networks* 10, no. 6 (August 2012): 1028–42.

29. M. Mushkin and I. Bar-David. Capacity and Coding for the Gilbert–Elliot Channels, *IEEE Transactions on Information Theory* 35, no. 6 (November 1989): 1277–90.

30. N. A. Pantazis, S. A. Nikolidakis and D. D. Vergados. Energy-Efficient Routing Protocols in Wireless Sensor Networks: A Survey. *IEEE Communications Surveys & Tutorials* 15, no. 2 (Second Quarter 2013): 551–91.

31. K. Sohrabi et al. Protocols for Self Organization of a Wireless Sensor Network. *IEEE Personal Communications* 7, no. 5 (October 2000): 16–27.

32. K. Akkaya and M. Younis. An Energy-Aware QoS Routing Protocol for Wireless Sensor Networks. *Proc. ICDCS '03*. Providence, RI (May 2003): 710–5.

33. T. He et al. SPEED: A Stateless Protocol for Real-Time Communication in Sensor Networks. *Proc. ICDCS '03*. Providence, RI (May 2003): 46–55.

34. S. C. Ergen and P. Varaiya. Energy Efficient Routing with Delay Guarantee for Sensor Networks. *Wireless Networks* 13, no. 5 (October 2007): 679–90.

35. E. Felemban, C. Lee and E. Ekici. MMSPEED: Multipath multi-SPEED protocol for QoS guarantee of reliability and timeliness in wireless sensor networks. *IEEE Transactions on Mobile Computing* 5, no. 6 (June 2006): 738–54.

36. M. A. Hamid, M. M. Alam and C. S. Hong. Design of a QoS-Aware Routing Mechanism for Wireless Multimedia Sensor Networks. *Proc. IEEE GLOBECOM 2008*. New Orleans, LA (November–December 2008): 1–6.

37. A. A. Ahmed and N. Fisal. A Real-Time Routing Protocol with Load Distribution in Wireless Sensor Networks. *Computer Communications* 31, no. 14 (September 2008): 3190–203.

38. A. A. Ahmed. An Enhanced Real-Time Routing Protocol with Load Distribution for Mobile Wireless Sensor Networks. *Computer Networks* 57, no. 6 (April 2013): 1459–73.

39. M. Chen et al. Directional Geographical Routing for Real-Time Video Communications in Wireless Sensor Networks. *Computer Communications* 30, no. 17 (November 2007): 3368–83.

40. Y. Lan, W. Wenjing and G. Fuxiang. A Real-Time and Energy Aware QoS Routing Protocol for Multimedia Wireless Sensor Networks. *Proc. 7th IEEE WCICA '08*. Chongqing, China (June 2008): 3321–6.

41. A. M. Zungeru, L.-M. Ang and K. P. Seng. Classical and Swarm Intelligence based Routing Protocols for Wireless Sensor Networks: A Survey and Comparison. *Journal of Network and Computer Applications* 35, no. 5 (September 2012): 1508–36.

42. Y. Sun et al. ASAR: An Ant-based Service Aware Routing Algorithm for Multimedia Sensor Networks. *Frontiers of Electrical and Electronic Engineering in China* 3, no. 1 (January 2008): 25–33.

43. L. Cobo, A. Quintero and S. Pierre. Ant-based Routing for Wireless Multimedia Sensor Networks Using Multiple QoS Metrics. *Computer Networks* 54, no. 17 (December 2010): 2991–3010.

44. S. Medjiah, T. T. Ahmed and F. Krief. GEAMS: A Geographic Energy-Aware Multipath Stream-based routing protocol for WMSNs. *Proc. IEEE GIIS 2009*. Hammamet, Tunisia (June 2009): 1–8.

45. B. Karp and H. T. Kung. GPSR: Greedy Perimeter Stateless Routing for Wireless Networks. *Proc. 6th ACM/IEEE MobiCom '00*. Boston (August 2000): 243–54.

46. T. S. Rappaport. *Wireless Communications: Principles & Practices*. Upper Saddle River, NJ: Prentice Hall (1996).

47. C. Wang et al. A Survey of Transport Protocols for Wireless Sensor Networks. *IEEE Network* 20, no. 3 (May–June 2006): 34–40.

48. M. Maimour, C. Pham and J. Amelot. Load Repartition for Congestion Control in Multimedia Wireless Sensor Networks with Multipath Routing. *Proc. 3rd ISWPC 2008*. Santorini, Greece (May 2008): 11–5.

49. M. Yaghmaee and D. Adjeroh. A New Priority based Congestion Control Protocol for Wireless Multimedia Sensor Networks. *Proc. WoWMoM 2008*. Newport Beach, CA (June 2008): 1–8.

50. M. O. Farooq, T. Kunz and M. St-Hilaire. Differentiated Services based Congestion Control Algorithm for Wireless Multimedia Sensor Networks. *Proc. IFIP Wireless Days 2011 IFIP*. Niagara Falls, ON, Canada (October 2011): 1–6.

51. S. Misra, M. Reisslein and G. Xue. A Survey of Multimedia Streaming in Wireless Sensor Networks. *IEEE Communications Surveys & Tutorials* 10, no. 4 (Fourth Quarter 2008): 18–39.

52. K. R. Rao, Z. S. Bojkovic and D. A. Milovanovic. *Multimedia Communication Systems: Technologies, Standards and Networks*. Upper Saddle River, NJ: Prentice Hall (2002).

53. S. Kwon, A. Tamhankar and K. R. Rao. Overview of H.264/MPEG-4 part 10. *Journal of Visual Communication & Image Representation* 17, no. 2 (April 2006): 186–216.

54. Z. Milicevic and Z. Bojkovic. H.264/AVC Standard: A Proposal for Selective Intra- and Optimized Inter-prediction. *Journal of Network and Computer Applications* 34, no. 2 (March 2011): 686–91.

55. B. Girod et al. Distributed Video Coding. *Proceedings of the IEEE* 93, no. 1 (January 2005): 71–83.

56. D. Slepian and J. Wolf. Noiseless Coding of Correlated Information Sources. *IEEE Transactions on Information Theory* 19, no. 4 (July 1973): 471–80.

57. A. Wyner and J. Ziv. The Rate-Distortion Function for Source Coding with Side Information at the Decoder. *IEEE Transactions on Information Theory* 22, no. 1 (January 1976): 1–10.

58. C. Yaacoub, J. Farah and B. Pesquet-Popescu. Joint Source-Channel Wyner-Ziv Coding in Wireless Video Sensor Networks. *Proc. 7th IEEE ISSPIT*. Cairo, Egypt (December 2007): 225–8.

59. Z. Xue et al. Distributed Video Coding in Wireless Multimedia Sensor Network for Multimedia Broadcasting. *WSEAS Transactions on Communications* 7, no. 5 (May 2008): 418–27.

60. J. J. Ahmad, H. A. Khan and S. A. Khayam. Energy Efficient Video Compression for Wireless Sensor Networks. *Proc. 43rd IEEE CISS 2009*. Baltimore (March 2009): 629–34.

61. R. Halloush, K. Misra and H. Radha. Practical Distributed Video Coding over Visual Sensors. *Proc. 27th IEEE PCS '09*. Piscataway, NJ (May 2009): 121–4.

62. F. Hu and S. Kumar. The Integration of Ad Hoc Sensor and Cellular Networks for Multi-Class Data Transmission. *Ad Hoc Networks* 4, no. 2 (March 2006): 254–82.

63. J. Zhang et al. Mobile Cellular Networks and Wireless Sensor Networks: Toward Convergence. *IEEE Communications Magazine* 50, no. 3 (March 2012): 164–9.

64. R. Rajbanshi, A. M. Wyglinski and G. J. Minden. An Efficient Implementation of the NC-OFDM Transceivers for Cognitive Radios. *Proc. 1st CROWNCOM*. Mykonos, Greece (June 2006): 1–5.

65. S. H. Lee et al. Wireless Sensor Network Design for Tactical Military Applications: Remote Large-Scale Environments. *Proc. IEEE MILCOM 2009*. Boston (October 2009): 911–7.

66. U. Lee et al. Dissemination and Harvesting of Urban Data Using Vehicular Sensing Platforms. *IEEE Transactions on Vehicular Technology* 58, no. 2 (February 2009): 882–901.

67. S. F. Midkiff. Internet-Scale Sensor Systems: Design and Policy. *IEEE Pervasive Computing* 2, no. 4 (October–December 2003): 10–3.

68. J. Ko et al. Wireless Sensor Networks for Healthcare. *Proceedings of the IEEE* 98, no. 11 (November 2010): 1947–60.

69. S.-L. Kim, W. Burgard and D. Kim. Wireless Communications in Networked Robotics [Guest Editorial]. *IEEE Wireless Communications* 16, no. 1 (February 2009): 4–5.

70. X. Li et al. Servicing Wireless Sensor Networks by Mobile Robots. *IEEE Communications Magazine* 50, no. 7 (July 2012): 147–54.
71. R. Falcon, X. Li and A. Nayak. Carrier-based Focused Coverage Formation in Wireless Sensor and Robot Networks. *IEEE Transactions on Automatic Control* 56, no. 10 (October 2011): 2406–17.

CHAPTER 7 SECURITY IN WIRELESS MULTIMEDIA COMMUNICATIONS

1. N. Doraswamy and D. Harkins. *IPsec: The New Security Standard for the Internet, Intranets and Virtual Private Networks.* Upper Saddle River, NJ: Prentice Hall (1999).
2. S. Parvin et al. Cognitive Radio Network Security: A Survey. *Journal of Network and Computer Applications* 35, no. 6 (November 2012): 1691–708.
3. A. G. Fragkiadakis, E. Z. Tragos and I. G. Askoxylakis. A Survey on Security Threats and Detection Techniques in Cognitive Radio Networks. *IEEE Communications Surveys & Tutorials* 15, no. 1 (First Quarter 2013): 428–45.
4. O. E. Muogilim, K.-K. Loo and R. Comley. Wireless Mesh Network Security: A Traffic Engineering Management Approach. *Journal of Network and Computer Applications* 34, no. 2 (March 2011): 478–91.
5. S. Glass, M. Portmann and V. Muthukkumarasamy. Securing Wireless Mesh Networks. *IEEE Internet Computing* 12, no. 4 (July–August 2008): 30–6.
6. J. Dong, R. Curtmola and C. Nita-Rotaru. Secure High-Throughput Multicast Routing in Wireless Mesh Networks. *IEEE Transactions on Mobile Computing* 10, no. 5 (May 2011): 653–68.
7. M. Guerrero-Zapata et al. The Future of Security in Wireless Multimedia Sensor Networks. *Telecommunication Systems* 45, no. 1 (September 2010): 77–91.
8. K. R. Rao, Z. S. Bojkovic and D. A. Milovanovic. *Introduction to Multimedia Communications: Applications, Middleware, Networking.* Hoboken, NJ: Wiley (2006).
9. D. Ma and G. Tsudik. Security and Privacy in Emerging Wireless Networks. *IEEE Wireless Communications* 17, no. 5 (October 2010): 12–21.
10. B. Bakmaz, M. Bakmaz and Z. Bojkovic. Elements of Security Aspects in Wireless Networks: Analysis and Integration. *International Journal of Applied Mathematics and Informatics* 1, no. 2 (2007): 70–5.
11. P. Sakarindr and N. Ansari. Security Services in Group Communications over Wireless Infrastructure, Mobile Ad Hoc, and Wireless Sensor Networks. *IEEE Wireless Communications* 14, no. 5 (October 2007): 8–20.
12. Y.-S. Shiu et al. Physical Layer Security in Wireless Networks: A Tutorial. *IEEE Wireless Communications* 18, no. 2 (April 2011): 66–74.
13. M. V. Bharathi et al. Node Capture Attack in Wireless Sensor Network: A Survey. *Proc. IEEE ICCIC 2012.* Tamilnadu, India (December 2012): 340–2.
14. D. R. Raymond et al. Effects of Denial-of-Sleep Attacks on Wireless Sensor Network MAC Protocols. *IEEE Transactions on Vehicular Technology* 58, no. 1 (January 2009): 367–80.

15. S. T. Zargar, J. Joshi and D. Tipper. A Survey of Defense Mechanisms against Distributed Denial of Service (DDoS) Flooding Attacks. *IEEE Communications Surveys & Tutorials.* 15, no. 4 (Fourth Quarter 2013): 2046–69.

16. C. Garrigues et al. Protecting Mobile Agents from External Replay Attacks. *The Journal of Systems and Software* 82, no. 2 (February 2009): 197–206.

17. Y.-C. Hu, A. Perrig and D. B. Johnson. Wormhole Attacks in Wireless Networks. *IEEE Journal on Selected Areas in Communications* 24, no. 2 (February 2006): 370–80.

18. C. S. R. Murthy and B. S. Manoj. *Ad Hoc Wireless Networks: Architectures and Protocols.* Upper Saddle River, NJ: Prentice Hall (2004).

19. J. Barros and M. R. D. Rodrigues. Secrecy Capacity of Wireless Channels. *Proc. IEEE ISIT 2006.* Seattle, WA (July 2006): 356–60.

20. C. Sperandio and P. G. Flikkema. Wireless Physical-Layer Security via Transmit Precoding over Dispersive Channels: Optimum Linear Eavesdropping. *Proc. IEEE MILCOM 2002.* Anaheim, CA (October 2002): 1113–7.

21. X. Li and E. P. Ratazzi. MIMO Transmissions with Information-Theoretic Secrecy for Secret-Key Agreement in Wireless Networks. *IEEE MILCOM 2005.* Atlantic City, NJ (October 2005): 1353–9.

22. A. A. Tomko, C. J. Rieser and L. H. Buell. Physical-Layer Intrusion Detection in Wireless Networks. *Proc. IEEE MILCOM 2006.* Washington, DC (October 2006): 1040–6.

23. S. Goel and R. Negi. Guaranteeing Secrecy using Artificial Noise. *IEEE Transactions on Wireless Communications* 7, no. 6 (June 2008): 2180–9.

24. B. Bakmaz, M. Bakmaz and Z. Bojkovic. Security Aspects in Future Mobile Networks. *Proc. IWSSIP 2008.* Bratislava, Slovak Republic (June 2008): 479–82.

25. K. R. Rao, Z. S. Bojkovic and D. A. Milovanovic. *Wireless Multimedia Communications: Convergence, QoS and Security.* Boca Raton, FL: CRC Press (2009).

26. 3GPP TS 33.401 V12.5.0. Technical Specification Group Services and System Aspects; 3GPP System Architecture Evolution (SAE); Security architecture (Release 12) (September 2012).

27. J. Cao et al. A Survey on Security Aspects for LTE and LTE-A Networks. *IEEE Communications Surveys & Tutorials.* 16, no. 1 (First Quarter 2014): 283–302.

28. L. Xiehua and W. Yongjun Security Enhanced Authentication and Key Agreement Protocol for LTE/SAE Network. *Proc. 7th WiCOM.* Wuhan, China (September 2011): 1–4.

29. A. A. Al Shidhani and V. C. M. Leung. Fast and Secure Reauthentications for 3GPP Subscribers during WiMAX-WLAN Handovers. *IEEE Transactions on Dependable and Secure Computing* 8, no. 5 (September/October 2011): 699–713.

30. B. Bakmaz, Z. Bojkovic and M. Bakmaz. Internet Protocol Multimedia Subsystem for Mobile Services. *Proc. IWSSIP & EC-SIPMCS 2007.* Maribor, Slovenia (June 2007): 353–6.

31. 3GPP TS 33.203 V12.2.0. Technical Specification Group Services and System Aspects; 3G security; Access security for IP-based services (Release 12) (June 2013).

32. 3GPP TS 33.320 V11.6.0. 3rd Generation Partnership Project; Technical Specification Group Services and System Aspects; Security of Home Node B (HNB)/Home evolved Node B (HeNB) (Release 11) (June 2012).

33. A. Golaup, M. Mustapha and L. B. Patanapongpibul. Femtocell Access Control Strategy in UMTS and LTE. *IEEE Communications Magazine* 47, no. 9 (September 2009): 117–23.

34. C. K. Han, H. K. Choi and I. H. Kim. Building Femtocell More Secure with Improved Proxy Signature. *Proc. IEEE GLOBECOM 2009*. Honolulu, HI (December 2009): 6299–304.

35. I. Bilogrevic, M. Jadliwala and J.-P. Hubaux. Security and Privacy in Next Generation Mobile Networks: LTE and Femtocells. *2nd Femtocell Workshop*. Luton, UK (June 2010).

36. 3GPP TS 22.368 V12.0.0. 3rd Generation Partnership Project; Technical Specification Group Services and System Aspects; Service Requirements for Machine-Type Communications (MTC) (Release 12) (September 2012).

37. J. Cao, M. Ma and H. Li. A Group-based Authentication and Key Agreement for MTC in LTE Networks. *Proc. IEEE GLOBECOM 2012*. Anaheim, CA (December 2012): 1017–22.

38. R. Chen et al. Toward Secure Distributed Spectrum Sensing in Cognitive Radio Networks. *IEEE Communications Magazine* 46, no. 4 (April 2008): 50–5.

39. X. Tan et al. Cryptographic Link Signatures for Spectrum Usage Authentication in Cognitive Radio. *Proc. 4th AMC WiSec '11*. Hamburg, Germany (June 2011): 79–90.

40. I. F. Akyildiz et al. A Survey on Spectrum Management in Cognitive Radio Networks. *IEEE Communications Magazine* 46, no. 4 (April 2008): 40–8.

41. T. C. Clancy and N. Goergen. Security in Cognitive Radio Networks: Threats and Mitigation. *Proc. 3rd CrownCom 2008*. Singapore (May 2008): 1–8.

42. C. N. Mathur and K. P. Subbalakshmi. Security Issues in Cognitive Radio Networks. In *Cognitive Networks: Towards Self-Aware Networks*, Q. Mahmoud, ed. Hoboken, NJ: Wiley (2007).

43. G. Jakimoski and K. P. Subbalakshmi. Towards Secure Spectrum Decision. *Proc. IEEE ICC '09*. Dresden, Germany (June 2009): 2759–63.

44. Y. Zhang, J. Zheng and H. Hu, eds. *Security in Wireless Mesh Networks*. Boca Raton, FL: CRC Press (2008).

45. C.-I. Fan, Y.-H. Lin and R.-H. Hsu. Complete EAP Method: User Efficient and Forward Secure Authentication Protocol for IEEE 802.11 Wireless LANs. *IEEE Transactions on Parallel and Distributed Systems* 24, no. 4 (April 2013): 672–80.

46. L. Lazos and M. Krunz. Selective Jamming/Dropping Insider Attacks in Wireless Mesh Networks. *IEEE Network* 25, no. 1 (January–February 2011): 30–4.

47. J. So and N. H. Vaidya. Multi-Channel MAC for Ad Hoc Networks: Handling Multi-Channel Hidden Terminals Using a Single Transceiver. *Proc. 5th ACM MobiHoc '04*. Tokyo, Japan (May 2004): 222–33.

48. S. Misra, M. Reisslein and G. Xue. A Survey of Multimedia Streaming in Wireless Sensor Networks. *IEEE Communications Surveys & Tutorials* 10, no. 4 (Fourth Quarter 2008): 18–39.

49. L. Tawalbeh, M. Mowafi and W. Aljoby. Use of Elliptic Curve Cryptography for Multimedia Encryption. *IET Information Security* 7, no. 2 (June 2013): 67–74.

50. J. Deng, R. Han and S. Mishra. Decorrelating Wireless Sensor Network Traffic to Inhibit Traffic Analysis Attacks. *Pervasive and Mobile Computing* 2, no. 2 (April 2006): 159–86.

CHAPTER 8 WIRELESS COMMUNICATIONS SYSTEMS IN THE SMART GRID

1. W. A. Wulf. Great Achievements and Grand Challenges. *The Bridge* 30, no. 3 & 4 (Fall/Winter 2000): 5–10.

2. H. Gharavi and R. Ghafurian. Smart Grid: The Electric Energy System of the Future. *Proceedings of the IEEE* 99, no. 6 (June 2011): 917–21.

3. G. W. Arnold. Challenges and Opportunities in Smart Grid: A position Article. *Proceedings of the IEEE* 99, no. 6 (June 2011): 922–7.

4. NIST Special Publication 1108. Framework and Roadmap for Smart Grid Interoperability Standards, Release 1.0 (January 2010).

5. Clean Energy Ministerial. Fact Sheet: International Smart Grid Action Network. Available at http://www.cleanenergyministerial.org, accessed 2011.

6. Final Report of the CEN/CENELEC/ETSI Joint Working Group on Standards for Smart Grids. Available at http://www.cen.eu (2011).

7. Available at http://www.smart-japan.org (2010).

8. Korea Smart Grid Institute. *Korea's Smart Grid Roadmap 2030.* Available at http://www.smartgrid.or.kr (2010).

9. State Grid Corporation of China. Framework and Roadmap for Strong and Smart Grid Standards. Available at http://www.sgstandard.org, accessed 2010.

10. M. Shargal and D. Houseman. Why Your Smart Grid Must Start with Communications. Available at http://www.smartgridnews.com, accessed February 2009.

11. V. Litovski and P. Petkovic. Why the Grid Needs Cryptography. *Electronics* 13, no. 1 (June 2009): 30–6.

12. Electric Power Research Institute. IntelliGrid[SM]: Smart Power for the 21st Century. Available at http://intelligrid.epri.com, accessed 2005.

13. Z. Bojkovic and B. Bakmaz. Smart Grid Communications Architecture: A Survey and Challenges. *Proc. 11th WSEAS ACACOS '12.* Rovaniemi, Finland (April 2012): 83–9.

14. A. Metke and R. Ekl. Security Technology for Smart Grid Networks. *IEEE Transactions on Smart Grid* 1, no. 1 (June 2010): 99–107.

15. R. Yu et al. Cognitive Radio Based Hierarchical Communications Infrastructure for Smart Grid. *IEEE Network* 25, no. 5 (September/October 2011): 6–14.

16. H. Wang, Y. Qian and H. Sharif. Multimedia Communications over Cognitive Radio Networks for Smart Grid Applications. *IEEE Wireless Communications* 20, no. 4 (August 2013): 125–32.

17. I. Koutsopoulos and L. Tassiulas. Challenges in Demand Load Control for the Smart Grid. *IEEE Network* 25, no. 5 (September/October 2011): 16–21.

18. D.-M. Han and J.-H. Lim. Design and Implementation of Smart Home Energy Management Systems based on ZigBee. *IEEE Transactions on Consumer Electronics* 56, no. 3 (August 2010): 1417–25.

19. H. Gharavi and B. Hu. Multigate Communication Network for Smart Grid. *Proceedings of the IEEE* 99, no. 6 (June 2011): 1028–45.

20. H. Gharavi and K. Ban. Multihop Sensor Network Design for Wide-band Communications. *Proceedings of the IEEE* 91, no. 8 (August 2003): 1221–34.

21. D. B. Johanson, D. A. Maltz and Y.-C. Hu. The Dynamic Source Routing Protocol (DSR) for Mobile Ad Hoc Networks for IPv4. *IETF RFC 4728* (February 2007).

22. C. Perkins, E. Belding and S. Das. Ad Hoc On-Demand Distance Vector (AODV) Routing. *IETF RFC 3561* (July 2003).

23. ITU-T. Ubiquitous Sensor Networks (USN). *ITU-T Technology Watch Briefing Report Service* no. 4 (February 2008).

24. ITU-T Rec. Y.2234. Open Service Environment Capabilities for NGN (September 2008).

25. A. Zaballos, A. Vallejo and J. M. Selga. Heterogeneous Communication Architecture for the Smart Grid. *IEEE Network* 25, no. 5 (September/October 2011): 30–7.

26. J.-S. Lee. Performance Evaluation of IEEE 802.15.4 for Low-Rate Wireless Personal Area Networks. *IEEE Transactions on Consumer Electronics* 52, no. 3 (August 2006): 742–9.

27. C. R. Stevenson et al. IEEE 802.22: The First Cognitive Radio Wireless Regional Area Network Standard. *IEEE Communications Magazine* 47, no. 1 (January 2009): 130–8.

28. N. Hatziargyriou et al. Microgrids: An Overview of Ongoing Research, Development and Demonstration Project. *IEEE Power and Energy Magazine* 5, no. 4 (July/August 2007): 78–94.

29. M. Erol-Kantarci, B. Kantarci and H. T. Mouftah. Reliable Overlay Topology Design for the Smart Microgrid Network. *IEEE Network* 25, no. 5 (September/October 2011): 38–43.

30. A. Perrig, J. Stankovic and D. Wagner. Security in Wireless Sensor Networks. *Communications of ACM* 47, no. 6 (June 2004): 53–7.

31. Z. Bojkovic, B. Bakmaz and M. Bakmaz. Security Issues in Wireless Sensor Networks. *International Journal of Communications* 2, no. 1 (December 2008): 106–15.

32. A. Tsikalakis and N. Hatziargyriou. Centralized Control for Optimizing Microgrids Operation. *IEEE Transactions on Energy Conversion* 25, no. 1 (March 2008): 241–8.

33. A. Jwayemi et al. Knowing When to Act: An Optimal Stopping Method for Smart Grid Demand Response. *IEEE Network* 25, no. 5 (September/October 2011): 44–9.

34. Z. M. Fadlullah et al. An Early Warning System against Malicious Activities for Smart Grid Communications. *IEEE Network* 25, no. 5 (September/October 2011): 50–5.

35. R. C. Qiu et al. Cognitive Radio Network for the Smart Grid: Experimental System Architecture, Control Algorithms, Security, and Microgrid Testbed. *IEEE Transactions on Smart Grid* 2, no. 4 (December 2011): 724–40.

CHAPTER 9 EVOLUTION OF EMBEDDED INTERNET

1. Available at http://www.autoidlabs.org (2013).

2. A. Gluhak et al. A Survey on Facilities for Experimental Internet of Things Research. *IEEE Communications Magazine* 49, no. 11 (November 2011): 58–67.

3. G. Marrocco. Pervasive Electromagnetics: Sensing Paradigms by Passive RFID Technology. *IEEE Wireless Communications* 17, no. 6 (December 2010): 10–7.

4. D. Pfisterer et al. SPITFIRE: Toward a Semantic Web of Things. *IEEE Communications Magazine* 49, no. 11 (November 2011): 40–8.

5. S. Hong et al. SNAIL: An IP-based Wireless Sensor Network Approach to the Internet of Things. *IEEE Wireless Communications* 17, no. 6 (December 2010): 34–42.

6. M. Zorzi et al. From Today's INTRAnet of Things to a Future INTERnet of Things: A Wireless and Mobility-related View. *IEEE Wireless Communications* 17, no. 6 (December 2010): 44–51.

7. R. K. Ganti, F. Ye and H. Lei. Mobile Crowdsensing: Current State and Future Challenges. *IEEE Communications Magazine* 49, no. 11 (November 2011): 32–9.

8. N. D. Lane. A Survey of Mobile Phone Sensing. *IEEE Communications Magazine* 48, no. 9 (September 2010): 140–50.

9. 3GPP. Systems Improvements for Machine-Type Communications. TR 23888, V.1.3.0 (June 2011).

10. 3GPP. General Packet Radio Service (GPRS) enhancements for Evolved Universal Terrestrial Radio Access Network (E-UTRAN) Access. TS 23.401, V.10.4.0 (June 2011).

11. F. Scioscia and M. Ruta. Building a Semantic Web of Things: Issues and Perspectives in Information Compression. *Proc. IEEE ICSC '09*. Berkeley, CA (September 2009): 589–94.

12. Z. Shelby. Embedded Web Services. *IEEE Wireless Communications* 17, no. 6 (December 2010): 52–7.

13. I. F. Akyildiz and J. M. Jornet. The Internet of Nano-Things. *IEEE Wireless Communications* 17, no. 6 (December 2010): 58–63.

14. I. F. Akyildiz and J. M. Jornet. Electromagnetic Wireless Nanosensor Networks. *Nano Communication Networks* 1, no. 1 (March 2010): 3–19.

15. I. F. Akyildiz, F. Brunetti and C. Blazquez. Nanonetworks: A New Communication Paradigm. *Computer Networks* 52, no. 12 (August 2008): 2260–79.

16. B. Sterling. *Shaping Things*. Cambridge, MA: MIT Press (2006).

17. A. Poolsawat, W. Pattara-Atikom and B. Ngamwongwattana. Acquiring Road Traffic Information through Mobile Phones. *Proc. 8th ITST 2008*. Phuket, Thailand (October 2008): 170–4.

18. P. Mohan, V. Padmanabhan and R. Ramjee. Nericell: Rich Monitoring of Road and Traffic Conditions Using Mobile Smartphones. *Proc. ACM SenSys*. Raleigh, NC (November 2008): 323–36.

19. I. Constandache et al. EnLoc: Energy-Efficient Localization for Mobile Phones. *Proc. IEEE INFOCOM 2009*. Rio de Janeiro, Brazil (April 2009): 2716–20.

20. R. Roman, P. Najera and J. Lopez. Securing the Internet of Things. *IEEE Computer* 44, no. 9 (September 2011): 51–8.

21. J. Krumm. A Survey of Computational Location Privacy. *Personal and Ubiquitous Computing* 13, no. 6 (August 2009): 391–9.

22. Available at http://tools.ietf.org/wg/6lowpan (2013).

23. Available at http://tools.ietf.org/wg/roll (2013).

24. Available at http://tools.ietf.org/wg/core (2013).

25. Z. Shelby and C. Bormann. *6LoWPAN: The Wireless Embedded Internet*. Chichester, UK: John Wiley and Sons, Ltd. (2009).

26. B. Bakmaz and Z. Bojkovic. Internet Protocol version 6 as Backbone of Heterogeneous Networks. *Proc. IWSSIP 2005*. Chalkida, Greece (September 2005): 255–9.

27. G. Montenegro et al. Transmission of IPv6 Packets over IEEE 802.15.4 Networks. *IETF RFC 4944* (September 2007).

28. N. Kushalnagar, G. Montenegro and C. Schumacher. IPv6 over Low-Power Wireless Personal Area Networks (6LoWPANs): Overview, Assumptions, Problem Statement, and Goals. *IETF RFC 4919* (August 2009).

29. T. Winter et al. RPL: IPv6 Routing Protocol for Low-Power and Lossy Networks. *IETF RFC 6550* (March 2012).

30. K. Jeonggil et al. Connecting Low-Power and Lossy Networks to the Internet. *IEEE Communications Magazine* 49, no. 4 (April 2011): 96–101.

31. C. Bormann, J. P. Vasseur and Z. Shelby. The Internet of Things. *IETF Journal* 6, no. 2 (October 2010): 1, 4–5.

32. Z. Shelby et al. Constrained Application Protocol (CoAP). *IETF draft-ietf-core-coap-09* (March 2012).

33. T. Berners-Lee, J. Hendler and O. Lassila. The Semantic Web. *Scientific American* 284, no. 5 (May 2001): 34–43.

34. W. Colitti, K. Steenhaut and N. De Caro. Integrating Wireless Sensor Networks with Web Applications. *Extending the Internet to Low Power and Lossy Networks (IP+SN 2011)*. Chicago (April 2011).

35. Available at http://www.tinyos.net (2013).

36. Available at http://www.contiki-os.org (2013).

37. R. Roman et al. Key Management Systems for Sensor Networks in the Context of the Internet of Things. *Computer & Electrical Engineering* 37, no. 2 (March 2011): 147–59.

38. M. Ha et al. Inter-MARIO: A Fast and Seamless Mobility Protocol to Support Inter-PAN Handover in 6LoWPAN. *IEEE GLOBECOM*. Miami, FL (December 2010).

39. C. Perkins, D. Johnson and J. Arkko. Mobility Support in IPv6. *IETF RFC 6275* (July 2011).

40. D. Mills. Network Time Protocol (Version 3) Specification, Implementation and Analysis. *IETF RFC 1305* (March 1992).

41. D. Mills. Simple Network Time Protocol (SNTP) Version 4 for IPv4, IPv6 and OSI. *IETF RFC 4330* (January 2006).

42. F. Sivrikaya and B. Yener. Time Synchronization in Sensor Networks: A Survey. *IEEE Network* 18, no. 4 (July/August 2004): 45–50.

43. M. Healy, T. Newe and E. Lewis. Analysis of Hardware Encryption versus Software Encryption on Wireless Sensor Network Motes. In *Smart Sensors and Sensing Technology*, vol. 20, Part I of *Lecture Notes in Electrical Engineering*, S. C. Mukhopadhyay and G. S. Gupta, eds. Berlin, Heidelberg: Springer (2008): 3–14.

44. R. H. Weber. Internet of Things—New Security and Privacy Challenges. *Computer Law & Security Review* 26, no. 1 (January 2010): 23–30.

45. C. Inhyok et al. Trust in M2M communication. *IEEE Vehicular Technology Magazine* 4, no. 3 (September 2009): 69–75.

46. R. Q. Hu et al. Recent Progress in Machine-to-Machine Communications [Guest Editorial]. *IEEE Communications Magazine* 49, no. 4 (April 2011): 24–6.

47. W. Geng et al. M2M: From Mobile to Embedded Internet. *IEEE Communications Magazine* 49, no. 4 (April 2011): 36–43.

48. K. Chang et al. Global Wireless Machine-to-Machine Standardization. *IEEE Internet Computing* 15, no. 2 (March/April 2011): 64–9.

49. S. L. Tompros, D. D. Vergados and N. P. Mouratidis. Harmonized Mobility Management in IMS Networks Based on the SSON Concept. *IEEE Wireless Communications* 18, no. 6 (December 2011): 74–81.

50. L. Foschini. M2M-Based Metropolitan Platform for IMS-Enabled Road Traffic Management in IoT. *IEEE Communications Magazine* 49, no. 11 (November 2011): 50–7.

51. B. Bakmaz, Z. Bojkovic and M. Bakmaz. Internet Protocol Multimedia Subsystem for Mobile Services. *14th IWSSIP*. Maribor, Slovenia (June 2007): 339–42.

52. S. Hiyama et al. Molecular Communication. *Proc. NSTI Bio-Nanotechnology Conference and Trade Show*, vol. 3. Anaheim, CA (May 2005): 391–4.

53. I. F. Akyildiz, J. M. Jornet and M. Pierobon. Nanonetworks: A New Frontier in Communications. *Communications of the ACM* 54, no. 11 (November 2011): 84–9.

54. J. M. Jornet and I. F. Akyildiz. Graphene-Based Nano-Antennas for Electromagnetic Nanocommunications in the TeraHertz Band. *Proc. 4th EuCAP*. Barcelona, Spain (April 2010): 1–5.

55. J. Weldon, K. Jensen and A. Zettl. Nanomechanical Radio Transmitter. *Physica Status Solidi (b)* 245, no. 10 (October 2008): 2323–5.

CHAPTER 10 FLEXIBLE FUTURE OF THE INTERNET

1. M. Conti et al. Research Challenges towards the Future Internet. *Computer Communications* 34, no. 18 (December 2011): 2115–34.
2. F. D. Cardoso et al. Physical Layer Aware Network Architecture for the Future Internet. *IEEE Communications Magazine* 50, no. 7 (July 2012): 168–76.
3. S. Eum et al. CATT: Cache Aware Target Identification for ICN. *IEEE Communications Magazine* 50, no. 12 (December 2012): 60–7.
4. J. Pan, S. Paul and R. Jain. A Survey of the Research on Future Internet Architectures. *IEEE Communications Magazine* 49, no. 7 (July 2011): 26–36.
5. B. Ahlgren et al. A Survey of Information-Centric Networking. *IEEE Communications Magazine* 50, no. 7 (July 2012): 26–36.
6. B. Ahlgren et al. Content, Connectivity, and Cloud: Ingredients for the Network of the Future. *IEEE Communications Magazine* 49, no. 7 (July 2011): 62–70.
7. N. M. M. K. Chowdhury and R. Boutaba. Network Virtualization: State of the Art and Research Challenges. *IEEE Communications Magazine* 47, no. 7 (July 2009): 20–6.
8. S. Balasubramaniam et al. Biological Principles for Future Internet Architecture Design. *IEEE Communications Magazine* 49, no. 7 (July 2011): 44–52.
9. H. Tuncer, Y. Nozaki and N. Shenoy. Virtual Mobility Domains—A Mobility Architecture for the Future Internet. *Proc. IEEE ICC 2012.* Ottawa, Canada (June 2012): 2774–9.
10. Available at http://www.4ward-project.eu (2013).
11. S. P. Sanchez and R. Bless. Network Design. In *Architecture and Design for the Future Internet: 4WARD Project*, L. M. Correia et al. eds. Dordrecht: Springer (2011).
12. T. Koponen et al. A Data-Oriented (and Beyond) Network Architecture. *ACM SIGCOMM Computer Communication Review* 37, no. 4 (October 2007): 181–92.
13. B. Ahlgren et al. Design Considerations for a Network of Information. *Proc. ACM CoNEXT '08.* Madrid, Spain (December 2008): Article no. 66.
14. V. Jacobson et al. Networking Named Content. *Proc. ACM CoNEXT '09.* Rome, Italy (December 2009): 1–12.
15. J. Choi et al. A Survey on Content-Oriented Networking for Efficient Content Delivery. *IEEE Communications Magazine* 49, no. 3 (March 2011): 121–7.
16. M. F. Bari et al. A Survey of Naming and Routing in Information-Centric Networks. *IEEE Communications Magazine* 50, no. 12 (December 2012): 44–53.
17. G. Zhang, Y. Li and T. Lin. Caching in Information Centric Networking: A Survey. *Computer Networks* 57, no. 16 (November 2013): 3128–41.
18. Y. Kim and I. Yeom. Performance Analysis of In-Network Caching for Content-Centric Networking. *Computer Networks* 57, no. 13 (September 2013): 2465–82.

19. V. Sourlas et al. Autonomic Cache Management in Information-Centric Networks. *Proc. IEEE/IFIP NOMS 2012*. Maui, HI (April 2012): 121–9.

20. W. K. Chai et al. Cache "Less for More" in Information-Centric Networks (extended version). *Computer Communications* 36, no. 7 (April 2013): 758–70.

21. Available at http://www.netinf.org (2013).

22. S. Farrell et al. Naming Things with Hashes. *IETF Internet-Draft* (August 2012).

23. N. Ramzan et al. Peer-to-Peer streaming of Scalable Video in Future Internet Applications. *IEEE Communications Magazine* 49, no. 3 (March 2011): 128–35.

24. H. Schwarz, D. Marpe and T. Wiegand. Overview of the Scalable Video Coding Extension of the H.264/AVC Standard. *IEEE Transactions on Circuits and Systems for Video Technology* 17, no. 9 (September 2007): 1103–20.

25. N. Ramzan, T. Zgaljic and E. Izquierdo. Scalable Video Coding: Source for Future Media Internet. In *Towards the Future Internet*, G. Tselentis et al. eds. Amsterdam: IOS Press (2010).

26. ITU-T Rec. H.264 (ISO/IEC 14496-10). Advanced Video Coding for Generic Audiovisual Services (April 2013).

27. G. J. Sullivan et al. Overview of the High Efficiency Video Coding (HEVC) Standard. *IEEE Transactions on Circuits and Systems for Video Technology* 22, no. 12 (December 2012): 1649–68.

28. M. Mrak et al. Performance Evidence of Software Proposal for Wavelet Video Coding Exploration Group. Tech. Rep. ISO/IEC JTC1/SC29/WG11/MPEG2006/M13146 (April 2006).

29. M. Alduan et al. System Architecture for Enriched Semantic Personalized Media Search and Retrieval in the Future Media Internet. *IEEE Communications Magazine* 49, no. 3 (March 2011): 144–51.

30. G. Bouabene et al. The Autonomic Network Architecture (ANA). *IEEE Journal on Selected Areas in Communications* 28, no. 1 (January 2010): 4–14.

31. M. Charalambides et al. Managing the Future Internet through Intelligent In-Network Substrates. *IEEE Network* 25, no. 6 (November/December 2011): 34–40.

32. G. Athanasiou et al. Multi-Objective Traffic Engineering for Future Networks. *IEEE Communications Letters* 16, no. 1 (January 2012): 101–3.

33. F. Francois et al. Optimizing Link Sleeping Reconfigurations in ISP Networks with Off-Peak Time Failure Protection. *IEEE Transactions on Network and Service Management* 10, no. 2 (June 2013): 176–88.

Index

Page numbers followed by f and t indicate figures and tables, respectively.

441